José Luis Millán

Mammalian Alkaline Phosphatases

Related Titles

Marangoni, AG

Enzyme Kinetics – A Modern Approach

248 pages
2002
Hardcover
ISBN 0-471-15985-9

Bisswanger, H.

Enzyme Kinetics

Principles and Methods

260 pages with 114 figures and 11 tables
2002
Hardcover
ISBN 3-527-30343-X

Copeland, R. A.

Enzymes

A Practical Introduction to Structure, Mechanism, and Data Analysis

414 pages
2000
Hardcover
ISBN 0-471-35929-7

Suelter, CH

Methods of Biochemical Analysis, Volume Thirty-Six
Bioanalytical Applications of Enzymes

280 pages
1992
Hardcover
ISBN 0-471-55880-X

José Luis Millán

Mammalian Alkaline Phosphatases

From Biology to Applications in Medicine
and Biotechnology

WILEY-VCH Verlag GmbH & Co. KGaA

Author

Prof. José Luis Millán
Burnham Institute for Medical Research
10901 North Torrey Pines Road
La Jolla, CA 92037

■ All books published by Wiley-VCH are carefully produced. Nevertheless, authors, editors, and publisher do not warrant the information contained in these books, including this book, to be free of errors. Readers are advised to keep in mind that statements, data, illustrations, procedural details or other items may inadvertently be inaccurate.

Library of Congress Card No.: applied for

British Library Cataloguing-in-Publication Data
A catalogue record for this book is available from the British Library.

Bibliographic information published by Die Deutsche Bibliothek
Die Deutsche Bibliothek lists this publication in the Deutsche Nationalbibliografie; detailed bibliographic data is available in the Internet at <http://dnb.ddb.de>.

© 2006 WILEY-VCH Verlag GmbH & Co. KGaA, Weinheim

All rights reserved (including those of translation into other languages). No part of this book may be reproduced in any form – nor transmitted or translated into machine language without written permission from the publishers. Registered names, trademarks, etc. used in this book, even when not specifically marked as such, are not to be considered unprotected by law.

Typesetting Kühn & Weyh, Satz und Medien, Freiburg
Printing betz-druck GmbH, Darmstadt
Bookbinding J. Schäffer GmbH i. G., Grünstadt

Printed in the Federal Republic of Germany.
Printed on acid-free paper.

ISBN-13: 978-3-527-31079-1
ISBN-10: 3-527-31079-7

Contents

Preface *IX*

Abbreviations *XI*

Glossary *XV*

Introduction *1*

Part I   Gene and Protein Structure *3*

1   **Gene Structure** *5*
1.1   Genomic Organization and Complexity *5*
1.2   Restriction Fragment Length Polymorphisms *11*

2   **Developmental Expression** *13*
2.1   The TNAP Gene *13*
2.2   The TSAP Genes *15*

3   **Gene Regulation** *19*
3.1   The TNAP Gene *19*
3.2   The TSAP Genes *22*

4   **Protein Structure and Functional Domains** *25*
4.1   The Three-dimensional Structure of PLAP *25*
4.1.1   Overview of the Structure *25*
4.1.2   The Active Site *28*
4.1.3   The Calcium Site *36*
4.1.4   The Disulfide Bonds *38*
4.1.5   The N-terminal Arm *38*
4.1.6   The Crown Domain *41*
4.1.7   The Monomer–Monomer Interface *46*
4.1.8   The Noncatalytic Peripheral Binding Site *46*
4.2   Genetic Polymorphism and Protein Variability *49*

Mammalian Alkaline Phosphatases: From Biology to Applications in Medicine and Biotechnology. J. L. Millán
Copyright © 2006 WILEY-VCH Verlag GmbH & Co. KGaA, Weinheim
ISBN: 3-527-31079-7

4.3	Post-translational Modifications 54
4.3.1	Glycosylation Sites 54
4.3.2	Ectoplasmic Localization of APs via a GPI Anchor 56
4.3.3	Nonenzymatic Glycation of APs 61
4.3.4	Quaternary Structure of APs 62
4.3.5	Subcellular Localization of APs 63

5	**Enzymatic Properties** 67
5.1	Catalytic Inhibition 67
5.1.1	Competitive and Noncompetitive Inhibitors of APs 69
5.1.2	Uncompetitive Inhibition 71
5.1.2.1	Mechanism of Inhibition in PLAP/GCAP 71
5.1.2.2	Inhibitor Binding in TNAP 76
5.2	Allosteric Behavior 80
5.3	Catalytic Efficiency of Mammalian APs 83
5.4	Substrate Specificities 85
5.5	APs as Members of a Superfamily of Enzymes 88

6	**Epitope Maps** 91
6.1	Epitopes in PLAP and GCAP 91
6.2	Epitopes in IAP 98
6.3	Discrimination Between Bone and Liver TNAP 99

Part II	*In Vivo* **Functions** 105

7	**The *In Vivo* Role of TNAP** 107
7.1	Function of TNAP in Bone 107
7.1.1	Hypophosphatasia 107
7.1.2	Hypophosphatasia Mutations 109
7.1.3	Variable Penetrance and Expressivity 123
7.2	Role of TNAP in Nonskeletal Tissues 124
7.3	Proposed Biological Functions of TSAPs 126
7.3.1	Proposed Functions of IAP 126
7.3.2	Putative Functions of GCAP and PLAP 128

8	**Knockout Mouse Models** 131
8.1	Phenotypic Abnormalities in $Akp2^{-/-}$ mice 131
8.1.1	Developmental and Skeletal Defects 131
8.1.2	Dental Abnormalities in $Akp2^{-/-}$ Mice 136
8.1.3	Deficient Mineralization by $Akp2^{-/-}$ Osteoblasts *In Vitro* 136
8.1.4	Metabolic Pathways Affected in $Akp2^{-/-}$ Mice 139
8.1.4.1	Neuro-physiological Abnormalities 139
8.1.4.2	The Function of TNAP in Bone Mineralization 141
8.1.4.3	Co-expression of TNAP and Fibrillar Collagens Restricts Calcification to Skeletal Tissues 148

8.1.4.4	Other Organs Affected in $Akp2^{-/-}$ Mice	152
8.2	Phenotypic Abnormalities in $Akp3^{-/-}$ Mice	154
8.3	Phenotypic Abnormalities in $Akp5^{-/-}$ Mice	160

Part III AP Expression in Health and Disease 165

9	**APs as Physiological and Disease Markers** 167	
9.1	Clinical Usefulness of TNAP	167
9.1.1	TNAP as a Marker of Bone Formation	167
9.1.2	TNAP and Bone Cancer or Bone Metastasis	173
9.1.3	TNAP Expression in Cholestasis	175
9.1.4	TNAP in Other Conditions	178
9.2	Clinical Usefulness of PLAP in Normal and Complicated Pregnancies 180	
9.3	IAP Expression in Relation to ABO Status, Fat Feeding and Other Pathologies 182	
9.4	Complexes of APs and Immunoglobulins	184
9.5	Hyperphosphatasia	185

10	**Neoplastic Expression of PLAP, GCAP, IAP (Regan, Nagao, Kasahara) and TNAP Isozymes** 187	
10.1	Some History	187
10.2	GCAP as Marker for Testicular Cancer	189
10.3	Usefulness of PLAP/GCAP in Ovarian Cancer	194
10.4	Other Tumors	196
10.5	Immunolocalization and Immunotherapy of Tumors Using PLAP/GCAP as Targets 197	
10.6	Tumoral Expression of IAP	204

Part IV Uses of APs in Industry and Biotechnology 207

11	**Applications of Recombinant APs** 209	
11.1	Expression of Recombinant APs	209
11.2	APs as *In Vitro* and *In Vivo* Reporters	213
11.3	APs as Molecular Biology and Diagnostic Reagents	218

12	**Use of APs in Prodrug Converting Strategies** 223	

13	**APs as Therapeutic Agents** 227	
13.1	In the Treatment of Hypophosphatasia	227
13.2	In the Treatment of CPPD Disease	231
13.3	Endotoxin Treatment	231
13.4	TNAP as Therapeutic Target for the Management of Ectopic Calcification 234	

14	**APs in the Food Industry** *237*	
15	**Veterinary Uses of AP Determinations** *239*	
16	**Methodologies** *241*	
16.1	Amperometric, Spectrophotometric and Potentiometric Assays	*241*
16.2	Using Inhibitors and Heat Inactivation *244*	
16.3	Electrophoretic Methods *245*	
16.4	Lectin-based Assays *245*	
16.5	High-performance Liquid Chromatographic Methods *247*	
16.6	Specific Immunoassays with Polyclonal and MAbs *249*	
16.7	mRNA-based Assays *253*	
16.8	Histochemical and Immunohistochemical Detection *254*	

References *257*

Index *315*

Preface

Determinations of serum levels of alkaline phosphatase (AP) isozymes are perhaps the most widely used biochemical measurements in the clinical laboratory but, despite their widespread clinical utility, the structure and biological functions of these widely known enzymes have remained elusive for decades. However, great progress is now being made in this area of research that justifies the writing of this monograph summarizing major recent advances. This book is not intended to be an exhaustive treatise on the subject. A superb compilation on APs, summarizing knowledge in this field up to the late 70s, was published by Drs Robert McComb, George Bowers and Solomon Posen (McComb et al., 1979) for references published before 1979, but a few papers published in the 1960s and 1970s are specifically cited here because of their unique relevance to some of our discussions. Two subsequent reports, much smaller in scope relative to McComb et al.'s impressive book, have also been published: the first (Human Alkaline Phosphatases, in *Progress in Clinical and Biological Research*, Vol. 166, Editors Torgny Stigbrand and William H. Fishman, Alan R. Liss, New York, 1984), reported the Proceedings of the First Symposium on Alkaline Phosphatases held in Umeå, Sweden, September 16–17, 1983. The second, published as a special issue of *Clinica Chimica Acta* [Vol 186 (2), pp. 125–320, 1990; Guest Editor José Luis Millán], presented selected papers from the Third Alkaline Phosphatase Symposium that took place at the La Jolla Cancer Research Foundation on February 1–3, 1989, on the occasion of Dr William H. Fishman's retirement. I will refer to individual papers in these two small volumes when needed.

The present work is intended to highlight and discuss key new information obtained in the last 25 years (1980–2005), with special emphasis on the structure of the proteins and genes, the function of the isozymes as revealed from the study of the individual mouse knockout strains and new uses of these isozymes in the clinical setting and in biotechnology. In this period, approximately 5000 papers have been published containing alkaline phosphatase in the title. If one searches for papers using alkaline phosphatase as a keyword, the number climbs to 36 000. Hence it would have been impossible to cite each of these references in a small treatise such as this and I apologize to those researchers whose work is either not cited or not presented in as much detail as I would have liked. Owing to space considerations, I have chosen to cite those references that, in my opinion, best illustrate new findings, a novel property or a function that advances the field.

Many confirmatory papers are not cited. Furthermore, this book is intended to discuss the structure and function of "mammalian" APs. Hence a considerable body of literature on bacterial APs is not covered here. Some papers on bacterial APs are cited only when there is a need to clarify a property or mechanism that has not been yet elucidated for the mammalian enzymes. The same is true for other nonmammalian APs. The book has been written with the aim of informing a wide spectrum of scientists and health care professionals who are not necessarily experts in this area of research. However, enough detail is provided so that investigators working on APs will find this to be an up-to-date account of the state-of-the-art in this field.

I would like to dedicate this book to the memory of two individuals who were responsible in large measure to my having dedicated most of my scientific career to the study of mammalian APs. To the late Dr Raúl Francisco Balado (1920–98), who mentored me through my first steps in clinical chemistry and in the use of APs in the clinical setting in my home town, Mar del Plata, Argentina, in the period 1974–77. His friendship, encouragement and support in obtaining a Rotary International Fellowship to travel abroad to initiate a research career represented a cornerstone of my scientific and personal life. That fellowship would take me, on August 4, 1977, to the laboratory of the late Professor William H. Fishman (1914–2001), who at that time had just moved from Tufts University in Boston to La Jolla, California, to found a new research institute, i.e. the La Jolla Cancer Research Foundation, now known as the Burnham Institute for Medical Research. His pioneering work on the Regan isozyme as a tumor marker had prompted him to create a research institution focused on oncodevelopmental biology and he was recruiting staff to build his new center. Fortunately, he did not turn away this young and inexperienced though aspiring trainee. Dr Fishman's mentoring and advice were paramount in my scientific career. He remained interested in advances in the field of APs to the day of his passing and continues to be a father figure to me and to many of the members of the institute that he created. I think he would have enjoyed this monograph.

I would like to thank Dr Andrea Pillmann of Wiley-VCH Verlag GmbH & Co., for suggesting writing this book after my lecture at the International Conference on Genes, Gene Families, and Isozymes, Berlin, 2003 and for her help and that of her staff in progressing the manuscript to final production. I want to thank my secretary Trixi Czink at the Burnham Institute for her invaluable help in compiling the bibliography for the book. I also want to express my sincere appreciation to all the members of my laboratory, past and present, who contributed much of the published data discussed and cited in this book. I have had the pleasure of working with many talented collaborators throughout the years. I would particularly like to thank Professor Marc F. Hoylaerts for the many years of extremely productive scientific interactions and friendship and for his many comments and suggestions on this book, particularly his input to the sections on enzyme kinetics and structure. I also want to thank Per Magnusson for his suggestions regarding the clinical sections of this book.

La Jolla, California
July 2005

José Luis Millán

Abbreviations

ADP	Adenosine diphosphate
AMP	Adenosine monophosphate
AP	Alkaline phosphatase
ATP	Adenosine triphosphate
ASO	Allele-specific oligonucleotide
bIAP	Bovine intestinal alkaline phosphatase
BsMAb	Bispecific monoclonal antibody
BMD	Bone mineral density
BMP	Bone morphogenetic protein
cAMP	Cyclic AMP
CAT	Chloramphenicol acetyltransferase
cDNA	Complementary deoxyribonucleic acid
CHO	Chinese hamster ovary
CIS	Carcinoma-*in-situ*
CMV	Cytomegalovirus
CPPD	Calcium pyrophosphate dihydrate
EAP	Embryonic alkaline phosphatase
ECAP	*Escherichia coli* alkaline phosphatase
ECM	Extracellular matrix
EDTA	Ethylenediaminetetraacetic acid
EIA	Enzyme immunoassay
ELISA	Enzyme-linked immunosorbent assay
EP	Etoposide phosphate
ES	Embryo-derived stem cells
FADP	Flavin adenine dinucleotide-3′-phosphate
GABA	γ-Aminobutyric acid
GCAP	Germ cell alkaline phosphatase
GPI	Glycosylphosphatidylinositol
GPI-PLC	GPI-specific phospholipase C
GPI-PLD	GPI-specific phospholipase D
HPAC	High-performance affinity chromatography
HPLC	High-performance liquid chromatography
Hyp	Hypophosphatasemic mice

Mammalian Alkaline Phosphatases: From Biology to Applications in Medicine and Biotechnology. J. L. Millán
Copyright © 2006 WILEY-VCH Verlag GmbH & Co. KGaA, Weinheim
ISBN: 3-527-31079-7

IAP	Intestinal alkaline phosphatase
IIAC	Idiopathic infantile arterial calcification
iPGM	Cofactor-independent phosphoglycerate mutase
IRMA	Immunoradiometric assays
k_{cat}	Catalytic rate constant
K_i	Inhibition constant
K_m	Michaelis constant
KO	Knockout
L-hArg	L-Homoarginine
LPS	Lipopolysaccharide
MAb	Monoclonal antibody
MALDI	Matrix-assisted laser desorption/ionization
MICA	Monoclonal antibody immunocatalytic assay
MOP	Mitomycin phosphate
MPLA	Monophosphoryl lipid A
mRNA	Messenger ribonucleic acid
MS	Mass spectrometry
MV	Matrix vesicle
NAD	Nicotinamide adenine dinucleotide
NADH	Nicotinamide adenine dinucleotide, reduced form
NPP1	Nucleotidetriphosphate pyrophosphohydrolase-1
OPN	Osteopontin
PAGE	Polyacrylamide gel electrophoresis
PCR	Polymerase chain reaction
PEA	Phosphoethanolamine
PGC	Primordial germ cells
P_i	Inorganic phosphate
PLAP	Placental alkaline phosphatase
PLP	Pyridoxal-5-phosphate
pNPP	*p*-Nitrophenylphosphate
PNPPate	*p*-Nitrophenylphosphonate
POMP	Phenol mustard phosphate
PP_i	Inorganic pyrophosphate
PSA	Prostate-specific antigen
QTL	Quantitative trait loci
RA	Retinoic acid
RFLP	Restriction fragment length polymorphism
RNA	Ribonucleic acid
RT-PCR	Reverse transcriptase polymerase chain reaction
SEAP	Secreted embryonic alkaline phosphatase
SLP	Surfactant-like particle
SNuPE	Single nucleotide primer extension
TGCTs	Testicular germ cell tumors
TNAP	Tissue-nonspecific alkaline phosphatase
TNFα	Tumor necrosis factor alpha

TOF	Time-of-flight
TSAPs	Tissue-specific alkaline phosphatases
UTR	Untranslated region
V_{max}	Maximum velocity
VSMC	Vascular smooth muscle cells
WGA	Wheat germ agglutinin
wt	Wild-type

Glossary

Allotypes	Allelic types. Synonymous with allozymes
Allozymes	Used as synonymous of allelic variants
Null	Used as synonymous of knockout or deficient, e.g. *Akp2* null mice, to refer to either $Akp2^{-/-}$ or $Akp2^{\beta geo/\beta geo}$ or TNAP-deficient mice
Biliary AP	An isoform of liver-derived TNAP
Bone AP	Used interchangeably with skeletal AP, bone-derived TNAP, bone-type TNAP, bone-specific TNAP isoform
Leukocyte AP	An isoform of TNAP
Neutrophil AP	Synonymous with leukocyte AP, an isoform of TNAP
Liver AP	Used interchangeably with liver-type TNAP or liver-specific TNAP isoform
Skeletal AP	Used interchangeably with bone AP, bone-derived TNAP, bone-specific TNAP isoform
Vitamer	Term used to refer to any of several chemical forms of a vitamin, e.g. pyridoxal is a vitamer of vitamin B_6

Introduction

Alkaline phosphatases (APs; EC 3.1.3.1) are ectoplasmic proteins that catalyze the following general reaction:

$$R-OP + H_2O \rightarrow R-OH + P_i$$

where the hydrolysis of R–OP gives rise to inorganic phosphate (P_i) and an alcohol, sugar, phenol, etc. (R–OH). As such, they are members of the class of enzymes known as phosphomonoesterases. They are unique in this class in the sense that they appear to be nonspecific and able to act on a wide variety of substrates, at least *in vitro*, and they are referred to as "alkaline" because of their ability to perform this reaction most efficiently at pH above neutral, e.g. pH 8–11. This propensity is somewhat surprising, since these enzymes are present *in vivo* in compartments not known to be particularly alkaline. Nevertheless, the properties of APs have made them useful for a variety of biotechnological applications ranging from dephosphorylating phosphoproteins and DNA fragments to using them for end-point detection in a wide variety of immunoassays and also as reporter molecules *in vivo*. In mammals, AP activity is found in a wide variety of tissues and organs and the enzymes are primarily ectoplasmic in location and attached to the plasma membrane via a glycosylphosphatidylinositol (GPI) anchor. Hence most of the artificial or macromolecular substrates used in the laboratory are never in contact with APs *in vivo* and in fact, as we will discuss, only two natural substrates have been unequivocally confirmed to date for one of the AP isozymes. The analysis of an inborn error of metabolism and of knockout mouse models will be a major focus in this book as these studies have clarified the biological role for some of the isozymes and for the others we now have good hypotheses to guide us in our future work. The clarification of the number and structure of the genes encoding human APs has also helped in our ability to detect abnormal alleles during prenatal diagnosis of hypophosphatasia and recent genotype/phenotype correlations of hypophosphatasia mutations have helped us predict the severity of phenotypic abnormalities in affected patients. Advances are also being made in experimental therapeutic approaches of hypophosphatasia and other illnesses. Although AP activity has long been used as indicator of health or disease in a variety of conditions, the potential use of APs itself as therapeutic target has only come into focus as a result of recent mechanistic studies. Efforts to develop pharmacological drugs able to modulate AP activity are being greatly helped by the wealth of new information gained in recent years on the three-dimensional structure and of the functional domains of mammalian APs. APs are also expressed in a wide variety of tumors and serve as tumor markers for immunodetection and immunolocalization of malignancies. Hence tremendous progress has been made in the last 25 years and this book will attempt to describe the most significant advances in each of these areas. The book is divided into four parts, i.e. Part I: Gene and Protein Structure; Part II: *In vivo* Functions; Part III: AP Expression in Health and Disease; and Part IV: Uses of APs in Industry and Biotechnology. Chapters 1–3 summarize current knowledge of the structure of the genes and their expression and regulation and serve as preparatory material for what I consider the core of this book, i.e. Chapters 4–8. Chapters 4–6 describe the structure of mammalian APs, their structural/functional domains and *in vitro* properties, and Chapters 6–8 elaborate on what we know about the biological function of each AP isozyme both in humans and in mice, through the study of hypophosphatasia and also of each mouse model of AP deficiency. Chapters 9 and 10 summarize a great deal of literature on the clinical utility of AP determinations in health and disease, particularly as cancer markers. Chapters 11–15 describe some exciting developments in the use of APs in industry and biotechnology, but also in the use of APs both as therapeutic agents and as therapeutic targets for some conditions. Finally, Chapter 16 gives an overview of the methodologies that have been improved or developed in the last 25 years. The reference list includes 1027 publications that are cited in this monograph. Readers are encouraged to consult the original sources for details beyond what I was able to include in this book, given the inherent limitations of space.

Part I
Gene and Protein Structure

1
Gene Structure

1.1
Genomic Organization and Complexity

In humans, alkaline phosphatases (APs) (EC 3.1.3.1) are encoded by four genes traditionally named after the tissues where they are predominantly expressed, although the gene nomenclature is now gaining wider use (Table 1). The tissue-nonspecific AP (TNAP) gene (*ALPL*) is expressed at highest levels in liver, bone and kidney (hence the alternative name "L/B/K" AP), in the placenta during the first trimester of pregnancy and at lower levels in numerous other tissues (McComb et al., 1979). The other three isozymes, i.e. placental AP (PLAP), placental-like or germ cell AP (GCAP) and intestinal AP (IAP), show much more restricted tissue expression, hence the general term tissue-specific APs (TSAPs). Within the short span of 2 years, several groups working independently published the sequences of the cDNAs of human PLAP (Kam et al., 1985; Millán, 1986; Henthorn et al., 1986), human IAP (Berger et al., 1987a; Henthorn et al., 1987) and human TNAP (Weiss et al., 1986; Terao and Mintz, 1987). The sequences of the four human AP genes were all published within months of each other in the following year, i.e. human TNAP (Weiss et al., 1988), IAP (Henthorn et al., 1988a), PLAP (Knoll et al., 1988) and GCAP (Millán and Manes, 1988).

The TNAP gene (*ALPL*; NM_000478), maps to the short arm of chromosome 1 (1p36–p34) (Swallow et al., 1986; Smith et al., 1988) and is at least five times larger that the TSAP genes, due in part to the first intron (at least 20 kb long) which separates exon I, which contains only 5′-untranslated (UTR) sequences, from exon II, which contains part of the 5′ UTR and also the start codon for translation. The *ALPL* gene is located on the distal short arm of chromosome 1, band 1p36.12, specifically at position chr1: 21581175–21650208, thus occupying a length of 69 034 bp. Subsequently it was demonstrated, first for the rat homolog (*Alp1*) (Toh et al., 1989) and subsequently for the human *ALPL* and the mouse *Akp2* genes (Matsuura et al., 1990; Toh et al., 1990; Studer et al., 1991), that the TNAP genes are really composed of 13 exons; the first two exons (Ia and Ib) are noncoding and are separated from one another and from the exon that contains the ATG translation initiation site (exon II) by relatively large introns. The last exon contains the termination codon and the 3′-untranslated region of the mRNA. The rest of the exonic sequences are interrupted by introns at positions analogous to those of the

Table 1 Nomenclature of the human, mouse, and rat AP isozymes and genes, including chromosomal location, gene size, and accession numbers.

Gene	Protein name	Common name	Chromosomal location	Accession No.
Human genes				
ALPL	TNAP	Tissue-nonspecific alkaline phosphatase; TNSALP; "liver–bone–kidney type" AP	chr1:21581174–21650208	NM_000478
ALPP	PLAP	Placental alkaline phosphatase; PLALP	chr2:233068964–233073097	NM_001632
ALPP2	GCAP	Germ cell alkaline phosphatase, GCALP	chr2:233097057–233100922	NM_031313
ALPI	IAP	Intestinal alkaline phosphatase, IALP	chr2:233146369–233150245	NM_001631
Mouse genes				
Akp2	TNAP	Tissue-nonspecific alkaline phosphatase; TNSALP; "liver–bone–kidney type" AP	chr4:136199753–136254338	NM_007431
Akp3	IAP	Intestinal alkaline phosphatase, IALP	chr1:87031694–87555136	NM_007432
Akp5	EAP	Embryonic alkaline phosphatase	chr1:86990248–86993641	NM_007433
Akp-ps1	N/a	AP pseudogene, pseudoAP	chr1:86968828–86972484	NG 001340
Akp6*	AKP6*	RIKEN sequence, new AP locus	chr1:87002298–87005230	AK008000
Rat genes				
Alp1	TNAP	Tissue-nonspecific alkaline phosphatase; TNSALP; "liver–bone–kidney type" AP	chr 5: 156511778–156568766	NM_013059
Alpi	IAPI	Intestinal alkaline phosphatase I	chr 9: 86076305–86079772	NM_022665
Alpi2	IAPII	Intestinal alkaline phosphatase II	chr9: 86107650–86110746	NM_022680

* tentative designation by the author

TSAP genes, but the introns are generally much larger (Fig. 1). Exons Ia in humans and rats have ~66% identical bases whereas sequence homology cannot be detected between the Ib exons in these two species (Toh et al., 1989). The major transcription start site and surrounding sequences of the 5′-most promoter Ia of the TNAP gene have been determined for humans (Weiss et al., 1988; Toh et al., 1989; Zernik et al., 1990) and mouse (Terao et al., 1990). Importantly, exons Ia and Ib are incorporated into the mRNA in a mutually exclusive fashion that results from the fact that each exon has its own promoter sequence. This results in two types of mRNAs, each encoding an identical polypeptide, but having different 5′-untranslated sequences (Kishi et al., 1989; Toh et al., 1989; Studer et al., 1991; Zernik et al., 1991). A 2127 bp cDNA encoding a functional feline TNAP of 524 amino acids has also been isolated (Ghosh and Mullins, 1995).

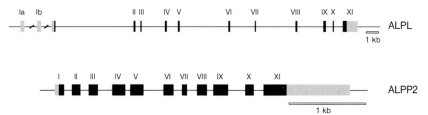

Fig. 1 Genomic organization of the human *ALPL* and *ALPP2* genes. Solid black boxes indicate coding exons and gray boxes mark the 5′ and 3′ untranslated regions (UTR).

The human TSAP genes, *ALPP*, *ALPP2* and *ALPI*, are clustered on human chromosome 2, bands q34–q37 (Griffin et al., 1987; Martin et al., 1987b) and are closely related to one another (Table 1). Their structures are almost identical, consisting of 11 exons interrupted by small introns (74–425 bp) at analogous positions all compressed in less than 5 kb of genomic DNA (Henthorn et al., 1988a; Knoll et al., 1988; Millán and Manes, 1988). Figure 1 shows the structure of the *ALPP2* gene as an example, in relation to the *ALPL* gene. The similarity in structure between all three TSAP genes suggests a divergent evolution for these genes since the chromosome mapping results show that the three related loci – *ALPP*, *ALPP2* and ALPI – are located in this order from centromere to telomere in the same region of the long arm of chromosome 2 (2q34–q37) (Griffin et al., 1987; Martin et al., 1987b). Each gene is comprised of 11 exons and 10 small introns contained within 4.5 kb of DNA. Specific regions of the introns, and also in the 3′-untranslated region of exon XI, show major differences in sequence and these regions have proven useful in the development of gene-specific probes. The PLAP gene also contains an Alu repeat sequence inserted in exon XI (Knoll et al., 1988). This Alu repeat sequence creates a new polyadenylation signal that may be responsible for alternative usage and consequently alternative size PLAP mRNA molecules approximately 300 bp shorter. Alternative PLAP mRNA molecules have already been observed in choriocarcinoma cells (Ovitt et al., 1986), colonic adenocarcinoma cells (Gum et al., 1987) and Hela cells (Chou and Takahashi, 1987). The intron–exon junctions in the coding region are, however, remarkably similar for all four genes. Knoll et al. have proposed that insertion/deletion events in the promoters of these three isozyme genes may partially explain evolution of promoter activity (Knoll et al., 1988).

In mice, besides the *Akp2* gene, which maps to chromosome 4 (Terao et al., 1988), four additional AP loci are found clustered on chromosome 1 (Table 1); the *Akp3* locus that encodes the IAP isozyme (Manes et al., 1990); the *Akp5* gene that encodes the embryonic AP (EAP) isozyme (Manes et al., 1990); the *Akp-ps1* non-transcribed, intron-containing, pseudogene (Manes et al., 1990); and finally a new locus recently identified in the databases after completion of the mouse genome project (Fig. 2). We tentatively have named this gene *Akp6* and have determined that it codes for an IAP-like isozyme albeit it has catalytic properties distinct from IAP (S. Narisawa et al., personal communication). Furthermore, with the caveat that negative data are seldom conclusive, to date there is no confirmation in the Mouse Genome Databases that the *Akp1* or *Akp4* loci actually exist. These loci were postulated based on hybrid maps and biochemical data (Wilcox et al., 1979), but no sequences have been posted to date for either of these two proposed loci. The *Akp3*, *Akp5* and *Akp-ps1* are around 5.0 kb in length and are composed of 11 exons interrupted by 10 small introns with an organization remarkably similar to that of the human tissue-specific TSAP genes. The smallest exon in all cases is exon VII (73 bp) and the largest in all cases is exon XI, which codes for the C-terminal end of the molecule and also contains the 3'-UTR (640 bp in length for *Akp5* and 1180 bp for *Akp3*). The introns are amongst the smallest reported with the largest one, splitting exons V and VI, being only 214 and 261 bp in *Akp5* and *Akp3*, respectively. The mRNA molecules for all AP isozymes, human or mouse, are of the order of 2.4–3.0 kb in length and encode peptides ranging from 518 to 535 amino acids. The rat genome harbors the TNAP gene (*Alp1*) (Toh et al., 1989) that produces an mRNA of around 2.5 kb (Noda et al., 1987b; Thiede et al., 1988) but reportedly also has two IAP genes, i.e. *Alpi* and *Alpi2*, based on cDNA (Lowe et al., 1990; Strom et al., 1991; Engle and Alpers, 1992) and genomic cloning (Xie

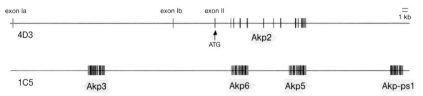

Fig. 2 Genomic organization of the murine AP loci. The mouse TNAP gene (*Akp2*) is located at 4D3 in chromosome 4. It stretches for 55 kb and consists of 12 exons and 11 introns including an alternative exon (exon 1b), located ~30 kb downstream of exon 1a. The mouse TSAP genes (*Akp3*, *Akp5*, *Akp6* and the *Akp-ps1* pseudogene) are closely linked at the 1C5 site in chromosome 1. The size of each TSAP gene is ~3.5 kb and they contain 11 exons and 10 introns. The direction of the *Akp3* gene and the *Akp-ps1* pseudogene is opposite to that of *Akp5* and *Akp6* genes. In the active AP genes, translation starts from the ATG site in the exon 2 and ends at the stop codon within the exon 11. Sequence numbers indicated beneath each gene are the actual location in the chromosome.

and Alpers, 2000). These genes produce mRNAs of 2.7 kb (IAP I) and 3.0 kb (IAP II) (Eliakim et al., 1990b).

After humans and rodents, *Bos taurus* is the species that has been studied the most in terms of its AP gene family. Besman and Coleman (1985) described two different IAP isozymes in the bovine intestine by C-terminal sequencing of chromatographically purified AP fractions. One of these was restricted to intestines of calves whereas the other was present in both calves and adults. Using the mouse IAP cDNA as probe, Weissig et al. (1993) cloned a gene that matched the amino terminus of the adult bovine AP isoform or bIAP I, and also an AP pseudogene. Subsequently, using a combination of protein sequencing and cDNA cloning, Manes et al. (1998) discovered an unprecedented level of complexity for the bovine AP family of genes, obtaining either full or partial cDNA evidence for up to seven IAP-like genes, i.e. bIAP II, III, IV, V, VI and VII. Garattini et al. (1987) cloned and sequenced a bovine kidney TNAP that displayed 90% homology with the human enzyme at both the nucleotide and amino acid levels. Manes et al. (1998) also found evidence for the existence of a second TNAP gene (TNAP-2). In the same year, another group reported the presence of two additional AP cDNAs present in bovine blastocysts (TSAP2 and TSAP3) (McDougall et al., 1998). Hence it seems that *Bos taurus* to date stands alone as the species with the highest degree of complexity in the AP family of genes.

Figure 3 shows a phylogenetic tree constructed from the homology of all the mammalian APs for which we have complete nucleotide or protein sequence. This cladogram was produced with the Clustal W algorithm of the DNASTAR software package. One can clearly see the clustering of the TNAP genes from different species as being considerably different from the TSAP isozymes that also cluster. Note that the duplication of the PLAP and GCAP genes appears to have occurred fairly recently based on the very limited number of substitutions. It has

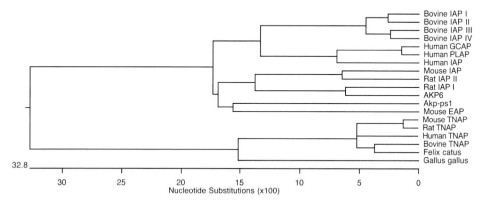

Fig. 3 Phylogenetic tree of all mammalian APs sequenced to date. The chicken TNAP (*Gallus gallus*) has also been included given that the chicken is a species used routinely for studies of bone mineralization. The sequences were aligned using the Clustal W method of the DNASTAR software package. The length of each pair of branches represents the distance between sequence pairs. The scale beneath the tree measures the distance between sequences.

been generally accepted that the appearance of the PLAP gene represents a late evolutionary event. Enzymes with the properties of PLAP were found expressed in the placenta of chimpanzee, orangutan and humans, but not in lower species (Doellgast and Benirschke, 1979; Goldstein and Harris, 1979). Doellgast et al. (1981) extended these observations to show that the placental tissue of the squirrel and spider monkey contain low but measurable levels of a heat-stable AP that resembled the human testis PLAP-like enzyme, that we now call GCAP.

In the late 1980s, we also had an opportunity to analyze genomic DNA samples from a number of Old and New World monkeys kindly provided by Dr Oliver A. Ryder (Center for Reproduction of Endangered Species, San Diego Zoo, San Diego, CA, USA) using the PLAP cDNA as a probe. As can be observed in Fig. 4, a complex hybridization pattern is observed in all the primate species indicating the probable existence of more than on PLAP-related gene. In order to attempt to quantify the number of genes present in each species, we rehybridized the blots using either a 144 bp *Sph–Sma*I or a 227 bp *Pst*I fragment containing the highly conserved exons V or IX, respectively, of the PLAP gene. The results enabled us to count the number of hybridizing bands and correlate them with the number of genes present in the different species. One duplication event seems to have occurred after the bifurcation of Prosimians and New World monkeys, very likely representing the duplication of the IAP gene. A second duplication event precedes the divergence of Old World monkeys and apes, where the precursor of the GCAP

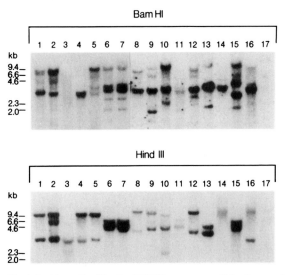

Fig. 4 Probing a Zoo blot for *ALPP*-like sequences. Southern blot analysis of genomic DNA from Old and New World monkey probed with the PLAP cDNA. The DNAs were digested with *Bam*HI and *Hind*III, respectively, and probed with the 2.0 kb *Eco*RI–*Kpn*I fragment of the PLAP cDNA. The lanes were loaded with digested genomic DNA from the following species: 1, human; 2, chimpanzee; 3, pygmy chimpanzee; 4, orang-utan; 5, lowland gorilla; 6, rhesus monkey; 7, lion tail macaque; 8, titi monkey; 9, red howler; 10, colobus monkey; 11, black spider monkey; 12, drill baboon; 13, douc langur; 14, ruffed lemur; 15, Allen's monkey; 16, siamanga; and 17, potto monkey.

gene may have duplicated to generate the PLAP gene (unpublished results). Doellgast and Wei (1984) had used extensively absorbed rabbit antisera to PLAP to investigate the presence of three distinct epitopes in old and New World monkeys. They concluded that the presence of two epitopes characteristic of the Nagao isozyme (see Chapter 10) in spider monkeys suggested that this gene product was closely related to the enzyme present in the primate placenta at the time of species divergence (humans and New World monkeys). Hence GCAP appears to be more ancestral than PLAP. However, definitive statements in this regard will have to await confirmation based on actual sequence data from the corresponding species.

1.2
Restriction Fragment Length Polymorphisms

Only two reports have been published concerning restriction fragment length polymorphisms (RFLPs) in the human *ALPL* gene (Weiss et al., 1987; Ray et al., 1988). *Bcl*I digestion of genomic DNA identified a two-allele polymorphism with either a single band of 7.4 kb or two bands of 4.3 and 3.1 kb (Weiss et al., 1987), whereas *Sst*I digestion identified two alleles of either 6.0 or 9.0 kb, respectively (Ray et al., 1988). In contrast, the orthologous mouse *Akp2* gene displays a considerably larger degree of genetic variability (Fröhlander and Millán, 2001). Differences between the *Akp2* alleles can be explained in some instances by simple mutations affecting a restriction site and in others by insertions or deletions affecting intron sizes. Single point mutations account for the *Apa*I 1.2 and 2.2 kb fragments in Balb/c and the *Pst*I site in intron 6 in 129/J and C57Bl/6J that generates 2.9 and 1.2 kb allelic fragments. Intron-size differences were also demonstrated through *Apa*I, *Bam*HI and *Hin*dII digestions. Sequencing of 17.5 kb of the *Akp2* gene also revealed an over-representation of repeats in the 5′ half of the gene, especially in introns 4 and 5, of possible relevance for the intron size differences. Di- and trinucleotide repeats, e.g. d(CA)n, d(GT)n, d(GT)n d(CCA)n, d(GGT)n, are most frequent but tetranucleotides, e.g. d(GGGC)n and d(GGGA)n, and also longer repeats are also present in the *Akp2* gene (Fröhlander and Millán, 2001). Finally, one RFLP has been reported for the bovine TNAP gene (Beever and Lewin, 1992) and a *Hin*dIII polymorphism for the porcine TNAP locus (Shalhevet et al., 1993).

Two RFLPs in the human *ALPP* gene were found using the restriction enzymes *Rsa*I (Martin, et al., 1987a) and *Pst*I (Tsavaler et al., 1987; Tsavaler et al., 1988). Both of these RFLPs correlated with electrophoretic polymorphisms (Beckman et al., 1989; Beckman et al., 1991). Further evidence, however, indicated that despite the correlation with PLAP electrophoretic types, *Pst*I(b) is an RFLP of the *ALPP2* locus, not *ALPP* (Beckman et al., 1992). Linkage disequilibria indicated close linkage between the *ALPP* and *ALPP2* loci, a situation now amply confirmed by the genome project data. *ALPP* and *ALPP2* RFLPs and haplotypes were found to show highly significant associations with spontaneous abortions in the Finnish and Swedish populations (Wennberg et al., 1995). The Finnish abortions were associated with the *ALPP2* allele *Pst*I(b)2 and the Swedish abortions with the *ALPP* allele *Pst*I(a)2. The authors discuss that a possible mechanism behind the associations may be linkage disequilibria with deleterious alleles within or close to the AP gene complex. The existence of only limited RFLPs in the *ALPP* gene contrast with the high degree of "electrophoretic" variability found at this locus (Section 4.2).

2
Developmental Expression

2.1
The TNAP Gene

TNAP is expressed in a multitude of tissues, hence the description tissue-nonspecific (McComb et al., 1979). However, this is an unfortunate misnomer since during embryonic development, TNAP shows a discreetly stage-specific and tissue-specific pattern of expression. Studies in mice have revealed that the *Akp2* gene is already expressed in one-cell stage embryos (Merchant-Larios et al., 1985; Hahnel et al., 1990). Detecting TNAP expression in embryo-derived stem (ES) cells can actually serve as a useful marker to ensure that ES cells are maintained in an undifferentiated state (Narisawa et al., 1997). At embryonic day 7.5 (E7.5), TNAP expression is observed in the primordial germ cells (PGCs) that appear close to the allantois (Ginsburg et al., 1990; Hahnel et al., 1990). Migrating PGC continue to be strongly positive for TNAP as they move into the endodermal epithelium of the hindgut and then into the mesentery to enter finally the genital ridges at E11–12 (MacGregor et al., 1995). At E8.5 of mouse development, TNAP expression also becomes prominent in the neuroepithelium. TNAP activity is homogeneously distributed in the neural tube and at E9.5, strongly AP-positive single cells appear in the brain and spinal cord area. At E10.5, strongly positive cells are observed between the mesencephalon and the rhombencephalon and also along the entire spinal cord. Cranial nerves emerging from the meylencephalon are also positive (Narisawa et al., 1994). At E11.5 and E12.5, the staining pattern is basically as in E10.5 except that a fiber-like appearance of the positive cells becomes more obvious after E11.5. At E13.5, homogeneous AP positivity is observed only in the basal telencephalon. At E14.5, AP expression becomes prominent in cartilage, whereas expression in the brain tissue is considerably reduced (Vorbrodt et al., 1986). TNAP is expressed in the peripheral nerves in developing mice (Narisawa et al., 1994), in the lumbosacral region of the spinal cord (Kwong and Tam, 1984) and especially enriched in nodal regions (nodes of Ranvier and paranodes) of diverse species including rodents, carnivores and humans (Scherer, 1996). The changes in TNAP activity also correlate with the extent of myelination (Ng and Tam, 1986). In the rat, most neurons in the cerebral cortex express AP activity on the plasma membrane of the nerve cell body and dendritic processes, from the thick trunk to the terminal postsynaptic ends (Mori and Nagano, 1985a). At E12,

Mammalian Alkaline Phosphatases: From Biology to Applications in Medicine and Biotechnology. J. L. Millán
Copyright © 2006 WILEY-VCH Verlag GmbH & Co. KGaA, Weinheim
ISBN: 3-527-31079-7

all layers of the prosencephalon display TNAP positivity that increases in the dorsal-to-ventral direction. Within the hypothalamic area a second rostro-ventral gradient exists from E14 onwards. At E18, both gradients have decreased. At E20, almost all TNAP positivity has disappeared from the hypothalamus, with the exception of some reaction product in the dorsal ventricular matrix of the hypothalamus (Lakke et al., 1988). TNAP activity is also present in early vascular sprouts and also throughout the subsequent maturation and differentiation of capillaries and arterial vessels in the rat cerebral cortex, whereas no TNAP activity is found in differentiated veins (Rowan and Maxwell, 1981). Recently, Fonta et al. (2004), studying marmosets and macaques, described that in addition to the presence of TNAP in endothelial cells, strong AP activity is found in the neuropile, matching the pattern of thalamo-cortical innervations in layer 4 of the primate sensory cortices (visual, auditory and somatosensory). Furthermore, TNAP activity was localized exclusively to the myelin-free axonal segments, including the node of Ranvier (Fonta et al., 2005). In humans, TNAP is expressed at high levels in the choroids plexus (Nishihara et al., 1994).

In all vertebrate species, however, TNAP expression is predominant in the developing skeleton. In mice, *Akp2* expression starts in the developing skeleton on E13 (MacGregor et al., 1995) and its expression mimics the developmental pattern described for the development of the ossified skeleton (Kaufman, 1992). TNAP immunoreactivity can be localized on the entire cell surface of preosteoblasts, and also the basolateral cell membrane of osteoblasts. It is also localized on some resting chondrocytes and most of the proliferative and hypertrophic cells in cartilage. In the incisor, cells of the stratum intermedium, the subodontoblastic layer, the proximal portion of secretory ameloblasts and the basolateral portion of odontoblasts show particularly strong expression. Immunoreactivity is observed also in other soft tissues, such as the brush borders of proximal renal tubules in kidney and on cell membranes of the biliary canalicula in the liver (Hoshi et al., 1997). TNAP activity is also found in cartilage matrix, and further among noncalcified collagen fibrils of osteoid tissue in bone (Bonucci et al., 1992) and on the cell surface of chondroblasts and osteoblasts, particularly on their shed matrix vesicles (MVs) (Ali et al., 1970; Bernard, 1978; Morris et al., 1992). In fact TNAP is enriched in MVs at least 2-fold compared with the plasma membrane and 10-fold compared with the cell layer in chondroblasts stimulated with 1,25-dihydroxyvitamin D_3 [1,25-$(OH)_2D3$] (Schwartz et al., 1988). TNAP is also present on odontoblasts and on dental pulp (Goseki et al., 1990). Peritubular cells in the testis (Chapin et al., 1987), pericytes (Mori and Nagano, 1985b) and myod cells in the intestinal wall are rich in TNAP as are endothelial cells (Rowan and Maxwell, 1981), although a study of skeletal muscle indicated that TNAP as an expression was prominent in arterial endothelium but not in venous endothelium (Grim and Carlson, 1990). TNAP is also expressed in pre-B and pre-T cells (Garcia-Rozas et al., 1982; Marquez et al., 1989, 1990). TNAP is further expressed in leukocytes, particularly neutrophils (Garattini and Gianni, 1996; Dotti et al., 1999), and in the kidney (McComb et al., 1979).

2.2
The TSAP Genes

Kim et al. (1989) suggested that multiples AP isozymes might be expressed in the early mouse embryo and Lepire and Ziomek (1989) identified an activity in mouse embryos similar to that of human PLAP, in addition to the well-defined expression of TNAP at that stage (Merchant-Larios et al., 1985). Subsequent cloning work revealed that indeed the so-called mouse embryonic AP (EAP) (Manes et al., 1990), encoded by the *Akp5* gene, is expressed in the two cell-to-blastocyst stage embryos at levels 10-fold higher that *Akp2* in these embryonic stages (Hahnel et al., 1990). EAP is not expressed after gastrulation but reappears in the adult testis, the thymus and also the intestine (Hahnel et al., 1990). In the human fetus, it is the GCAP isozyme, and not TNAP as in mouse, that is expressed in migrating PGCs and in the fetal gonads (Heath, 1978; Hustin et al., 1987), but it remains to be established if GCAP is also expressed in human ES cells and in the preimplantation human embryo. GCAP is also found in trace amounts in the testis, lung, cervix and thymus (Chang et al., 1980; Goldstein et al., 1982). Both humans and monkeys also express GCAP in type I pneumocytes in the lung. Nouwen et al. (1990), using cytochemistry and immunocytochemistry at the light- and electron-microscopic level of resolution, observed PLAP-like expression in the apical and basal plasma membrane, in apical and basal caveolae and in the underlying basement membrane of type I pneumocytes. The level of heat-stable PLAP-like expression in the human lung was 10-fold lower than in the monkeys. In human fetal lung, the onset of heat-stable PLAP-like expression was associated with the development of the alveolar epithelium from 17–20 weeks gestation onwards (Nouwen et al., 1990). It is now clear that in the lung, it is the GCAP isozyme that is expressed by type I pneumocytes (Hirano et al., 1990; Koshida et al., 1990; Nouwen et al., 1990) and this expression explains the basal serum levels of GCAP detectable in all normal individuals (Millán et al., 1985a). It is unclear what cell type is responsible for GCAP expression in the thymus. During embryogenesis, germ cells originate from extra-embryonic PGCs that migrate from the allantois to the developing gonad. During this process, some PGCs are arrested and survive in the pineal gland and in the thymus and may, in adult life, give rise to germ cell tumors of extra-gonadal origin. How these PGCs survive and what function they serve in these sanctuaries is at present unknown. It is possible that the trace expression of GCAP observed in the thymus is derived from these trapped PGCs. GCAP is localized at the cell membrane of PGCs and is first expressed during their early migration towards the genital ridges (Hustin et al., 1987). Even when the gonads are not yet identifiable (first trimester of pregnancy), scattered GCAP-positive PGCs can be seen along the genital ridges. At this stage, PGCs express GCAP strongly and are the only cells expressing this isozyme in the entire human embryo. In older embryos obtained during the second trimester of pregnancy, most of the germ cells are trapped within sex cords. Here, GCAP expression decreases progressively and is completely negative in term specimens. Subsequently, GCAP is never produced in infantile and prepubertal testes, but is weakly re-expressed by spermatogonia in the adult male (Hustin et al., 1990). However, peritubular cells are the richest source of AP activity in adult testis owing to expression of the TNAP gene. In the mouse testis, besides the strong TNAP expression by peritubular cells, EAP is expressed in M-phase spermatocytes (Narisawa et al., 1992).

PLAP is expressed in large amounts in the syncytiotrophoblast cells of the placenta from about the eighth week of gestation throughout pregnancy (Fishman et al., 1976), where it shows a clustered distribution (Jemmerson et al., 1985a). Povinelli and Knoll (1991) further documented that even though expression of the *ALPP* gene is predominant in the human placenta, transcripts from the *ALPP2* are also detectable at about 2% of the level of *ALPP* transcripts. Whether this indicates that both PLAP and GCAP are functional during pregnancy or that the GCAP gene is also expressed owing to its proximity to the active PLAP gene is unclear at this point. However, Ovitt et al. (1986) suggested that transformation of normal to malignant trophoblast is associated with a switch from *ALPP* to *ALPP2* expression. Trace amounts of PLAP are also produced by the epithelium lining the female genital tract, i.e. oviduct, endometrium, endometrial glands, endocervix and endocervical glands (Nozawa et al., 1980; Davies et al., 1985a; Nouwen et al., 1985; van de Voorde et al., 1985b; Nouwen et al., 1987; Hamilton-Dutoit et al., 1990). Trace amounts of PLAP and/or GCAP have also been detected in the normal breast (McLaughlin et al., 1984), ovary and oviduct (Nouwen et al., 1987). Also, using high-performance liquid chromatography (HPLC) for separation and a specific immunoassay for detection, Garattini et al. (1985) demonstrated that PLAP is also present in extracts of liver and intestine in appreciable amounts. PLAP is also present in the neonatal intestine (Behrens et al., 1983) and in the adult colon (Wada et al., 2005).

Intestinal AP (IAP), as the name implies, is expressed in the small intestine of many species. In humans, IAP is detectable already in amniotic fluid at 16–18 weeks gestation but owing to the altered electrophoretic mobility compared with adult IAP it is referred to as fetal IAP (Moss and Whitaker, 1987). The properties of the fetal IAP isoform appear to be generally similar to those of adult IAP except for a difference in the sialic acid content (Miki et al., 1978; Mulivor et al., 1978a). Sequencing of the N-terminal amino acids (Hua et al., 1986), peptide map analysis (Behrens et al., 1983; Mueller et al., 1985; Vockley et al., 1984a) and monoclonal antibody (MAb) binding studies (Vockley et al., 1984b) revealed differences between fetal and adult IAP, which suggested differences in the peptide structure in addition to carbohydrate differences. The accumulated evidence appeared to indicate that fetal IAP consists of heterodimers of IAP and PLAP (Behrens et al., 1983), but this interpretation was challenged by Vockley et al. (1984a) and also by Mueller et al. (1985), who suggested that the two proteins might be encoded by separate gene loci or, alternatively, may arise from differential processing at the RNA or protein level of the product of a single gene. Given that the human genome project has not provided evidence for the existence of another IAP-like gene in humans, it remains to be seen whether alternative splicing may account for some of the differences between fetal and adult IAP.

Adult IAP is found associated with the brush border of the intestinal epithelium and enriched in surfactant-like particles (SLP) (Eliakim et al., 1989; Alpers et al., 1994, 1995; Eliakim et al., 1997; Goetz et al., 1997). Fat feeding *in vivo* or transfection of an IAP cDNA into intestinal-like Caco-2 cells increases SLP secretion (Eliakim et al., 1991; Tietze et al., 1992) and, after fat feeding, an IAP-containing membrane surrounds the lipid droplet in the enterocyte (Yamagishi et al., 1994b; Zhang et al., 1996). These findings suggest a role for SLP in fat absorption involving the induction of IAP by fat, which in turn results in IAP-induced SLP secretion. Comparing the two mRNAs encoding IAP in the rat (IAP I and II), IAP II is the most active in inducing SLP secretion (Engle and Alpers, 1992; Tietze et al., 1992; Xie et al., 1997). In the rat, IAP I mRNA can be found in both the duodenal and jejunal mucosa, but IAP II mRNA is only expressed in the duodenum (Calhau et al., 2000). The co-expression of two distinct IAP mRNAs can give rise to three isoforms of IAP, i.e. two homodimeric forms and one heterodimeric form, that can be found distributed in a gradient along the length of the small intestine (Shidoji and Kim, 2004). Furthermore, while expression of the IAP I mRNA appears restricted to the intestine, the IAP II mRNA is also expressed in the liver and fat feeding increases the expression of IAP I mRNA in the intestine and that of IAP II mRNA both in the intestine and in the liver (Goseki-Sone et al., 1996). The levels of IAP II mRNA in the jejunum appear to be induced by ingestion of a lactose-containing diet (Sogabe et al., 2004)

Both IAP I mRNA and TNAP are expressed by type II pneumocytes in the rat (Harada et al., 2002). Heat shock induces the elevation of IAP I mRNA and IAP protein synthesis in rat IEC-18 intestinal epithelial cells (Harada et al., 2003). In humans, IAP and PLAP are co-expressed in the neonatal intestine (Behrens et al., 1983) and also expressed in the adult colon (Wada et al., 2005). IAP is also expressed in the human kidney and Verpooten et al. (1989) reported that ~25% of the total AP content of renal tissue at the transition between cortex and medulla is of the intestinal type with the remainder of the activity corresponding to TNAP. Immunoperoxidase staining using specific MAbs against liver-type TNAP and IAP revealed that TNAP is present throughout the different segments of the proximal tubule, whereas IAP is found exclusively in tubulo-epithelial cells of the S3 segment of the proximal tubule (Verpooten et al., 1989). Hirano et al. (1989) reported that the biophysical properties of the kidney IAP were more similar to those of fetal meconial IAP than to adult IAP and that those differences were most probably related to differences in the carbohydrate chains (Hirano et al., 1989). In mice, both the *Akp3* and *Akp5* genes are expressed in the small intestine (Hahnel et al., 1990). More recently, Narisawa et al. (2005) we have found expression of the newly identified *Akp6* gene throughout the small intestine (S. Narisawa et al., personal communication).

3
Gene Regulation

3.1
The TNAP Gene

The expression of the TNAP gene is regulated by the presence of two leader exons, Ia and Ib, resulting in the synthesis of two alternatively spliced mRNAs that are different only in part of their 5′-untranslated region (Terao et al., 1990; Studer et al., 1991). The promoter usage has primarily been inferred by examination of the structure of TNAP cDNAs isolated from various tissues. In humans and rats, the upstream promoter (Ia) is preferentially utilized by osteoblasts (bone-type mRNA) and the downstream promoter (Ib) preferred in kidney and liver (liver-type mRNA) (Zernik et al., 1991). The bone-type mRNA is also predominantly transcribed in peripheral neutrophils and in neutrophils cultured *in vitro* (Sato et al., 1994). The first promoter is also active in ES cells, whereas the second promoter is silent under basal conditions. In the whole animal, the transcript driven by the first promoter is found in most tissues, albeit at different levels, whereas the one driven by the second promoter is specifically expressed at high levels only in the heart (Studer et al., 1991).

Although the 5′-flanking regions of human, rat and mouse TNAP have been analyzed in transient transfection assays (Kiledjian and Kadesch, 1990, 1991), more work is needed to elucidate the mechanism of tissue-specific regulation. The promoter of the rat *ALPL* gene displays features of a "housekeeping" gene promoter: an atypical TATA-box (TTCATAA); 3 potential Sp1 binding sites; high GC content (82% in positions −134 to −14); and a high CpG to GpC ratio (60:89 in the 0.85 kb promoter region), indicating an abundance of potential methylation sites. Transient transfection of chloramphenicol acetyltransferase (CAT) fusion genes into ROS 17/2.8 rat osteosarcoma-derived cells revealed a weak expression from the promoter and proximal 5′-flanking sequences, which could be enhanced by an SV40 enhancer. The homologous human *ALPL* promoter demonstrates a similar combination of tissue-specific and housekeeping characteristics (Zernik et al., 1990).

Studies using agents that increase TNAP activity have provided evidence for both a transcriptional and post-transcriptional regulation of TNAP expression (Rodan and Rodan, 1983; Sato et al., 1994). Retinoic acid, dexamethasone and granulocyte colony-stimulating factor (G-CSF) all promote transcription of TNAP from the upstream Ia promoter and there is evidence that TNAP transcription in nonexpressing tissues of mice is repressed *in vivo* by methylation at the Ia promoter (Zernik et al., 1990; Scheibe et al., 1991; Heath et al., 1992; Escalante-Alcalde et al., 1996). Nevertheless, the Ib promoter is activated by dexamethasone (Zernik et al., 1991), dibutyryl cAMP and also by all-*trans*-retinoic acid (Gianni et al., 1993; Orimo and Shimada, 2005). Regulation of TNAP expression by the forkhead transcription factor (FKHR) via a forkhead response element in the TNAP promoter has been demonstrated (Hatta et al., 2002). The transcription factor Sp3 was found to activate the TNAP gene in hematopoietic cells (Yusa et al., 2000). Also, the TNAP promoter appears to be repressed by Smad-interacting protein 1 (Tylzanowski et al., 2001), whereas Smad3 promotes TNAP expression and mineralization in osteoblastic cells lines (Sowa et al., 2002a,b). There is also evidence for regulation of TNAP at the translational or post-translational stages (Kobayashi and Robinson, 1991; Leboy et al., 1991; Studer et al., 1991; Sato et al., 1994; Kyeyune-Nyombi et al., 1995). G-CSF specifically accumulates TNAP mRNA without showing a substantial increase in the rate of transcription of the TNAP gene in polymorphonuclear neutrophils isolated from chronic myelogenous leukemia or chronic myelomonocytic leukemia patients. Once increased by G-CSF, TNAP mRNA is very stable, showing a half-life of more than 4 h in the presence of actinomycin D (Rambaldi et al., 1990).

Since the expression of TNAP in osteoblasts is linked to differentiation, some agents which cause elevated enzyme levels in these cells may do so by increasing expression of osteoblast-specific transcription factors, by relieving osteogenic suppression rather than acting directly on the TNAP gene and even by stabilizing the TNAP mRNA. However, the choice of cell line for these kinds of studies is important. Rat osteosarcoma cells are a reasonably good experimental model for bone formation since they retain osteoblast properties including the ability to form bone and to mineralize and possess specific cytoplasmic receptors for 1,25-$(OH)_2$D3 that respond by enhancing TNAP activity (Manolagas et al., 1981). Osteosarcoma-fibroblast cell hybrids regain elevated TNAP when treated with a combination of 1,25-$(OH)_2$D3 and TGF, suggesting that this combination diminishes repression of the TNAP gene (Johnson-Pais and Leach, 1996). However, basic fibroblast growth factor (bFGF) has been shown to reduce the level of TNAP in ROS 17/2.8 cells (Rodan et al., 1989). Nuclear runoff analysis indicates that the transcription rate of the *ALPL* gene is stimulated 5-fold by 1,25-$(OH)_2$D3-treated osteosarcoma cells and that the treatment also increased *ALPL* mRNA stability (Kyeyune-Nyombi et al., 1991). Ascorbic acid and 1,25-$(OH)_2$D3 are both required to achieve maximum TNAP activity in both ROS 17/2.8 rat and MG-63 human osteosarcoma cells (Franceschi and Young, 1990). Treatment of MG-63 cells with 1,25-$(OH)2$D3 rapidly stimulated Type I collagen synthesis and acid-precipitable hydroxyproline production and this stimulation was further increased by ascorbic acid (Franceschi et al., 1988). These results suggest that induction of TNAP is directly or indirectly coupled to collagen matrix synthesis and/or accumulation. Dexamethasone also increases TNAP levels up to 7-fold in rat ROS 17/2.8 cells (Majeska et al., 1985). These increases in TNAP activity are detectable after about 5 h and are inhibited by both actinomycin D and cycloheximide, indicating that glucocorticoids increase *de novo* enzyme synthesis in ROS 17/2.8 cells. Sato et al. (1987) studied the effects of thyroid hormone on TNAP activity also in ROS 17/2.8 cells and found that they express T3 nuclear receptors and that TNAP activity is stimulated by thyroid hormone. Murray et al. (1987) reported that the basal TNAP activity of human SAOS-2 osteosarcoma cells was 100–1000 times greater than that of other established human osteogenic cell lines, TE-85 or SAOS-1. However, in contrast to those other cell lines, in which TNAP activity is stimulated several-fold by steroid hormones, 1,25-$(OH)_2$D3 and hydrocortisone, the TNAP activity of SAOS-2 cells was not affected by 1,25-$(OH)_2$D3 treatment despite the presence of classical receptors for this hormone. Schwartz et al. (1991) have shown that the regulation of TNAP and phospholipase A_2 activity by vitamin D_3 metabolites in cartilage cells is mediated by changes in calcium influx. A mouse osteoblast-like cell line MC3T3-E1 cells can also mimic the process of differentiation and mineralization *in vitro* provided that ascorbic acid and β-glycerophosphate are supplied in the culture (Quarles et al., 1992).

Other reagents may act indirectly on TNAP expression by stimulating osteogenic pathways. For example, ascorbic acid has been shown to increase TNAP mRNA in MC3T3-E1 cells by a slow mechanism requiring increased extracellular collagen synthesis which, in turn, promotes increased osteoblast differentiation (Franceschi et al., 1994). The transcriptional activity of exon Ia promoter is upregulated by retinoic acid (RA) through a putative RA-responsive element (Heath et al., 1992; Escalante-Alcalde et al., 1996), although Zhou et al. (1994) have also shown that retinoic acid achieves its marked induction of TNAP gene expression through a post-transcriptional effect in the nuclei of rat pre-osteoblastic UMR 201 cells in addition to its transcriptional effect. They concluded that RA treatment leads to stabilization of nascent TNAP mRNA chains and suggested that regulation of mRNA processing occurs independently of gene transcription (Zhou et al., 1994). Vitamin A is also an important regulator of TNAP expression in fetal rat small intestine and in IEC-6 intestinal crypt cells (Nikawa et al., 1998). Studies in transgenic mice using lacZ reporter constructs revealed the presence of negative regulatory elements within 8.5 kb of the TNAP exon Ia promoter. Sequences in the endogenous TNAP gene promoter, and also in the 8.5 kb lacZ transgene, in nonexpressing tissues were found to be highly methylated, suggesting that TNAP expression is being negatively regulated by methylation affecting either specific regulatory sequences or chromatin structure (Escalante-Alcalde et al., 1996). Kobayashi et al. (1998) identified an enhancer sequence in the 5′-flanking region of exon Ia in the murine *Akp2* gene. Using luciferase assays, they identified a segment containing a 9 bp inverted repeat (nt −1212 to −1179) with enhancer activity and an E-box at nt −234 that did not have prominent promoter activity but was required for the enhancer activity (Kobayashi et al., 1998). The cellular levels of cAMP-dependent protein kinase (PKA) activity have also been shown to regulate the expression of the TNAP gene both at the level of mRNA and protein (Uhler and Abou-Chebl, 1992). Distinct mitogen-activated protein kinase (MAPK) pathways seem to modulate independently osteoblastic cell proliferation and differentiation, with Erk playing an essential role in cell replication, whereas p38 is involved in the regulation of TNAP expression during osteoblastic cell differentiation (Suzuki et al., 1999, 2002). Smad3 promotes TNAP activity and mineralization in osteoblastic MC3T3-E1 cells (Sowa et al., 2002a) and activations of ERK1/2 and JNK by transforming growth factor beta (TGFβ) negatively regulate Smad3-induced TNAP activity and mineralization in mouse osteoblastic cells (Sowa et al., 2002b). TNAP expression is stimulated by bone morphogenetic protein-2 (BMP-2) treatment, the activation of BMP receptors and R-Smads and the expression of the transcription factors Dlx5 and Runx2. Dlx5 transactivates TNAP expression directly by binding to its cognate response element and/or indirectly by stimulating Runx2 expression and Msx2 counteracts the direct transactivation of Dlx5 (Kim et al., 2004). The Wnt autocrine loop has also been found to mediate the induction of TNAP and mineralization by BMP-2 in pre-osteoblastic cells (Rawadi et al., 2003). The PI 3-kinase and mTOR pathways contribute to the induction of TNAP activity induced by osteogenic protein-1 (OP-1) and the synergistic effect of OP-1 and IGF-I on TNAP activity in fetal rat calvarial cells (Shoba and Lee, 2003). Insulin growth factor binding protein-2 (IGFBP-2), at nearly equimolar concentration with IGF-II, plays a potentiating role of IGF-II action on rat tibial osteoblast differentiation *in vitro* (Palermo et al., 2004). Noda et al. (2005) recently found that IGF-I induces the phosphorylation of Akt in MC3T3-E1 cells but that the Akt inhibitor LY294002 significantly suppressed the IGF-I-stimulated TNAP activity suggesting that phosphatidylinositol 3-kinase/Akt plays a role in the IGF-I-stimulated TNAP activity in osteoblasts.

3.2
The TSAP Genes

Expression of the GCAP gene is a highly regulated process associated with the malignant transformation of human placental cells, since human choriocarcinoma cells (malignant trophoblasts) express primarily the *ALPP2* gene and only low or nondetectable levels of the *ALPP* gene normally found expressed in the human placenta. Butyrate, a natural product of colonic bacterial flora, has been reported to increase the activities of a number of enzymes, including APs. Some characteristics of the butyrate induction include its reversiblity and the variability of the induction period. Also, as early as 1961, it was known that corticosteroids added to the medium of HeLa cells increased AP activity (McComb et al., 1979). Like prednisolone and hyperosmolarity treatments, butyrate arrests cells in the G1 phase of the cell cycle (Xue and Rao, 1981). It is believed that the butyrate effect is exerted by a combination of inhibition of phosphorylation and hyperacetylation of histones and alterations in chromatin structure, although the two mechanisms may not necessarily be dependent on each other. The induction of AP expression by hyperosmolarity is for the most part restricted to PLAP/GCAP, while the TNAP and IAP isozymes are not affected. As Herz (1984) has pointed out, each inducer exerts an independent effect on the expression of individual isozymes and these effects can be either additive or synergistic. Administration of sodium butyrate to choriocarcinoma cells greatly increased the transcription rate of the *ALPP2* gene, resulting in an increase in GCAP mRNA expression and enzyme biosynthesis. The butyrate-modulated induction of AP is blocked by cycloheximide, suggesting that a mediator protein may be involved (Pan et al., 1991). When optimal concentrations of both sodium butyrate and dibutyryl cAMP are added simultaneously to cells, they cause a synergistic induction of activity, suggesting that these compounds use separate mechanisms to induce GCAP activity and that it is the cAMP moiety of dibutyryl cAMP that induces enzyme activity (Telfer and Green, 1993a). Both prednisolone and butyrate increased the steady-state levels of PLAP mRNA in HeLa S3 cells (Chou and Takahashi, 1987). PLAP mRNA induction by prednisolone was also shown in a uterine cervical epidermoid cancer cell line SKG-IIIa (Nozawa et al., 1989). Estradiol increased PLAP activity 2–3-fold over control in human endometrial tumor Ishikawa cells (Albert et al., 1990). The mechanism for this increase appeared to be at the level of transcription, at least in part, since there was an increase in the concentration of PLAP mRNA.

Deng et al. (1992a) examined the *ALPP2* gene promoter using transient transfection experiments. The 5'-flanking region of the gene was found to have positive regulatory elements in nucleotides –1 to –170 and –363 to –512 (relative to the start of transcription). A negative control element was also found to be present in the region between nucleotides –170 and –363. Mobility shift electrophoresis indicated that a nuclear factor bound to the promoter between bases –182 and –341. Furthermore, the activity of the *ALPP2* promoter was found to be inducible by sodium butyrate. In contrast, the closely related PLAP gene promoter (*ALPP*) exhibited almost no response to this agent (Deng et al., 1992a). In agreement with those data, Wada and Chou (1993) showed that nucleotides –156 to –1 region relative to the gene transcription start site (+1) contain cis-acting DNA elements that direct GCAP expression in choriocarcinoma cells. Three nuclear protein-binding sites, I (–63/–44), II (–87/–67) and III (–136/–103), were identified by DNase I footprinting analysis. Sites I and II contain a sequence known to bind the transcription factor AP-2; the AP-2 site in site II overlaps a consensus motif for the transcription factor Sp1. Gel retardation experiments showed that similar nuclear protein factor(s) in JEG-3 choriocarcinoma cells bind to all three sites, with highest affinity to sites I and II (Wada and Chou, 1993). Subsequently, Park et al. (1996) showed that the first 100 bp upstream of the GCAP gene, which contains sites I and II, constitutes a minimal GCAP promoter. The simultaneous presence of both sites I and II is necessary for GCAP expression and its induction by sodium butyrate. The PLAP promoter directs only a very low level of gene expression in choriocarcinoma cells and the expression does not respond to butyrate. The –100/–1 DNA regions between the GCAP and PLAP promoters differ by only eight base pairs. However, the GC-rich regions in sites I and II of the GCAP promoter are disrupted in the corresponding PLAP promoter. This disruption blocks or markedly reduces the binding of choriocarcinoma nuclear factors to the PLAP promoter, leading to a reduction in expression and a loss of butyrate response. The authors further showed that nucleotides –75 to –58 in both AP promoters, which bind a human Y-box binding protein, appear to down-regulate GCAP expression (Park et al., 1996).

IAP expression is dramatically regulated by a variety of dietary, developmental and hormonal factors, including 1,25-(OH)$_2$D3 (Strom et al., 1991) and cortisone or cortisone plus thyroxine (Yeh et al., 1994). In turn, fasting in the adult rat causes a dramatic decrease in IAP levels. These alterations in IAP levels likely have important physiological consequences in regard to the absorption of dietary fat. The 5'-untranslated regions of the two rat IAP genes differ completely in their primary sequence. Transfection of these promoter regions (1.7 and 1.2 kb, respectively for IAP I and IAP II) into a kidney cell line, COS-7, produced a different response to oleic acid that would have been expected from previous *in vivo* studies, i.e. a 3-fold increase response to oleic acid using the IAP II region and little response to IAP I (Xie et al., 2000), possibly because the 1.7 bp promoter did not contain all the necessary sequence to duplicate *in vivo* response. Kim et al. (1999) used sodium butyrate-treated HT-29 cells as an *in vitro* model system to study the molecular mechanisms underlying human IAP gene activation. Transient transfection assays using human IAP-CAT reporter genes along with DNase I footprinting were used to identify a novel, Sp1-related cis-regulatory element in the human IAP gene that appears to play a role in its transcriptional activation during differentiation *in vitro*.

Among the known transcriptional regulators of IAP expression are the thyroid hormone receptor (TR) and the gut-enriched Kruppel-like factor (KLF4) (Hinnebusch et al., 2004; Malo et al., 2004). Siddique et al. (2003) found that the TR complex and KLF4 synergistically activate the IAP gene, suggesting a previously unrecognized interrelationship between these two transcription factor pathways. These authors also examined the promoter region of *ALPI* and concluded that ALPI transactivation by KLF4 is likely mediated through a critical region located within the proximal IAP promoter region (Hinnebusch et al., 2004). Cdx1 and Cdx2 are members of the caudal-related homeobox family and also appear to play critical roles in gut differentiation, proliferation and neoplasia. Alkhoury et al. (2005) have identified a previously unrecognized interaction between two other important gut transcription factors, Cdx1 and Cdx2, in the context of IAP gene regulation. Cdx1 activates the IAP gene via a novel cis-element, whereas Cdx2 inhibits the Cdx1 effects. Transcriptional activation of the IAP gene is also associated with changes in the acetylation state of histone H3 as detected in the butyrate-treated HT-29 *in vitro* model of enterocyte differentiation (Hinnebusch et al., 2003). The authors concluded that butyrate-induced differentiation is associated with specific and localized changes in the histone acetylation state within the IAP promoter (Hinnebusch et al., 2003). Olsen et al. (2005) conducted a computer-assisted cis-element search of the proximal human *ALPI* promoter sequence. A putative recognition site for the transcription factor hepatocyte nuclear factor (HNF)-4 was predicted at the positions from −94 to −82 in relation to the translational start site. The ability of HNF-4a to stimulate the expression from the *ALPI* promoter was investigated in the nonintestinal Hela cell line. Cotransfection with an HNF-4a expression vector demonstrated direct activation of the *ALPI* promoter through this −94 to −82 element.

4
Protein Structure and Functional Domains

4.1
The Three-dimensional Structure of PLAP

4.1.1
Overview of the Structure

APs occur widely in nature and are found in many organisms from bacteria to humans (McComb et al., 1979). Irrespective of their origin, most APs are homodimeric enzymes and each catalytic site contains three metal ions, two Zn and one Mg, necessary for enzymatic activity. The enzymes catalyze the hydrolysis of monoesters of phosphoric acid and also catalyze a transphosphorylation reaction in the presence of high concentrations of phosphate acceptors. While the main features of the catalytic mechanism are conserved between bacterial and mammalian APs, mammalian APs have higher specific activity and K_m values, have a more alkaline pH optimum, display lower heat stability, are membrane-bound and are inhibited by L-amino acids and peptides through an uncompetitive mechanism. These properties, however, differ considerably among the different mammalian AP isozymes.

For many years the crystallographic coordinates of the *Escherichia coli* AP (ECAP) (Kim and Wyckoff, 1991) molecule provided the only source of structural information on APs, but now the three-dimensional structure of the first mammalian AP, i.e. human PLAP (1EW2; http://pdbbeta.rcsb.org/pdb/; Le Du et al., 2001) has been solved (Figure 5). As had been predicted from sequence comparisons the central core of PLAP, consisting of an extended β-sheet and flanking α-helices, is very similar to that of ECAP. The overall structure of PLAP is a dimer and each monomer contains 484 residues, four metal atoms, one phosphate ion and 603 water molecules. The two monomers are related by a 2-fold crystallographic axis. The surface of PLAP is poorly conserved with that of ECAP with only 8% residues in common. PLAP possesses additional secondary structure elements, comprising an N-terminal α-helix (residues 9–25), an α-helix and a β-strand in a highly divergent region (residues 208–280) and a different organization of the small β-sheet in domain 365–430. In the active site, only the residues that are essential for catalysis are preserved, i.e. the catalytic Ser, the three metal ion sites, M1 (occupied by Zn^{2+}, also called Zn1), M2 (occupied by Zn^{2+}; also called

Mammalian Alkaline Phosphatases: From Biology to Applications in Medicine and Biotechnology. J. L. Millán
Copyright © 2006 WILEY-VCH Verlag GmbH & Co. KGaA, Weinheim
ISBN: 3-527-31079-7

Zn2) and M3 (occupied by Mg^{2+}) in addition to their ligands, while most of the surrounding residues are different. Half of the enzyme surface corresponds to three clearly identifiable regions whose sequences vary widely among human APs and are lacking in nonmammalian enzymes, i.e. the long N-terminal a-helix, an interfacial flexible loop known as the "crown domain" and a fourth metal-binding domain (M4). The availability of the PLAP structure facilitated modeling the human GCAP, IAP and TNAP isozymes, revealing that all the novel features discovered in PLAP are conserved in those human isozymes also (Le Du and Millán, 2002). The model of GCAP, IAP and TNAP, listed in order of structural similarity, was built from the PLAP structure using the sequence alignment shown in Figure 6. The PLAP and GCAP molecules display 98% identity, with no insertion or deletion relative to PLAP. The IAP and PLAP molecules show 87% identity and 91% homology, with no insertion or deletion relative to PLAP. The TNAP and PLAP molecules, however, display 57% identity and 74% homology and TNAP has four insertions of one residue, one insertion of three residues and one deletion of two residues relative to PLAP. As expected, the overall structure of each model is very close to that of PLAP. We will now discuss each of the structural elements of PLAP and compare them with the corresponding structures of ECAP where appropriate or of other mammalian APs and discuss mutagenesis and kinetic data that support structure/function assignments of individual residues in the structure.

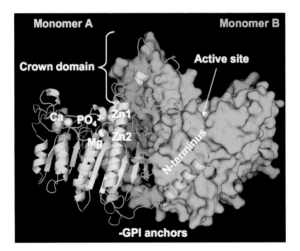

Fig. 5 Three-dimensional structure of PLAP. Overview of the structure of human PLAP from the coordinates determined by Le Du et al. (2001). Monomer A is shown in ribbon representation and in cyan, while monomer B is shown in surface representation in yellow. Indicated are the active site metals, Zn1, Zn2 and Mg, the novel fourth metal site occupied by Ca, the crown domain and the N-terminal arm. The relative location of the GPI anchor on the processed enzyme is also indicated.

```
PLAP:   1  IIPVEEENPDFWNREAAEALGAAKKLQPAQT-AAKNLIIFLGDGMGVSTVTAARILKGQK   59
GCAP:   1  IIPVEEENPDFWNRQAAEALGAAKKLQPAQT-AAKNLIIFLGDGMGVSTVTAARILKGQK   59
IAP :   1  VIPAEEENPAFWNRQAAEALDAAKKLQPIQK-VAKNLILFLGDGLGVPTVTATRILKGQK   59
TNAP:   1  LVPEKEKDPKYWRDQAQETLKYALELQKLNTNVAKNVIMFLGDGMGVSTVTAARILKGQL   60
              <--------------->           <*****>     <------------>

PLAP:  60  KDKLGPEIPLAMDRFPYVALSKTYNVDKHVPDSGATATAYLCGVKGNFQTIGLSAAARFNQ 120
GCAP:  60  KDKLGPETFLAMDRFPYVALSKTYSVDKHVPDSGATATAYLCGVKGNFQTIGLSAAARFNQ 120
IAP :  60  NGKLGPETPLAMDRFPYLALSKTYNVDRQVPDSAATATAYLCGVKANFQTIGLSAAARFNQ 120
TNAP:  61  HHNPGEETRLEMDKFPFVALSKTYNTKAQVPDSAGTATAYLCGVKANEGTVGVSAATERSR 121
                                <****>     <-------->

PLAP: 121  CNTTRGNEVISVMNRAKKAGKSVGVVTTTRVQHASPAGTYAHTVNRNWYSDADVPASARQE 181
GCAP: 121  CNTTRGNEVISVVNRAKKAGKSVGVVTTTRVQHASPAGTYAHTVNRNWYSDADVPASARQE 181
IAP : 121  CNTTRGNEVISVMNRAKQAGKSVGVVTTTRVQHASPAGTYAHTVNRNWYSDADMPASARQE 181
TNAP: 122  CNTTQGNEVTSILRWAKDAGKSVGIVTTTRVNHATPSAAYAHSADRDWYSDNEMPPEALSQ 182
              <----->    <*******>

PLAP: 182  GCQDIATQLISNM-DIDVILGGGRKYMFRMGTPDPEYPDDYSQGGTRLDGKNLVQEWLA-- 239
GCAP: 182  GCQDIATQLISNM-DIDVILGGGRKYMFPMGTPDPEYPDDYSQGGTRLDGKNLVQEWLA-- 239
IAP : 182  GCQDIATQLISNM-DIDVILGGGRKYMFPMGTPDPEYPADASQNGIRLDGKNLVQEWLA-- 239
TNAP: 183  GCKDIAYQLMHNIRDIDVIMGGGRKYMYPKNKTDVEYESDEKARGTRLDGLDLVDTWKSFK 243
              <----->    <****>                              <----->

PLAP: 240  -KRQGARYVWNRTELMQASLDP-SVTHLMGLFEPGDMKYEIHRDSTLDPSLMEMTEAALRL 298
GCAP: 240  -KHQGARYVWNRTELLQASLDP-SVTHLMGLFEPGDMKYEIHRDSTLDPSLMEMTEAALLL 298
IAP : 240  -KHQGAWYVWNRTELMQASLDQ-SVTHLMGLFEPGDTKYEIHRDPTLDPSLMEMTEAALRL 298
TNAP: 244  PRHKHSHFIWNRTELL--TLDPHNVDYLLGLFEPGDMQYELNRNNVTDPSLSEMVVVAIQI 302
              <**> <------->       <***>                <---------->

PLAP: 299  LSRNPRGFFLFVEGGRIDHGHHESRAYRALTETIMFDDAIERAGQLTSEEDTLSLVTADHS 359
GCAP: 299  LSRNPRGFFLFVEGGRIDHGHHESRAYRALTETIMFDDAIERAGQLTSEEDTLSLVTADHS 359
IAP : 299  LSRNPRGFYLFVEGGRIDHGHHEGVAYQALTEAVMFDDAIERAGQLTSEEDTLTLVTADHS 359
TNAP: 303  LRKNPKGFFLLVEGGRIDHGHHEGKAKQALHEAVEMDRAIGQAGSLTSSEDTLTVVTADHS 363
             >   <*****> <-----> <-------------------->  <*****>

PLAP: 360  HVFSFGGYPLRGSSIFGLAPGKA-RDRKAYTVLLYGNGPGYVLKDGARPDVTESESGSPEY 419
GCAP: 360  HVFSFGGYPLRGSSIFGLAPGKA-RDRKAYTVLLYGNGPGYVLKDGARPDVTESESGSPEY 419
IAP : 360  HVFSFGGYTLRGSSIFGLAPSKA-QDSKAYTSILYGNGPGYVFNSGVRPDVNESESGSPDY 419
TNAP: 364  HVFTFGGYTPRGNSIFGLAPMLSDTDKKPFTAILYGNGPGYKVVGGERENVSMVDYAHNNY 424
              <*>                          <***>              <--->

PLAP: 420  RQQSAVPLDEETHAGEDVAVFARGPQAHLVHGVQEQTFIAHVMAFAACLEPYTA-CDLAPP 479
GCAP: 420  RQQSAVPLDGETHAGEDVAVFARGPQAHLVHGVQEQTFIAHVMAFAACLEPYTA-CDLAPP 479
IAP : 420  QQQAAVPLSSETHGGEDVAVFARGPQAHLVHGVQEQSFVAHVMAFAACLEPYTA-CDLAPP 479
TNAP: 425  QAQSAVPLRHETHGGEDVAVFSKGPMAHLLHGVHEQNYVPHVMAYAACIGANLGHCAPASS 484
              <*>      <****>            <*>       <-------->
```

Fig. 6 Sequence alignment of human PLAP, GCAP, IAP and TNAP. The alignment shows the secondary structures of PLAP (<– – –>, α-helix; <***>, β-sheet). Residues highlighted in yellow are buried by more than 10 Å upon dimerization; residues in blue have an accessibility between 10 and 100 Å^2; residues in red have an accessibility higher than 100 Å^2; and residues underlined are located in the 12 Å sphere around the phosphate group in the active site. Taken from Le Du and Millán (2002) and reproduced with permission from the *Journal of Biological Chemistry*.

4.1.2
The Active Site

Only some of the active site residues are strictly conserved on comparing ECAP and mammalian and other APs. The active site Ser is conserved in all species where an AP has been sequenced to date. Ligands to the crucial active site metal ions also have conserved functions, as will be described below. The catalytic mechanism has been elucidated for ECAP through the analysis of numerous crystal structures. The most recent revised mechanism is that of Stec et al. (2000) that builds upon the original mechanism proposed by Kim and Wyckoff (1991) but takes into account the participation of not only the Zn1 and Zn2 active site metals but also of the Mg ion in the reaction. According to Stec et al., "In the free enzyme (E, Fig. 7a, top left), three water molecules fill the active site and the Ser102 hydroxyl group participates in a hydrogen bond with the Mg-coordinated hydroxide ion. Upon binding of the phosphomonoester (ROP) to form the Michaelis enzyme–substrate complex (E•ROP, Fig. 7b, top right), the Ser102 O^γ becomes fully deprotonated for nucleophilic attack with the concomitant transfer of the proton to the Mg-coordinated hydroxide group to form a Mg-coordinated water molecule. Coordination of Zn2 stabilizes Ser102 O^γ in its nucleophilic state. In the first in-line displacement, the activated hydroxyl group of Ser102 attacks the phosphorus center of the substrate in the enzyme–substrate complex (E•ROP) to form a covalent serine–phosphate intermediate (E-P, Figure 7d, bottom right). Zn1 participates in this step by coordinating the bridging oxygen atom of the substrate and facilitating the departure of the alcohol leaving group (RO⁻). In the second in-line displacement step, a nucleophilic hydroxide ion coordinated to Zn1 attacks the phosphorus atom, hydrolyzing the covalent serine–phosphate intermediate to form the noncovalent enzyme-phosphate product complex (E•P_i, Fig. 7c, bottom left) and regenerate the nucleophilic Ser102. Zn1 lowers the pK_a of the coordinated water molecule to form effectively the nucleophilic hydroxide ion, while the Mg-coordinated water molecule acts as a general acid to reprotonate O^γ of Ser102. Protonation of Ser102 may facilitate departure of the phosphate product from the noncovalent E•P_i complex. Alternatively, the Mg-coordinated water molecule may directly protonate the phosphate group for its release. The release of phosphate from the E•P_i complex to give the free enzyme (E, Fig. 7, top left) may be facilitated also by the increased mobility of the Arg166 side-chain ... For the first time, this study offers evidence for a direct role of the Mg ion, the third metal ion of the catalytic triad, in the mechanism. The Mg occupancy dependence of the M3 site on the Ser102 conformation strongly implicates a well-positioned, Mg-bound water molecule as the general base in the generation of the Ser102 nucleophile and as a general acid in the regeneration of the Ser102 hydroxyl group. The octahedral geometry established by an Mg ion in the M3 site is necessary to position the water molecule for its function".

Structure–function studies comparing PLAP and the ECAP structure have found a conserved function for those residues that stabilize the active site Zn1, Zn2 and Mg metal ions (Kozlenkov et al., 2002). Figure 8, shows the residues ser-

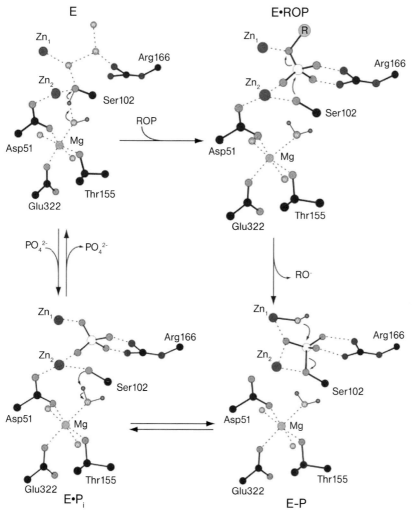

Fig. 7 The catalytic mechanism of APs elucidated using the *Escherichia coli* AP structure (Stec et al., 2000). Hydrogen atoms have been included only at catalytically relevant sites. In the free enzyme (E), the phosphate-binding site is filled with three water molecules. The Ser102 hydroxyl group participates in a hydrogen bond with a Mg-coordinated hydroxide ion. Formation of the enzyme–substrate complex (E·ROP) involves coordination of the ester oxygen atom to Zn1 and additional interactions between the nonbridging oxygen atoms of the substrate with Zn2 and the guanidinium group of Arg166. Ser102 occupies the position opposite the leaving group. Upon phosphomonoester binding, the Mg-coordinated hydroxide ion acting as the general base deprotonates Ser102 O^γ for nucleophilic attack on the phosphorus atom. The formation of the covalent enzyme–phosphate intermediate (E-P) results in inversion of the phosphorus center and the loss of the leaving group (RO$^-$). A nucleophilic hydroxide ion coordinated to Zn1 attacks the covalent E-P intermediate, forming the noncovalent enzyme–phosphate complex (E·P$_i$) and causing a second inversion of configuration at the phosphorus center. The water molecule coordinated to Mg now acts as a general acid, donating a proton to O^γ of Ser102 or, alternatively, inorganic phosphate. Most of the hydrogen atoms and ligands to Zn1 and Zn2 are not shown. The figure was kindly provided by Professor Evan R. Kantrowitz and is reproduced here with permission from the *Journal of Molecular Biology*.

ving as ligands to the three catalytically important metal ions, Zn1, Zn2 and Mg, i.e. Asp42, His153, Ser155, Glu311, Asp316, His320, Asp357, His358, His360 and His432 in mammalian APs. In the PLAP numbering the active site residue is Ser92 (equivalent to Ser102 in ECAP). Residues serving as ligands to catalytically important metal ions, i.e. Asp42, His153, Ser155, Glu311, Asp316, His320, Asp357, His358, His360 and His432 were mutagenized into Ala. Two residues in the vicinity of the Mg-binding site in PLAP that are not conserved in ECAP, i.e. His153 and His317, were mutagenized into Ala or to Asp and Lys, respectively. Some active site mutations were studied both in the context of wild-type PLAP, i.e. [E429]PLAP, and in [G429]PLAP, since this substitution has been shown to have profound influences on the behavior of the enzyme by conferring GCAP characteristics (Hummer and Millán, 1991) (as will be discussed in Sections 4.1.6–4.1.8, 4.2 and 5.1.2.1). Figure 9 shows a detailed comparison of the structure of the active site region of PLAP and ECAP that will help the reader follow the discussion of changes induced in the active site by mutagenizing these residues. Table 2 presents the kinetic data of all the PLAP mutants discussed in this section.

	Zn2/Mg	PO$_4^{3-}$				Mg				Mg	Zn1			Zn1	Zn2					Zn1		
	42	91	92	101	121	153	155	166	183	311	316	317	319	320	357	358	360	367	429	432	467	474
E. Coli	D	D	S	T	H	D	T	R	G	E	D	K	D	H	D	H	H	P	S	H	-	-
chTNAP	D	D	S	C	C	H	T	R	C	E	D	H	H	H	D	H	H	Y	Q	H	C	C
cTNAP	D	D	S	C	C	H	T	R	C	E	D	H	H	H	D	H	H	Y	H	H	C	C
bTNAP	D	D	S	C	C	H	T	R	C	E	D	H	H	H	D	H	H	Y	H	H	C	C
rTNAP	D	D	S	C	C	H	T	R	C	E	D	H	H	H	D	H	H	Y	H	H	C	C
mTNAP	D	D	S	C	C	H	T	R	C	E	D	H	H	H	D	H	H	Y	H	H	C	C
hTNAP	D	D	S	C	C	H	T	R	C	E	D	H	H	H	D	H	H	Y	H	H	C	C
bIAP I	D	D	S	C	C	H	S	R	C	E	D	H	H	H	D	H	H	Y	S	H	C	C
bIAP II	D	D	S	C	C	H	S	R	C	E	D	H	H	H	D	H	H	Y	S	H	C	C
bIAP III	D	D	S	C	C	H	S	R	C	E	D	H	H	H	D	H	H	Y	S	H	C	C
bIAP IV	D	D	S	C	C	H	S	R	C	E	D	H	H	H	D	H	H	Y	S	H	C	C
rIAP I	D	D	S	C	C	H	S	R	C	E	D	Q	H	H	D	H	H	Y	S	H	C	C
rIAP II	D	D	S	C	C	H	S?	C	E	D	R	H	H	D	H	H	Y	S	T	C	C	
mIAP	D	D	S	C	C	H	S	R	C	E	D	R	H	H	D	H	H	Y	S	H	C	C
mEAP	D	D	S	C	C	H	S	R	C	E	D	H	H	H	D	H	H	Y	S	H	C	C
hIAP	D	D	S	C	C	H	S	R	C	E	D	H	H	H	D	H	H	Y	S	H	C	C
hGCAP	D	D	S	C	C	H	S	R	C	E	D	H	H	H	D	H	H	Y	G	H	C	C
hPLAP	D	D	S	C	C	H	S	R	C	E	D	H	H	H	D	H	H	Y	E	H	C	C
Mutated to:	A	A/N	G	S	S	A	A/D/T	S	A	A	A	A/K		A	A	A	A		G/F	A	S	S

Fig. 8 Comparison across species of critical functional residues. The sequences compared include all mammalian APs known to date and also the chicken TNAP sequence in comparison with the E. coli AP sequence. They are: E. coli AP (ECAP); chicken TNAP (chTNAP); cat TNAP (cTNAP); bovine TNAP (bTNAP); rat TNAP), mouse TNAP (mTNAP); human TNAP (hTNAP); bovine IAP I isozyme (bIAP I); bovine IAP II, III and IV (bIAP II, bIAP III and bIAP IV); rat IAP I and II (rIAP I and rIAP II); mouse IAP and EAP (mIAP and mEAP); human IAP (hIAP); human GCAP (hGCAP); and human PLAP (hPLAP). A colored box over the residue number indicates that it is a ligand to the active site Mg (yellow), Zn1 (green), Zn2 (purple) or both Zn2 and Mg (blue). A red box marks the active site Ser92 that covalently binds phosphate during catalysis. The disulfide bonds between C121–C183 and C467–C474 are shown by a thin line.

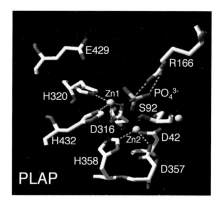
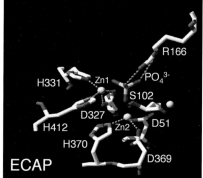
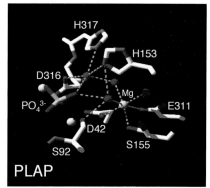
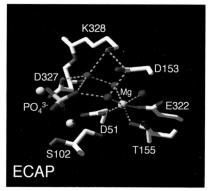

Fig. 9 Comparison of the residues coordinating to the active site metals in PLAP and ECAP. The upper panels focus on the environment of the Zn1 and Zn2 metal sites and their ligands and the lower panels display the environment of the Mg metal site and its ligands. Water molecules are shown as red spheres. Green dotted lines denote metal–ligand interactions and hydrogen bonds. Taken from Kozlenkov et al. (2002) and reproduced with permission from the *Journal of Biological Chemistry*.

When substituting the Zn1 ligands in PLAP, i.e. Asp316, His320 and His432, two of the three mutants, i.e. [A316]PLAP and [A432]PLAP, retained significant activity. The k_{cat} and K_m of [A316]PLAP showed a 2.8- and a 2.25-fold decrease, respectively, relative to wild-type (wt) PLAP. Thus, the catalytic efficiency (k_{cat}/K_m) of the [A316]PLAP mutant remains comparable to that of wt PLAP. The k_{cat} of [A432]PLAP was also reduced 2.7-fold but its K_m increased 3.7-fold for a resulting 5.8-fold reduction in catalytic efficiency (Table 2). In contrast, the introduction of the H320A mutation reduced the specific activity of PLAP by more than 200-fold. Saturation of each of the mutants with concentrations of Zn^{2+} up to 10 mM did not result in any increase in activity. It should be noted that analogous mutations in ECAP were reported to have very different consequences. Notably, the D327A substitution in ECAP (analogous to Asp316 in PLAP) resulted in a 3000-fold decrease in k_{cat} and a 2000-fold increase in K_m for a 10^7-fold decrease in catalytic efficiency that was not reversible by the addition of Zn^{2+} (Xu and Kantrowitz,

Table 2 Kinetic parameters of PLAP mutants. Reproduced from Kozlenkov et al. (2002) with permission from the *Journal of Biological Chemistry*.

PLAP mutant	k_{cat} (s^{-1})	K_m (mM)	k_{cat}/K_m (s^{-1}mM^{-1})
wt PLAP	460 ± 11	0.36 ± 0.03	1288
[G429]PLAP	344 ± 14	0.10 ± 0.005	3400
GCAP	277 ± 16	0.11 ± 0.005	2518
Conserved active-site residues			
[A42]PLAP	6.4 ± 0.8	0.68 ± 0.04	9
[A91]PLAP	39 ± 4	1.1 ± 0.07	35
[N91]PLAP	30 ± 15	0.62 ± 0.09	48
[G92]PLAP	<1.0	n.d.	n.d.
[A92]PLAP	<1.0	n.d.	n.d.
[A155]PLAP	5.1 ± 0.5	0.38 ± 0.02	14
[T155]PLAP	529 ± 37	0.18 ± 0.01	2813
[A166]PLAP	20 ± 9	4.4 ± 0.41	4.5
[A311]PLAP	2.8 ± 0.4	1.26 ± 0.08	2
[A316]PLAP	193 ± 6	0.16 ± 0.01	1073
[A320]PLAP	1.8 ± 0.9	n.d.	n.d.
[A357]PLAP	18 ± 3.7	0.21 ± 0.02	78
[A358]PLAP	< 1.0	n.d.	n.d.
[A432]PLAP	170 ± 6	1.36 ± 0.18	221
Nonconserved active-site residues			
[D153]PLAP	313 ± 14	0.71 ± 0.03	442
[A153]PLAP	989 ± 53	1.22 ± 0.08	825
[A153, G429]PLAP	546 ± 35	0.31 ± 0.03	1761
[K317]PLAP	906 ± 64	0.80 ± 0.03	1097
[A317]PLAP	999 ± 31	1.13 ± 0.19	884
[A317, G429]PLAP	797 ± 38	0.25 ± 0.02	3188
[D153, K317]PLAP	400 ± 18	1.2 ± 0.1	340
Active-site neighboring residues			
[A319]PLAP	6.5 ± 1.5	n.d.	n.d.
[A319, G429]PLAP	12.8 ± 2.2	n.d.	n.d.

Table 2 Continued.

PLAP mutant	k_{cat} (s^{-1})	K_m (mM)	k_{cat}/K_m (s^{-1}mM^{-1})
[A360]PLAP	552 ± 23	1.4 ± 0.1	608
[A367]PLAP	195 ± 11	0.35 ± 0.02	557
[F367]PLAP	178 ± 14	0.27 ± 0.02	659
[A367, G429]PLAP	202 ± 6	0.22 ± 0.01	918
[F367, G429]PLAP	200 ± 11	0.17 ± 0.01	1176
Cysteine residues			
[S101]PLAP	489 ± 13	0.41 ± 0.05	1193
[S467]PLAP	206 ± 15	0.36 ± 0.05	572
[S474]PLAP	221 ± 13	0.33 ± 0.03	670
[S467, S474]PLAP	244 ± 11	0.40 ± 0.04	610
[S121]PLAP & [S183]PLAP	n.d.	n.d.	n.d.

1992). In contrast, the activity of the H412A mutant in ECAP (analogous to the H432A in PLAP) was responsive to 0.2 mM Zn^{2+}, reaching values of k_{cat} and K_m only 2-fold lower than those of wt ECAP (Ma and Kantrowitz, 1994). The H331A mutation (analogous to H320A in PLAP) has not been studied in ECAP. These results indicate that there are significant differences in the environment of Zn1 in the PLAP structure compared with the ECAP structure and that substitutions of the PLAP Zn1 ligands are better tolerated than in ECAP. This may reflect the fact that the top flexible loop or crown domain, which harbors Glu429 in PLAP, appears to provide additional stabilization to the active site environment, so that Zn^{2+} cannot easily diffuse in or out of the PLAP molecule as has been noted previously (Hoylaerts et al., 1997). Therefore, even though the state of coordination of Zn1 is affected by the mutations, the Zn^{2+} ion remains in place and is able to function in catalysis.

Alanine substitutions of the Zn2 ligands in PLAP, i.e. Asp42, Asp357 and His358, resulted in significant decreases in specific activity, ranging from more than 25-fold (D357A) to undetectable levels (H358A). None of these values changed in response to the addition of Zn^{2+}. Whereas the K_m of [A42]PLAP nearly doubled, the K_m of [A357]PLAP slightly decreased. Thus, the catalytic efficiencies of [A42]PLAP and [A357]PLAP were reduced by 130- and 16-fold, respectively. Although no studies have been performed in ECAP on residues analogous to D357 and H358, the [A51]ECAP mutant, analogous to [A42]PLAP, was shown to be more than 800-fold less active than the wt ECAP (Tibbitts et al., 1996). Ala42 is a bidentate ligand, coordinating not only to Zn2 but also to Mg. Alanine substitution of the other two Mg ligands, i.e. Ser155 and Glu311, reduced the specific activity of PLAP approximately 100- and 200-fold, respectively. The K_m for [A155]PLAP did not change, but it increased about 4-fold for [A311]PLAP. A similar pattern was seen for the corresponding E322A mutation in ECAP (Xu and Kantrowitz, 1993). Interestingly, the S155T substitution hardly affects the activity of the resulting mutant and even doubles its catalytic efficiency (Table 2).

Whereas most of the AP active site residues are perfectly conserved throughout evolution, some important differences exist in the neighborhood of the Mg ion (Fig. 9) in PLAP and other mammalian APs. His153 and His317 in PLAP are homologous to Asp153 and Lys328, respectively, in ECAP. The substitution of D153H and K328H in ECAP produced enzymes with kinetic properties similar to those of mammalian APs. For example, the D153H/K328H double-ECAP mutant displayed a 5.6-fold higher k_{cat} and a 30-fold higher K_m, a decrease in heat stability and a shift in pH optimum to alkaline pH values (Murphy et al., 1995). Kozlenkov et al. (2002) constructed the reciprocal mutations, i.e. H153D and H317K, in PLAP and also the double mutation (H153D/H317K). The authors also introduced a H153A and H317A mutation both in PLAP and in [G429]PLAP. The expectation was that by reverting to the residues found in ECAP, one would confer ECAP-like properties to PLAP. Surprisingly, however, no decrease in K_m was observed in any of the mutants. Instead, the K_ms consistently increased for all the mutants. The effects on k_{cat} were variable. No significant change in the pH dependence or heat stability of the mutants was observed with reference to wt PLAP. In the ECAP active site, the environment of the Mg ion is one of octahedral coordination, including three amino acids and three water molecules (Kim and Wyckoff, 1991). These water molecules are further coordinated and stabilized by other amino acid residues including Asp153. Hung and Chang (2001a) found that in PLAP, the Mg site also has an octahedral configuration and that the binding at this M3 site is a slow binding process with a low binding affinity (K_{app} = 3.32 mM). They also reported that the Zn^{2+} ion has a high affinity for this site (K_{app} = 0.11 mM) and that Zn^{2+} can act as a time-dependent inhibitor of the enzyme. Their structural data revealed that the M3 site is converted to a distorted tetrahedral coordination when Zn^{2+} ion substitutes for Mg^{2+} ion at the M3 site. In another study, Tian et al. (2003) reported on the refolding kinetics of calf IAP using 3.0 M guanidine hydrochloride. They concluded that for this isozyme, Mg^{2+} was a more efficient inducer of reconstitution of the active site than Zn^{2+}.

In ECAP, the D153H mutation was shown to destabilize the octahedral Mg coordination in favor of a tetrahedral one and resulted in an enzyme that had reduced Mg^{2+} affinity, increased Zn^{2+} affinity and was activated significantly by Mg^{2+} (Murphy et al., 1995). This is strongly reminiscent of the behavior of the IAP isozyme (McComb et al., 1979) but not of human PLAP, which binds Mg^{2+} tightly and where the further addition of Mg^{2+} does not increase activity. A possible explanation for these data comes from the analysis of the structure of PLAP around the Mg^{2+} ion (Fig. 9). In PLAP, His153 and His317 are positioned so that they can serve the same purpose as the corresponding Asp and Lys in ECAP, i.e. they are direct ligands to active site water molecules and indirect ligands to the Mg ion and the noncovalently bound phosphate group. Kozlenkov et al. (2002) proposed that His153 and His317 in PLAP are already well positioned to stabilize these water-mediated interactions and that introducing different residues at these positions would result in a decrease in affinity for phosphate, thus increasing K_m and k_{cat} rather than decreasing these parameters. Because the IAP isozyme is more dependent on Mg^{2+} activation (McComb et al., 1979), one could speculate that the structure of IAP in the immediate environment of the Mg ion would be more similar to and could be better modeled by the structure of ECAP than by the structure of PLAP. Upon mutagenizing H153A or H317A in PLAP, both mutant enzymes displayed about a 2-fold increase in k_{cat} and about a 3-fold increase in K_m compared with wt PLAP. This can be explained by the disruption of their water-mediated interactions with the phosphate group (Fig. 9) via the same water molecule. Disruption may lead to an enzyme with lower affinity for both substrate and product, thus having higher k_{cat} and K_m. Interestingly, by combining the H153A and H317A mutations with the Gly429 mutation, it was possible to engineer mutant enzymes with increased catalytic efficiency compared with wt PLAP but not compared with [G429]PLAP. The [A153, G429]PLAP and [A317, E429G]PLAP mutants had restored K_m values to those of wt PLAP while largely preserving the increase in k_{cat}. This is especially true of the [A317, G429]PLAP, in which both specific activity and catalytic efficiency (k_{cat} and k_{cat}/K_m values) were increased 2-fold compared with wt PLAP and 5-fold compared with [G429]PLAP (Kozlenkov et al., 2002).

It is clear that preserving the Zn and Mg ion environments is essential to maintaining the integrity of the catalytic activity of the enzymes. However, work by Bortolato et al. (1999) on the bIAP isozyme has indicated that these metal sites may also be responsible for maintaining the quaternary structure of mammalian APs. They showed that demetalated apobIAP, prepared using ion-chelating agents, exhibited a dramatic decrease in hydrolase activity, concomitant with conformational changes in its quaternary structure. By rate-zonal centrifugation and electrophoresis, they demonstrated that the loss of divalent ions leads to a monomerization process for the metal-depleted bIAP. For wt bIAP, three steps of temperature-induced changes were exhibited, whereas for apobIAP, only one step was exhibited at 55 °C. Their work on bIAP showed two main differences from ECAP, i.e. the loss of the divalent ions induces protein monomerization and the total recovery of enzyme activity by divalent ion addition to apobIAP is obtained. In the case of ECAP, monomers can be obtained by heat and pH denaturation and reactivation is possible but this involves the dimerization process in which Zn^{2+} ions are involved (Harris and Coleman, 1968). Yan et al. (2003) studied the effect of the addition of exogenous Zn^{2+} on calf IAP. Under conditions of slow binding of Zn^{2+} to calf IAP, increasing Zn^{2+} first inhibited enzymatic activity and further increases in Zn^{2+} resulted in an increase in activity. For quick reversible binding of Zn^{2+}, the effect on calf IAP activity changed at lower concentrations of substrate, indicating a complex cooperativity between Zn^{2+} and substrate. Tian et al. (2003) also studied the reconstitution of calf IAP by first denaturing it in 3.0 M guanidine hydrochloride and then diluting it 20-fold in a buffer solution containing Mg^{2+}, Zn^{2+} and nucleotide phosphate. The authors concluded that for calf IAP, Mg^{2+} is a more efficient inducer of reconstitution of the active site than Zn^{2+}.

Two other active site residues, His360 and His319, perfectly conserved in mammalian APs, were also investigated by Kozlenkov et al. (2002). In ECAP, His372 is 3.8 Å away from Zn1 in the active site and correspondingly His360 is located within 4.4 Å from Zn1 ion in PLAP. The side-chain of His360 forms a hydrogen bond to the side-chain of Asp316, which is a direct ligand to Zn1. The [A372]ECAP showed changes in catalytic behavior, such as a 20% reduction in k_{cat} and a 4-fold reduction in K_m in the presence of a phosphate acceptor (Xu et al., 1994). In contrast, [A360]PLAP displayed an increase in specific activity 20% over that of wt PLAP and a 3.8-fold increase in K_m. Judging from its three-dimensional positioning, residue His319 should mainly have a structural role, being involved in hydrogen bonds with Thr48 and Tyr393. The H319A and H319A/E429G mutations, however, displayed 70- and 37-fold decreases in specific activity, respectively, indicating that the interactions formed by His319 are important for maintaining an optimal conformation in the active site.

4.1.3
The Calcium Site

An additional noncatalytic metal-binding site M4 that appears to be occupied by calcium and that is not present in ECAP was revealed upon solving the PLAP 3D structure (Le Du et al., 2001; Mornet et al., 2001). This fourth metal site is conserved in all human and mouse APs and presumably represents a novel feature common to all mammalian APs. However, the structural and functional significance of this new metal site remains to be established. The overall binding domain of this M4 site comprises 76 residues (209–285) and is folded into two β-strands flanked by two α-helices. It includes the glycosylation site at Asn249, stabilized by a stacking interaction with Trp248 and a metal ion (Fig. 10). In the original structural data set, the peak for this metal corresponds to about 10 electrons. Its height and the crystallization conditions suggested the presence of an Mg^{2+}, but its coordination by carboxylates from Glu216, Glu270 and Asp285, by the carbonyl of Phe269 and a water molecule, appeared to suggest that the site is occupied by a Ca^{2+} ion. Although the ligands to this site have not yet been mutagenized experimentally, there are a number of hypophosphatasia mutations affecting this site in TNAP (Mornet et al., 2001; Section 7.1) that give a clear indication as to the significance of the integrity of this site. The structure of this M4 site appears to be conserved in all mammalian APs. The realization of the existence of this novel M4 calcium-containing site amply confirms early studies indicating that in cartilage, TNAP is a Ca^{2+}-binding glycoprotein (De Bernard et al., 1986).

4.1 The Three-dimensional Structure of PLAP | 37

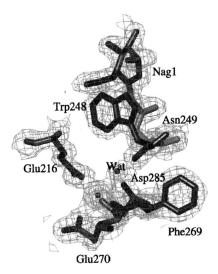
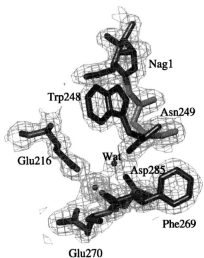

Fig. 10 Stereo view of the electron density for the M4 metal ion. The carbohydrate connected to Asn249 is in stacking interaction with Tr248 and the M4 metal is in coordination with Glu216, Phe269–CO, Glu270–Oε2, Asp285 and one water molecule. In each case, the $2F_o$–F_c map is shown in blue and is contoured at the 1.2σ level. The F_o–F_c map is shown in green and contoured at the 10σ level. Taken from Le Du et al. (2001) and reproduced with permission from the authors and from the *Journal of Biological Chemistry*.

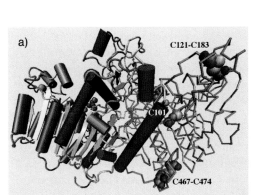
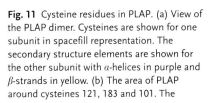
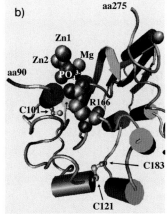

Fig. 11 Cysteine residues in PLAP. (a) View of the PLAP dimer. Cysteines are shown for one subunit in spacefill representation. The secondary structure elements are shown for the other subunit with α-helices in purple and β-strands in yellow. (b) The area of PLAP around cysteines 121, 183 and 101. The catalytically important Arg166 and phosphate group in the active site of the enzyme are also shown. The polypeptide segment between Arg166 and Cys183 is shown in green. Taken from Kozlenkov et al. (2002) and reproduced with permission from the *Journal of Biological Chemistry*.

4.1.4
The Disulfide Bonds

As was shown in Fig. 8, all mammalian APs have five cysteine residues (Cys101, Cys121, Cys183, Cys467 and Cys474 in PLAP) per subunit, not homologous to any of the four cysteines in ECAP. They form two disulfide bonds, Cys121–Cys183 and Cys467–Cys474 (Fig. 11a), whereas the Cys101 residue remains in free form. The sequence contained within the Cys121–Cys183 disulfide bond (Fig. 11b) harbors three important elements of secondary structure: (a) a stretch of the central β-strand; (b) the Arg166 residue known to be crucial for catalysis (Hoylaerts et al., 1992a) and (c) the ligands stabilizing the M4 metal ion. Kozlenkov et al. (2002) found that the cysteine PLAP mutants C101S, C467S, C474S and C467S/C474S were appropriately expressed by transfected COS-1 cells and retained residual activity. [S467]PLAP, [S474]PLAP and [S467, S474]PLAP enzymes all displayed k_{cat} values 2-fold lower than the wt PLAP (Table 2). In contrast, disrupting the disulfide bond between the Cys121 and Cys183 completely prevented the formation of the active enzyme. The substitution of the free Cys101 did not significantly affect the properties of the enzyme as shown in Table 2. The authors examined the possibility that Cys101 would be available for covalent modification by a fluorescent probe. Only when wt PLAP was incubated with guanidinium chloride did the Cys101 side-chain become available for covalent modification. This confirms that Cys101 is not likely to play a role in regulating activity or stability of the enzyme (Kozlenkov et al., 2002).

4.1.5
The N-terminal Arm

The N-terminus of PLAP has an additional α-helix not present in the ECAP structure and its position is removed from the rest of the monomer, but it interacts with the second monomer with a buried surface area of 555 Å^2, suggesting an involvement in enzyme dimerization (as shown on Fig. 5). Residues of the N-terminal extremity interact with Arg370 from loop 366–375 and with Asn106 and Arg117 from the second monomer. Figure 12a identifies two residues within the first nine amino acid sequence, capable of structural stabilization through side-chain interactions, i.e. Glu6 (of the A subunit) interacts with Arg370 (of the A subunit) and Glu7 (A subunit) interacts with Arg117 (B subunit). The subsequent 16 amino acid residues form an α-helix and are capable of interacting with monomer B at several points of contact, as shown in detail in Fig. 12b and c, that show a left view and a right view of the α-helix, respectively. These figures reveal potential interactions between Pro9 and Val129 (B, 3.94 Å), Trp12 and Lys104 (B, 3.73 Å), Asn13 ands His460 (B, 3.2 Å), Ala19 and Phe457 (B, 3.79 Å), Leu20 and Tyr471 (B, 3.86 Å), Ala23 and His450 (B, 3.46 Å) and Lys24 and Phe467 (B, 3.9 Å). Recently, Hoylaerts et al. (2005) have shown that the integrity of the N-terminal arm and in particular of the α-helix comprising residues 10–25 represents a structural requirement for the active site to execute the intramolecular transition required during enzyme catalysis in both PLAP and TNAP. Deletion and mutagenesis analysis revealed that the progressive deletion of the first nine N-terminal amino acid residues in PLAP caused some fluctuations in the observed catalytic rate constants, measured at pH 9.8, but resulted in a considerable decrease in k_{cat} for [Δ1–9]PLAP. Surprisingly, the decrease in k_{cat} was not accompanied by a decrease in K_m. The more drastic deletion of 25 N-terminal amino acid residues, including the α-helix, almost abrogated enzyme activity at pH 9.8. However, the K_m for the residual activity was not affected by the N-terminal deletion. These results were also valid at pH 7.5, even when at this pH catalysis occurs with a lower K_m since deletion of the nine N-terminal amino acid residues had no effect on K_m whereas k_{cat} decreased as a consequence of the deletion. Substitution in PLAP of E7K (as found in TNAP) had little impact on catalysis, at the level of both k_{cat} and K_m. Mutation of the potential amino acid ligand for Glu6 in PLAP (Arg370, Fig. 12) likewise had little impact on catalysis. These findings illustrate that the N-terminal α-helix of one PLAP monomer plays a crucial role in the control of PLAP active site function but that the structural destabilization of [Δ1–9]PLAP did not result from elimination of N-terminal amino acid residues directly involved in structural stabilization, but rather suggested that the propensity of α-helix formation was affected in [Δ1–9]PLAP, a conclusion supported by computational analysis of α-helix formation of the PLAP peptide 1–25 and 10–25. Defective folding of the new N-terminus in [Δ1–9]PLAP would affect positioning of Trp12, a highly conserved amino acid in all mammalian AP isozymes.

The corresponding analysis for TNAP revealed an equivalent qualitative, but more pronounced, response. The progressive deletion of N-terminal amino acid residues in TNAP did not affect the K_m for p-nitrophenylphosphate (pNPP), but strongly reduced k_{cat}. Thus, already deleting five N-terminal amino acid residues in TNAP reduced k_{cat} more than 10-fold, both at pH 9.8 and 7.5, without affecting K_m. Substituting Lys7 in TNAP for the residue found in PLAP, i.e. K7E, had little effect on catalysis, and also substitution of Arg374 in TNAP (which stabilizes Glu6) had little effect on catalysis, ruling out an important role for both Glu6 and Lys7 of TNAP in the control of enzyme integrity. However, [Δ1–9]TNAP was totally inactive, illustrating that the further deletion of the four N-terminal amino acid residues of [Δ1–5]TNAP suffices to abolish TNAP activity. Correspondingly, [Δ1–25]TNAP is also an inactive enzyme. Studies using a panel of epitope-mapped anti-TNAP MAbs (Section 6.3) as probes indicated that the loss of TNAP catalysis did not result from the loss of critical residues in stretch 1–9, but that deletion of this stretch affected the role of the α-helix in controlling the TNAP structure, findings also supported by the computed α-propensity for the TNAP peptide 1–25 and 10–25.

Measurements of residual AP activity during inhibition by P_i, a competitive inhibitor of APs, revealed that the N-terminal deletion of five amino acids as in [Δ1–5]TNAP had no effect on the IC_{50} for P_i, when analyzed at pH 9.8, in the presence of 2 mM pNPP. Likewise, [Δ1–9]PLAP and PLAP were similarly inhibited by P_i, indicating that the active site entrance was not affected in these deletion mutants. This evidence indicates that k_1 and k_{-1} for substrate positioning in the

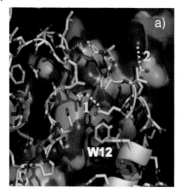 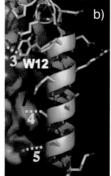 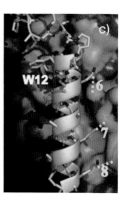

Fig. 12 N-terminal amino acid residues positioning and interactions in PLAP. Detailed interactions between N-terminal α-helical residues of monomer A (shown in wireframe and ribbon representation) and backbone residues of monomer B (shown as spacefilled surface residues). (a): detailed interactions between N-terminal aa residues of A with the Y367 harboring loop of A and distance between potentially interacting residues; 1: Gu6 (A) and Arg370 (A), 2.86 Å; 2: Glu7 (A) and Arg117 (B), 2.56 Å. (b and c): detailed interactions between terminal α-helical residues of A and backbone residues of B, viewed either from a front or back position and distance between potentially interacting residues; 3: Trp12 (A) and Lys104 (B), 3.73 Å; 4: Ala19 (A) and Phe457 (B), 3.79 Å; 5: Ala23 (A) and His450 (B), 3.46 Å; 6: Pro9 (A) and Val129 (B), 3.94 Å; 7: Asn13 (A) and His460 (B), 3.2 Å; 8: Leu20 (A) and Tyr471 (B), 3.86 Å; 9: Lys24 (A) and Phe467 (B), 3.9 Å. Stabilizing bonds are shown via dashed yellow lines, against a background of monomer B; W12 is specified as an internal reference. Taken from Hoylaerts et al. (2005).

active site are not modified by the deletions. Likewise, inhibition of PLAP and [Δ1–9]PLAP by the uncompetitive inhibitor L-Phe did not differ and inhibition of TNAP and [Δ1–5]TNAP by L-homoarginine (L-hArg) were identical, when investigated at both pH 9.8 and pH 7.5. The higher IC_{50}s for both inhibitors at pH 7.5 are compatible with the uncompetitive mechanism by amino acids that requires the unprotonated amino function of the compounds for inhibition (Section 5.1.2.1). These experiments suggested that in the expression for K_i either the ratio k_3/k_2 was not modified during inhibition by L-Phe and L-hArg or alternatively that $k_3 \ll k_2$, in which case $K_i = K_{EPI}$, for inhibition by these amino acids (see schemes and equations in Section 5.1). In view of the fact that the ratio k_{cat}/K_m, apart from k_1 and k_{-1}, only depends on k_2, these data imply that N-terminal amino acid residue deletions reduce the rate of covalent phospho-complex formation, i.e. k_2. This explains why k_{cat} decreases. Since both the expression for k_{cat} and that for K_m contain a term k_2/k_3 and K_m is not affected by the N-terminal amino acid deletions, it follows that either $k_2 \ll k_3$ or k_2/k_3 is not affected, implying that the rate of phospho-complex hydrolysis/transphosphorylation k_3 is reduced comparably to k_2. The uncompetitive inhibition data confirm that the second option is correct; the expression for K_m contains a term k_3/k_2, which is identical for deletion mutants and wt enzymes. In view of a modified k_2 in the mutants, these findings would require that $k_3 \ll k_2$, in contradiction with the assumption made above that that

$k_2 \ll k_3$. Therefore, N-terminal deletions in TNAP and PLAP affect folding of the N-terminal α-helix and this change is responsible for the measurable structural destabilization that is translated in the active site through a reduced catalytic efficacy slowing down phospho-complex formation and processing (Hoylaerts et al., 2005).

4.1.6
The Crown Domain

The top flexible loop that constitutes the crown domain of PLAP (Fig. 5) is formed by the insertion of a 60-residue segment (366–430) from each monomer. It consists of two small interacting β-sheets, each composed of three parallel strands and surrounded by six large and flexible loops containing a short α-helix (Le Du et al., 2001). This region displays the least degree of sequence conservation among mammalian APs. As will be discussed in detail here and in subsequent sections, isozyme-specific properties, such as the characteristic uncompetitive inhibition (Hoylaerts and Millán, 1991; Hoylaerts et al., 1992a; Hummer and Millán, 1991; Kozlenkov et al., 2002, 2004), their variable heat-stability (Bossi et al., 1993) and their allosteric behavior (Hoylaerts et al., 1997) have been attributed to residues located on this crown domain unique to mammalian APs.

This domain may also mediate the interaction between APs and extracellular matrix (ECM) proteins. Tsonis et al. (1988) identified a sequence within the crown domain of PLAP that had homology to cartilage matrix protein, Von Willebrand factor and MAC-1 among others, suggesting that this domain may mediate an interaction of APs with ECM components. In fact, Vittur et al. (1984) and also Wu et al. (1991) documented that TNAP is able to bind to collagen and proposed that this interaction may be a step in the process leading to skeletal calcification. Bossi et al. (1993) determined that the crown domain of TNAP was in large part responsible for binding of TNAP to collagen. The authors introduced unique restriction sites at identical positions in the mouse TNAP and human PLAP cDNAs to allow the homologous exchange of the loop domain of the TNAP (T domain) and PLAP (P domain) isozymes and the generation of the reciprocally chimeric molecules PLAP-T and TNAP-P. The introduction of the T loop into PLAP reduced the heat stability of PLAP-T almost to that of TNAP. The domain substitution was accompanied by a conformational change that resulted in the loss of immune reactivity with four of 17 epitope-mapped anti-PLAP MAbs (Section 6.1). The T and P loops provided stabilization to the side-chains of specific uncompetitive AP inhibitors and the introduction of the T domain also conferred collagen-binding properties to PLAP-T accounting for half of the binding affinity of TNAP for collagen, while not affecting PLAP binding to IgG (Bossi et al., 1993). Two residues in this crown domain, Glu429 and Tyr367, are of particular structural and functional significance and will be referred to repeatedly throughout Part I of this book.

The identity of residue 429 is isozyme-specific in human APs and also in other species. PLAP has a Glu at this position, GCAP a Gly, IAP a Ser and TNAP a His. The placement of this residue at the very entrance of the active site pocket of APs

(Fig. 13) has profound implications for the stability of the active site of PLAP as described above and also for the isozyme-specific inhibition by L-amino acids (Section 5.1.2.1). The single most important difference between PLAP and GCAP is the E429G substitution (PLAP numbering) that effectively converts PLAP into an enzyme with the kinetic, inhibition and heat stability properties of GCAP, as will be detailed later (Hummer and Millán, 1991; Watanabe et al., 1991; Hoylaerts et al., 1992b). Glu429 is located in the immediate neighborhood of the active site Zn1 (Fig. 13) and the nature of this residue confers completely different ionic and steric properties to the immediate surrounding of the active site, i.e. Glu429 in PLAP has a theoretical pK_a around 4.3, Gly429 in GCAP has no side-chain and provides important flexibility to the neighboring loop, Ser429 in IAP is neutral and polar; and His434 in TNAP has a theoretical pK_a of 6.0. The close proximity of this residue to the active site suggests that it is directly involved in substrate binding. The change in side-chain and pK_a can therefore selectively affect the nature of the substrate favored to bind to the active site of each AP isozyme.

Although substitutions at position 429 will greatly affect the conformation of the surface loop of PLAP and the inhibition properties (Hoylaerts and Millán, 1991), they do not introduce any major structural changes in the active site pocket itself, as revealed experimentally by inhibition kinetic experiments using free phosphate as a competitive inhibitor (Hoylaerts et al., 1992a). At physiological pH, PLAP is extremely stable, resisting a temperature of 65°C for 60 min (McComb et al., 1979). Using urea as a denaturant, the red shifts of fluorescence spectra show a complex unfolding process involving multiple equilibrium intermediates indicating differential stability of the subdomains of the enzyme. PLAP does not lose its phosphate-binding ability after substantial tertiary structure changes, suggesting that the substrate-binding region is more resistant to chemical denaturant than the other structural domains (Hung and Chang, 2001b). A biphasic denaturation process was observed for human PLAP using guanidinium chloride as denaturant (Hung and Chang, 1998) and an unfolding intermediate state was also observed between 1.6 and 2.0 M guanidinium chloride for the calf IAP isozyme (Zhang et al., 2003) with the refolding process proceeding also in two phases (Zhang et al., 2002b, 2003). Therefore, these intermediate denaturation stages are most likely mediated by changes in the surface domains of PLAP, particularly the crown domain and/or the N-terminal arm. In contrast, a single E429G substitution results in a >60% reduction in activity (Hoylaerts et al., 1992b). It is clear from these results that the conformational change that occurs in PLAP when Glu429 is substituted for Gly429 has a major impact on the general stability of the isozyme.

Another interesting structural feature of PLAP, with no counterpart in the ECAP structure, is Y367 (Fig. 13). This residue is part of the subunit interface in the PLAP dimer, where it protrudes from one subunit and its hydroxyl group is located 6.1 Å from the phosphate and 3.1 Å from His432, which in turn chelates the zinc atom Zn1 in the active site of the other subunit. This structural feature could not have been postulated from the ECAP structure, as the loop containing residues 366–375 in PLAP is an insertion with respect to ECAP. Chang and

Chang (1984) documented that PLAP is inactivated by treatment with tetranitromethane in a biphasic process and that the inactivation was due to the conversion of a tyrosine residue to 3-nitrotyrosine. Since phosphate prevented inactivation of the enzyme, the authors suggested the involvement of tyrosyl residues at the phosphoryl site of the phosphorylated enzyme. It seems very likely that the tyrosyl residue referred to by these authors is indeed Y367. The location of Y367 in the structure and the fact that this residue is perfectly conserved in mammalian APs (Fig. 8) implicate this residue in an important structural/functional role. Kozlenkov et al. (2002) mutated Tyr367 to Phe as the closest possible structural analogue of Tyr, and also to Ala (Fig. 8). Both mutations were done in the context of wt PLAP and of [G429]PLAP. Residue 429 is situated close to Y367, so it was of interest to check the effect of the double substitutions. The effects of the substitutions at Y367 on the kinetic properties of PLAP were very similar for all four recombinant mutants studied, i.e. [A367]PLAP, [F367]PLAP, [F367, G429]PLAP and [A367, G429]PLAP (Table 2). The k_{cat} values were in the range 39–45% of the wt PLAP value. The K_m values were not significantly changed from the value of wt PLAP (0.35 mM). However, the Y367A and Y367F substitutions significantly compromised the heat stability of the mutant PLAP enzymes. In addition, whereas [F367]PLAP, like wt PLAP, displayed a monophasic inactivation curve, [A367]PLAP displayed instead a biphasic inactivation mechanism. Interestingly, the [A367]PLAP mutant was more stable than the [F367]PLAP enzyme after a prolonged incubation, despite a higher initial rate of inactivation. As will be described in Section 5.1.2.1, both Glu429 and Y367 play a crucial role in determining the uncompetitive inhibition properties of the mammalian APs.

Residues in the crown domain participate in forming the roof of the active site cleft in mammalian APs. At the active site cleft, Le Du and Millán (2002) defined a fingerprint characteristic of each AP isozyme. This fingerprint is compatible with the hypothesis of isozyme-specific specialization for a phosphate donor or phosphate acceptor in the case of the transphosphorylation reaction. On the upper part of the active site cleft, the authors observed a large cluster composed of Arg314, Tyr276, Arg326, Arg323 and Arg420, surrounded by Glu321 and Glu418. This cluster, located 11–17 Å from the phosphate group, exists in PLAP and GCAP, but not in IAP or TNAP. On the lower part of the active site cleft, Lys87, Phe107, Glu108, Arg166, Asn167, Tyr169 and Glu429 from one monomer and Tyr367 from the second monomer form a hydrophobic pocket in PLAP. This hydrophobic pocket, which involves residues from both monomers, is conserved in GCAP and IAP, except at position 87 in IAP, which displays the conservative substitution K87R. In TNAP, among the eight residues of this pocket, there are five substitutions: K87A, F107E, Q108G, N167D and E429H (PLAP numbering). These substitutions remove the hydrophobic character, converting it into strongly ionic in TNAP. Therefore, the properties of this pocket in the case of TNAP are completely different from those of PLAP, GCAP or IAP. These findings correlate well with the differential behavior of the AP isozymes towards uncompetitive inhibitors, i.e. PLAP, GCAP and IAP are inhibited by L-Phe but not by L-hArg, while TNAP is inhibited by L-hArg but not by L-Phe (Sections 5.1.2.1 and 5.1.2.2). The

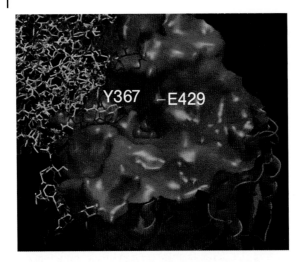

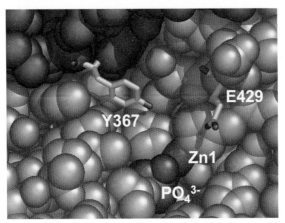

Fig. 13 Location and environment of the Y367 residue. (A) Top view of the entrance to the active site, showing the position of the Y367 residue from one subunit (wireframe representation) in the immediate vicinity of the E429 residue of the other subunit (space-fill representation). Taken from Kozlenkov et al. (2002) and reproduced with permission from the *Journal of Biological Chemistry*. (B) Detail of the active site entrance in spacefill representation showing the location of E429 relative to Zn1 and the active site PO_4^{3-} in one subunit and the relative location of Y367 of the other subunit.

location and orientation of this pocket with regard to the phosphoseryl intermediate during catalysis suggest that it may participate in stabilizing the phosphate donor at the first step of the reaction. Therefore, the substrate of the TSAPs or of TNAP must display ionic properties compatible with the highly divergent ionic properties of the corresponding pocket. The only difference between PLAP and GCAP, within a 12 Å sphere around the phosphate moiety of the covalent phosphoseryl intermediate P-Ser92, occurs at position 429. At a structural level, this substitution results in a larger pocket in GCAP and may affect the local pK_a in close proximity of the phosphate moiety of the phosphoserine. In IAP, position 429 is a serine and three additional substitutions, G93A, D428S and A433G, characterize the intestinal enzyme. G93A and A433G are conservative and are unlikely to affect the enzymatic properties, but D428S may also alter the local pK_a in the same manner as E429S. In TNAP, there are 10 substitutions, including a histidine substitution at position 429. The substitutions, mainly towards ionic residues, cluster around the hydrophobic pocket, causing it to lose its hydrophobic character and to gain hydrophilic properties. This modification of the overall character of the pocket is consistent with L-hArg, an ionic amino acid being the preferred inhibitor of this isozyme (Section 5.2.2.2). These studies have defined three regions at the active site cleft which characterize each human AP. PLAP is characterized by residue Glu429, the Arg cluster at the roof and the hydrophobic pocket at the floor of the cleft; GCAP is defined by Gly429, the Arg cluster and the hydrophobic pocket; IAP has Ser429, no Arg cluster but has the hydrophobic pocket; and TNAP contains His429, no Arg cluster and a highly ionic pocket at the floor of the cleft (Le Du and Millán, 2002). These structural fingerprints suggest that the TSAPs isozymes and TNAP are likely to have very different substrate specificities, at least *in vivo*.

Recently, Harada et al. (2005) constructed structural models of rat IAP I and IAP II based on the 3D crystal structure of PLAP. Both rat isozymes displayed the expected $\alpha\beta$ topology, but the crown domain of IAP I contained an additional β-sheet, whereas the embracing N-terminal arm of rat IAP II lacked the α-helix, when each model was compared with PLAP. The representations of surface potential in the rat IAPs were predominantly positive at the base of the active site but the coordinated metal at the active site was predicted to be a zinc triad in IAP I, whereas the typical combination of two Zn atoms and one Mg atom was predicted for IAP II. The authors found no differences among the residues that directly interact with the metal at the M3 site on comparing IAP I and IAP II, but observed a difference in a ligand with indirect interaction i.e. Gln317 in IAP I and an Arg317 in IAP II. As shown in Fig. 9, the His317 of PLAP is indirectly associated with the M3 site via a water molecule. Comparative stereoviews of the M3 site residues in the active site of rat IAP I and IAP II indicated that the Mg^{2+} at the M3 site was coordinated octahedrally with three residues and three water molecules, i.e. the OD2 oxygen atom of Asp42, the OG2 oxygen atom of Ser155 (Thr155 in ECAP) and the OE1 oxygen atom of Glu311 in PLAP. The positions of Asp42, Ser155 and Glu311 did not vary significantly from their positions in the side-chains of PLAP. His153 was clearly closer to the metal position in rat IAP than in PLAP. A comparison of the distances between the residues and the metal revealed that His153 was the closest residue to the metal at 2.40 Å in rat IAP I and was closer to the metal than Glu311 in rat IAP II. Because of these differences in the side-chain structure, the position of the water-interacting atom, i.e. a nitrogen atom of Gln317 in rat IAP I and a nitrogen atom of Arg317 in rat IAP II, was also different from that in PLAP. In support of these modeling data, Harada et al. (2005) used metal-depleted extracts from rat duodenum or jejunum and PLAP and performed enzyme assays under restricted metal conditions. With the duodenal and jejunal extract, but not with PLAP, enzyme activity was restored by the addition of zinc, whereas in nonchelated extracts, the addition of zinc inhibited duodenal IAP and PLAP but not jejunal IAP. Western blotting revealed that nearly all IAP activity in the jejunum extracts was IAP I, whereas in duodenum the percentage of IAP I (55%) correlated with the degree of AP activation (60% relative to that seen with jejunal extracts). These data are consistent with the presence of a triad of zinc atoms at the active site of rat IAP I, but not IAP II or PLAP, and suggest that the side-chain position of His153 and the alignment of Q317 might be the major determinants for activation of the zinc triad in rat IAP I.

4.1.7
The Monomer–Monomer Interface

The monomer–monomer interface in PLAP has a strong hydrophobic character, in contrast to that of ECAP. Less than 30% of the residues are involved in hydrogen-bonding interactions, which confer flexibility on the interface. The surface buried in the dimer interface is only slightly larger in PLAP (4150 Å^2) than in ECAP (3900 Å^2), but the residues involved in dimerization are mostly different. In ECAP, 36 out of 82 total residues at the dimer interface are involved in hydrogen bonds, whereas in PLAP only 24 out of 83 residues are involved in hydrogen bonding interactions and only nine of these are conserved between the two. According to Le Du et al. (2001), many hydrogen bond interactions have been replaced in PLAP by less specific van der Waals contacts, which are more likely to allow rearrangement of the two monomers. Both the N-terminal arm and the crown domain take part in stabilizing the AP dimeric structure. Modeling of the structure of TNAP, GCAP and IAP (Le Du and Millán, 2002) revealed that the overall surface buried at the interface varies between 4134 and 4244 Å^2 per monomer for these isozymes, which corresponds to about 25% of the overall protein surface and comprises about 90 residues per monomer. The comparison of the interfaces reveals high conservation in the TSAPs compatible with the fact that PLAP–GCAP or PLAP–IAP heterodimers form readily in nature (Behrens et al., 1983; Imanishi et al., 1990a,b; Kodama et al., 1994). In TNAP, however, a number of charged substitutions lead to repulsive forces at the interface that do not favor the formation of heterodimers between TNAP and any of the tissue-specific APs (Le Du and Millán, 2002), although one paper has documented the existence of abnormal IAP/TNAP heterodimers (Vergnes et al., 2000).

4.1.8
The Noncatalytic Peripheral Binding Site

Recently, Llinas et al. (2005) identified a novel, noncatalytic peripheral site on the surface of PLAP. This newly identified peripheral site is 28 Å from the active site but only 13 Å from the calcium binding site (Fig. 14a). The residues that constitute this site are located on two α-helices, 250–257 and 287–297. The loop between

Ser257 and Ser287 contains the three residues that coordinate the Ca^{2+} ion, namely Phe269, Glu270 and Asp285, located only two residues from Ser287. Arg250 is located two residues away from Trp248 that interacts with the Ca^{2+} through a water molecule. Therefore, the large 250–297 loop that belongs to the peripheral domain of PLAP includes both the M4 site and the peripheral binding site. The location of the Ca^{2+} ion in this loop suggests that the function of this novel M4 site could be related to the conformational stabilization of the two α-helices that form the peripheral site. By crystallizing PLAP in the presence of different ligands, i.e. AMP, pNPP and L-Phe, Llinas et al. observed that the strongest electron density corresponded to the rings of L-Phe, pNPPate or 5′-AMP all in the same orientation. Considering the diversity of the compounds that are able to bind, this site shows a strong affinity for hydrophobic rings without any particular selectivity. In the native enzyme, the side-chain of Met254 occupies the position that the hydrophobic ring occupies in the complexes (Fig. 14b). Ligand binding involves the displacement of the side-chain of Met254, opening of a pocket in which the ring of L-Phe, pNPPate or 5′-AMP are able to fit, followed by stabilization of the side-chain of Met254 (Fig. 14c). The identification of this peripheral site helps to explain the results obtained by Chang et al. (1990), who reported that 1,N^6-ethenoadenosine monophosphate (epsilon AMP) was both a substrate and a partial mixed-type inhibitor of PLAP. Their results indicated that the reagent and substrate could combine with the enzyme simultaneously and the authors proposed that in addition to the catalytic site, a hydrophobic region exists for additional substrate-binding.

The architecture of the peripheral site in PLAP is conserved in IAP but not in GCAP or TNAP, suggesting that this site might contribute to tissue specific functions. The sequences of GCAP and PLAP isozymes differ at only nine positions; one is position 429 in the active site and two at the peripheral site, 254 and 297. Thus, Met254 and Arg297, which in PLAP sandwich the ring of the ligand, are substituted by two leucine residues in GCAP, suggesting that whatever functional role this peripheral site might have in PLAP it is not shared by GCAP. However, the peripheral site is fully conserved in IAP but not in TNAP. In TNAP, residues Arg250, Leu253 and Glu290 at the peripheral site that interact directly with the ligand in PLAP correspond to Arg255, Leu258 and Glu294. These residues belong to a cluster of mutations: R255L, L258P, E294K, W253X and D289V around the peripheral site associated with a severe phenotype of the inherited disease hypophosphatasia (Mornet et al., 2001; Section 7.1). Although there is no conservation of the peripheral site between PLAP and TNAP, the residues Trp253, Arg255, Leu258, Asp289 and Glu294 appear to be involved in some yet undefined important structural and/or physiological function in TNAP. Hence it appears likely that this newly defined noncatalytic binding site in APs may be of physiological relevance in all mammalian APs.

48 | *4 Protein Structure and Functional Domains*

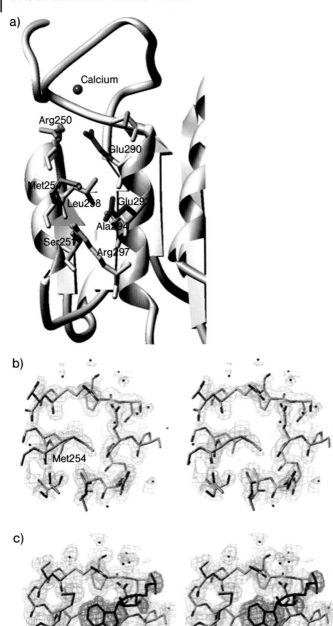

Fig. 14 A noncatalytic peripheral binding site in PLAP. Ribbon representation of the region of the peripheral binding site of PLAP, in ball-and-stick form, showing the residues of the peripheral binding site of PLAP (a). Stereo view of an F_o-F_c omit map contoured at the 3.5σ level (green) and a $2F_o-F_c$ omit map contoured at the 1.5σ level (gray) in the native active site of PLAP (b) or co-crystallized with 5'-AMP (c). Wn, nucleophilic water molecule.

4.2
Genetic Polymorphism and Protein Variability

The AP gene family has long captured the interest of biochemical geneticists. The degree of genetic polymorphism varies considerably between AP isozymes: whereas PLAP has been called the most-polymorphic enzyme system know to date (Harris, 1980), no electrophoretic polymorphism has been described for the IAP or TNAP isozymes. However, the GCAP isozyme has been shown to display a fairly high degree of genetic polymorphism by the use of MAbs (Section 6.1).

Briefly, and to summarize some older but pertinent literature, there are three common alleles of PLAP, accounting for the six common electrophoretically distinguishable phenotypes of this isozyme, i.e. S, FS, F, I, SI and FI, according to their electrophoretic mobility on starch gel (S = slow, F = fast, I = intermediate) (McComb et al., 1979). In the nomenclature introduced by Donald and Robson (1974a,b), these common phenotypes are called 1, 2-1, 2, 3, 3-1 and 3-2, respectively. This pattern is consistent with the expression of three common autosomal alleles, Pl^1, Pl^2 and Pl^3. Heterozygous individuals show triple bands, in agreement with the fact that the enzyme is a dimer and that hybrid enzymes result from a random combination of the monomers (Robson and Harris, 1965; Beckman et al., 1966). Since the three common alleles are responsible for only 97.5% of the placental phenotypes, the frequency of rare alleles is unusually high (2.5%)(Harris, 1980). At least 18 rare alleles at the *ALPP* locus have been identified, which give rise to hybrid phenotypes in combination with one of the common alleles. Figure 15 shows a schematic representation of the electrophoretic pattern of the common and some rare variants of PLAP as resolved by starch gel electrophoresis at pH 7.4 and 8.9. Altogether, 48 phenotypes of PLAP have been described in a survey of 5000 placentas (Donald and Robson, 1974a,b). One of these rare alleles, D, was first thought to be restricted to Africans (Boyer, 1961), in whom it has been found at a frequency of about 0.8%. The D allele is apparently absent in Asian populations (Beckman, 1970) but has been found in most European ethnic groups, with the highest frequency (1.4%) found in Finns (Beckman et al., 1994). Population studies had suggested that PLAP types of both the F and D alleles may be involved in intrauterine selection (Beckman et al., 1972a,b; Beckman and Beckman, 1975) and therefore there was, and still is, considerable interest in understanding the potential differential properties of these two allozymes during pregnancy. A study also suggested a relation between the deleterious effects of smoking on birth weight depending on PLAP genotype and maternal age (Magrini et al., 2003).

Early structural studies on allelic variants of PLAP were performed on the S, F and I allozymes of PLAP purified to homogeneity from homozygous individuals (Holmgren and Stigbrand, 1978). The allozymes were shown to have similar molecular weight, sedimentation coefficient, amino acid composition, pH optimum, sensitivity to the uncompetitive inhibitors L-Phe and L-Leu and similar K_m values with different substrates. Using anti-PLAP antisera extensively cross-absorbed with purified S, F or I allozymes, Doellgast et al. established that the D allozyme was identical with allozyme 18 (Doellgast et al., 1980). Comparison of the amino acid sequences of the type 1 (S) and 3 (I) allozymes deduced from their respective cDNAs showed seven amino acid differences, each due to a single base substitution, and three additional silent substitutions in the coding regions and three base differences in the 1.0 kb 3'-untranslated region (Henthorn et al., 1986). These substitutions were P-19L, M44V, R241H, Q255R, T263A, Y367C and S372G. This high degree of difference between the type 1 and 3 allozymes was surprising, as the authors themselves discussed (Henthorn et al., 1986; Harris, 1990), and some of these substitutions are likely to be due to cloning artifacts or sequencing errors. In particular, given the critical role played by residue Y367 that is perfectly conserved in mammalian APs, the Y367C substitution reported for the type 3 (I) allozyme is not likely to be correct. The type 2 (F) allelic variant cloned by Millán (1986) contained a Pro209 residue, whereas the type 1 allozyme cloned by Hen-

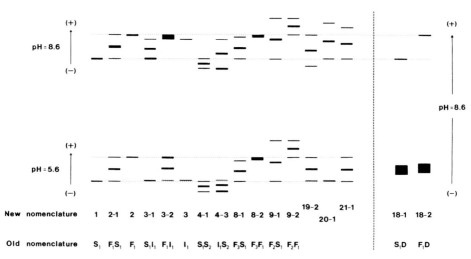

Fig. 15 Schematic representation of the electrophoretic mobility of representative genetic variants of PLAP. The allelic variants 1–9 correspond to those identified by Donald and Robson (1974) and 19, 20 and 21 as reported later by Millán et al. (1982). The allelic D variant described initially by Beckman and Beckman (1968) was later renamed variant 18 (Doellgast et al., 1980). Only the isozyme bands migrating fast towards the anode are shown. The relative activity of the components is indicated by the widths of the bands. (a) Electrophoretic mobility of the allozymes on starch gels at both pH 8.6 and 5.6. (b) Electrophoretic mobility at pH 8.6 of two samples containing the D variant of PLAP display a characteristic slow mobility on starch gels. Taken from Millán et al. (1982) and reproduced with permission from *Human Genetics*.

thorn et al. (1986) contained a P209R substitution. The replacement of a neutral by a positively charged amino acid agrees with the "slow" electrophoretic mobility of the S PLAP phenotype in starch gel electrophoresis and polyacrylamide gel electrophoresis (PAGE). The use of allozyme-specific MAbs had indicated that the S and the D PLAP allozymes shared the epitope that distinguished them from the F allozyme (Hoylaerts and Millán, 1991). In fact, later studies revealed that similarly to the S and I alleles of PLAP (Henthorn et al., 1986), the D allele encodes an Arg at codon 209 and at codon 429, the Glu found in the three common alleles is replaced by a Gly in the D variant (Wennberg et al., 2002). These authors also used a single nucleotide primer extension (SNuPE) assay to detect the presence of the F and D allele in genomic DNA samples. The assay is based on the allelic polymorphism at codons 209 (692C>G) and 429 (1352A>G). By analyzing these two substitutions on each sample, the authors were able to distinguish the PLAP F and D alleles from the PLAP S/I alleles in genomic DNA samples. The S and I alleles both have the Arg209 and the Glu429 residues, so they cannot be distinguished in this assay. Although seven amino acid substitutions were reported for the PLAP I allele compared with the PLAP F cDNA sequence, as presented above, the authors could not identify any consistent mutation within the 1.9 kb PCR fragment of interest that could be useful in identifying the I allele from the S allele by this SNuPE assay. These kinds of assays on DNA samples, rather than placental tissue samples, will enable future large-scale population studies to re-examine the possible association between allelic variability of PLAP and the incidence of spontaneous abortions. A comparison of the biochemical properties of purified recombinant S, F and D allozymes, i.e. [R209]PLAP, [P208]PLAP and [R209; G429]PLAP, confirmed a number of distinctive properties of the D allozymes. Thus, the sensitivity of the D allozyme to uncompetitive inhibition by L-Leu was significantly higher than that of the S and F allozymes, whereas the enzymes from SD and FD phenotypic variants, consisting of mixtures of homodimers and heterodimers, responded in an intermediate fashion to inhibition. The same pattern was repeated in the thermostability experiments, where the recombinant D allozyme displayed an increased heat sensitivity, losing half of its activity at 61 °C (Hoylaerts and Millán, 1991; Wennberg et al., 2002). These properties, all explained by the E429G substitution, are reminiscent of those of the GCAP isozyme and these shared characteristics between the D allozyme of PLAP and the GCAP isozyme explain the initial controversial results linking the re-expression of the rare D variant of PLAP with tumorigenesis (Section 10.1).

The genetic polymorphism of GCAP was initially demonstrated by the use of MAbs that allowed the classification of nine allotypes based on their reactivity pattern (Millán et al., 1982a–c; Wahren et al., 1986; Hendrix et al., 1990). The nucleotide sequences obtained from genomic DNA (Millán and Manes, 1988) and also from three independent cDNAs isolated from JEG3 choriocarcinoma cells (Watanabe et al., 1989), BeWo cells (Lowe and Strauss, 1990) and LS174T cells (Gum et al., 1990) confirmed the existence of allelic polymorphism at this locus. The following substitutions were found in the GCAP sequences: M38I, V133N or V133M, T159A, L254M, L297R, V361L, P479R and A512T. Most of these substitutions were subsequently analyzed via mutagenesis experiments and reactivity with MAbs (Hoylaerts et al., 1992b), as will be described in Section 6.1. It is of interest that the GCAP allozymes are not well resolved by starch gel or agarose gel electrophoresis, largely owing to the presence of Gly429 in these allozymes (Hoylaerts et al., 1992b). The conformational changes in the crown domain when Gly429 is present lead to a very diffuse migration on different electrophoretic supports. This also explains the difficulty in demonstrating genetic variability in the PLAP-like isozymes expressed by tumors in the early days before the advent of cloning.

Although no electrophoretic variability has been demonstrated for human TNAP, a number of single nucleotide polymorphisms have been documented. Table 3 (taken from the web site maintained by Etienne Mornet; http://www.sesep.uvsq.fr/database_hypo/Polymorphism.html) details the polymorphic mutations in the human *ALPL* gene identified to date. Most of the substitutions either reside in introns or do not change an amino acid in TNAP. However, three substitutions do result in point substitutions, i.e. R135H (Mumm et al., 2002), Y246H (Henthorn et al., 1992) and V505A (Greenberg et al., 1993). Although the phenotypic consequences of the R135H polymorphism have not been investigated experimentally, Goseki-Sone et al. (2005) showed a significantly association between the 787T>C (Y246H) polymorphism of the *ALPL* gene and bone mineral density (BMD) among 501 postmenopausal women. Furthermore, the Y246H TNAP mutant was examined kinetically and whereas the specific activity was similar to that of the control enzyme, the K_m of [H246]TNAP was 0.29 compared with 0.5 for wt TNAP, indicating an increase in the affinity for substrate for this allozyme (Goseki-Sone et al., 2005). The authors suggested that genetic variability in the *ALPL* gene might be a factor contributing to age-related bone loss in humans. In a somewhat related and relevant mouse study, Srivastava et al. (2005) analyzed two inbred strains of mice, MRL/MpJ and SJL, which exhibit a 90% difference in total serum AP activity (268 ± 26 versus 140 ± 15 U L^{-1}, respectively, $p < 0.001$) and reported a genome-wide scan for cosegregation of genetic marker data with serum AP activity. Three major quantitative trait loci (QTL), one each on chromosomes 2, 6 and 14, were identified in addition to one additional suggestive QTL on chromosome 2. Together, these QTLs explained 22.5% of the variance in serum AP between these two strains. Serum AP showed a moderate but significant correlation with body weight-adjusted total body bone mineral density and periosteal circumference at midshaft tibia in F2 mice. The chromosome 6 region harboring the major serum AP QTL also contains a major BMD and bone size QTL and in addition is also close to the locus that regulates IGF-I levels in C3HB6 F2 mice. The authors concluded that common QTLs indicate that the observed difference in AP and BMD or bone size may be regulated by the same loci.

Table 3 Polymorphisms in the human *ALPL* gene. Nucleotide numbering is given according to Weiss et al. (1988) and the Nomenclature Working Group (Antonarakis et al., 1988): the first nucleotide (+1) corresponds to the A of the ATG initiation codon. Amino acid numbering is given according to both the nonstandardized and standardized nomenclatures.
Nonstandardized nomenclature is according to Weiss et al. (1988) and takes into account the 17-residue signal peptide, i.e. the ATG initiation codon is numbered as residue −17. The standardized nomenclature follows the recommendations of the Human Genome Variation Society (http://www.hgvs.org/), i.e. the first codon is the ATG initiation codon.
p, Protein numbering; c, nucleotide numbering of the coding sequence. This table was taken from the site maintained by Etienne Mornet (http://www.sesep.uvsq.fr/database_hypo/Polymorphism.html) and is reproduced here with his permission.

Exon	Base change	Non-standard	Standard	Frequency	No. tested	Reference
IVS5	c.472+12delG			0,07	35	Mornet (2000)
5	c.330C>T	S93S	p.S110S	0,05	30	Greenberg et al. (1993)
5	c.455G>A	R135H	p.R152H	0,02	168	Mumm et al. (2002)
6	c.534C>T	Y161Y	p.Y178Y	0,05	20	Goseki-Sone et al. (2005)
7	c.787T>C	Y246H	p.Y263H	0,33	73	Henthorn et al. (1992)
IVS7	c.793–31C>T			0,21	21	Mornet (2000)
IVS8	c.862+20G>T			0,07	23	Mornet (2000)
IVS8	c.862+51G>A		n.d.			
IVS8	c.862+58C>T			0,12	20	Mornet (2000)
IVS8	c.863–7T>C		n.d.			
IVS8	c.863–12C>G			0,06	42	Orimo et al. (1997)
9	c.876A>G	P275P	p.P292P	0,47	71	Henthorn et al. (1992)
12	c.1565T>C	V505A	p.V522A	0,26	47	Greenberg et al. (1993)

Electrophoretic variability exists in the mouse TNAP isozyme. The existence of allelic variation in the mouse *Akp2* gene, consistent with two common alleles designated *Akp2*[a] and *Akp2*[b], was initially demonstrated by gel electrophoresis (Wilcox and Taylor, 1981). MAbs distinguishing this variation were subsequently developed (Dairiki et al., 1989). Two *Akp2* cDNA sequences have been reported, derived from Icr random-bred mice (designated mouse placental AP) and the 129/J derived cell line NULLI-SCC1 (Terao and Mintz, 1987; Hahnel and Schultz, 1990). The transcribed region of the *Akp2* gene from C57Bl/6J was subsequently reported in exact agreement with that of the mouse placental TNAP cDNA (Terao et al., 1990). The two cDNAs differed at only two amino acid positions, Ser for Leu in the signal peptide and Arg for Pro at position 504. Neither of these substitutions is present in the fully processed anchored enzyme and, in fact, both the C57Bl/6J and 129/J correspond to the $Akp2^b$ phenotype, as determined by cellulose acetate electrophoresis (Wilcox and Taylor, 1981). Hence the amino acid difference(s) that may account for the differential electrophoretic mobility and MAb reactivity between the $Akp2^a$ and $Akp2^b$ allozymes remains to be elucidated.

4.3
Post-translational Modifications

4.3.1
Glycosylation Sites

PLAP and GCAP have two glycosylation sites, at Asn122 and Asn249, that vary in their degree of glycosylation depending of the placental sample. Initial data had suggested that only Asn249 was actually glycosylated in PLAP (Millán et al., 1985b; Millán, 1986). Endo et al. (1988), using a combination of sequential exoglycosidase digestion and methylation analysis, reported that the structure of the *N*-linked glycan of PLAP is NeuNAcα2–3Galβ1–4GlcNAcβ1–2Manα1–3 (NeuNAcα2–3Galβ1–4GlcNAcβ1–2Manα1–6)Manβ1–4GlcNAcβ1–4($\pm$Fucα1–6)GlcNAc (Fig. 16). More recently, in a structure paper, Le Du et al. (2001) reported that evidence of *N*-linked glycosylation was observed at both Asn122 and Asn249, depending on the allelic variant that was crystallized. As PLAP and GCAP have identical putative glycosylation sites, Watanabe et al. (1992) investigated GCAP mutants, lacking one or both these putative glycosylation sites, expressed in COS-1 cells and concluded that both sites were glycosylated. Importantly, the glycan moieties are not required either for dimerization of the subunits or for the catalytic activity of GCAP (Pan et al., 1991).

Concerning other APs, Hua et al. (1985) estimated that the *N*-linked carbohydrates of human adult IAPs and also of the bovine isozyme contribute 16.7 and 8%, respectively, to the total mass of the proteins. Garattini et al. (1986) found 33% of the mass of the liver TNAP isoform to be derived from glycosylation. TNAP is also known as liver/bone/kidney AP since these are tissues known to express high levels of TNAP. These different isoforms differ in their post-translational modifications of the carbohydrate side-chains. Fractionation of bone and liver TNAP isoforms present in serum by serial lectin affinity chromatography demonstrated differences in the sugar chain structure with a lower content of high-mannose or hybrid-type sugar chains and a higher content of biantennary complex-type chains that similar isoforms isolated from the corresponding tissues

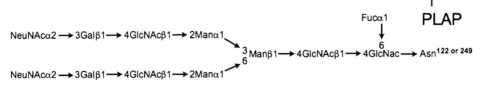
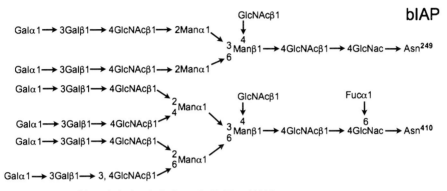

Fig. 16 Structure of the N-linked carbohydrates in PLAP and bIAP.

(Kuwana et al., 1991). The amino acid sequence suggests five putative N-glycosylation sites in TNAP, i.e. Asn123, Asn213, Asn254, Asn286 and Asn413 (Weiss et al., 1986). Nosjean et al. (1997) have confirmed that both the bone and liver TNAP isoforms are N-glycosylated and found N-glycosylation to be absolutely essential for TNAP activity but not for the activity of PLAP or IAP. However, the authors did not determine the number of N-glycosylation sites that actually are linked to oligosaccharides. Their data further suggested that the main difference between the bone and liver isoforms is due to a difference in O-glycosylation, i.e. bone-derived TNAP has some O-linked oligosaccharides, whereas liver-derived TNAP does not, but that differences in immunoreactivity between the bone and liver isoforms are mainly due to differences in N-glycosylation (Nosjean et al., 1997). It has also been reported that bone-derived TNAP contains more fucose (Kuwana et al., 1991) and sialic acid residues (Nosjean et al., 1997) compared with the liver isoform. Of interest is the fact that the degree of N-glycosylation of TNAP and also the mRNA and protein levels of this enzyme were found to be affected by retinoic acid treatment using the P19 cell line as a paradigm (Mueller et al., 2000). However, to date, the structures of the N- and O-linked oligosaccharides and other oligosaccharide differences of bone and liver TNAP have not been reported. Nevertheless, carbohydrate differences are the basis for the discrimination of up to seven isoforms of bone and liver TNAP detectable in serum by HPLC in children, healthy adults and patients with different pathologies (Section 9.1.1).

Whereas neither the nature nor the degree of O-linked glycosylation has been established for any of the IAP isozymes, the N-linked oligosaccharides were studied for the tumor-derived (IAP-like) Kasahara isozyme (Section 10.1), purified from FL amnion cells (Endo et al., 1990). Almost all of the oligosaccharides (98%) were acidic components that could be converted to neutral oligosaccharides upon sialidase digestion. Structural analysis of the oligosaccharides by sequential exoglycosidase digestion in combination with methylation analysis revealed that the Kasahara isozyme contains sialylated mono-, bi-, tri- and tetraantennary complex-type sugar chains with Galβ1–4GlcNAcβ1 outer chains. Some of the tetraantennary sugar chains contain a single Galβ1–4GlcNAcβ1–3Galβ1–4GlcNAcβ1 outer chain on their Manα1–6 arm. Both fucosylated and nonfucosylated trimannosyl cores were found in the sugar chains. However, it is of interest that the core portion of monoantennary oligosaccharide was not fucosylated and that of the tetraantennary oligosaccharides with a tetrasaccharide outer chain was completely fucosylated. Bublits et al. (2001) determined the compositions and structures of the major glycans linked to the two putative glycosylation sites, N249 and N410, of calf IAP. Using IAP purified from chyme and mucosa from the intestines of 12–15 calves, they extracted one anchorless fraction (Fraction I) and four GPI–IAP fractions (Fractions II–V) exhibiting increasing hydrophobicity caused by different fatty acid contents of their anchors. They found both glycosylation sites to be occupied. There were at least eight different types of glycans linked to Asn249, mainly nonfucosylated, bi- or triantennary structures with a bisecting N-acetylglucosamine. The glycans linked to Asn410 were also a mixture of at least nine, mainly tetraantennary, fucosylated structures with a bisecting N-acetylglucosamine. The most abundant glycopeptide (40 and 35%) at each glycosylation site is presented in Fig. 16. The majority of the glycans were capped at their nonreducing termini by α-galactose residues. In contrast to the glycans linked to other AP isozymes, no sialylation was observed in calf IAP. The glycopeptide mass fingerprints of both glycosylation sites and glycan contents did not differ between AP from mucosa and chyme (Bublits et al., 2001). It is possible, however, that some of the heterogeneity in the glycan structures uncovered by Bublits et al. may result in part from the presence of more than one bIAP isozyme in their purified enzyme preparations, due to the difficulty in obtaining a pure bIAP isozyme from the bovine intestine given the large number of very similar IAP genes expressed in the gut in this species (Manes et al., 1998).

4.3.2
Ectoplasmic Localization of APs via a GPI Anchor

Whereas ECAP is located in the periplasmic space of the bacterium, mammalian APs are ectoenzymes bound to the plasma membrane via a glycosylphosphatidylinositol (GPI) anchor (Low and Saltiel, 1988). The GPI-glycan moiety is added post-translationally to the nascent PLAP peptide chain in a transamidation reaction that involves the removal of a stretch of 29 hydrophobic amino acids from the C-terminus and the concomitant transfer of the preassembled GPI anchor to the new C-terminal amino acid. Sequencing and amino acid analysis in PLAP showed that the new C-terminus of the mature polypeptide chain was Asp484 of the proenzyme as deduced from the corresponding cDNA (Micanovic et al., 1988). Further analysis of the peptide showed it to be a peptidoglycan containing one residue of ethanolamine, one residue of glucosamine and two residues of neutral hexose. The inositol glycan is linked to the α-carboxyl group of the aspartic acid through the ethanolamine. Location of the inositol glycan on Asp484 of the proenzyme indicates that a 29-residue peptide is cleaved from the nascent protein during the post-translational condensation with the phosphatidylinositol-glycan. Despite the fact that all mammalian APs are thought to be anchored via GPI, PLAP remains to date the only isozyme where the anchoring site has been unequivocally defined.

PLAP was in fact used as a model for studying the biosynthesis of the PGI-protein linkage in intact cells and in cell-free systems. Interruption of the C-terminal 29-amino acid stretch by a charged residue prevented GPI-glycan addition and resulted in a protein that was secreted into the medium (Lowe, 1992). Also, deletion of amino acids in this hydrophobic region prevented the addition of GPI-glycan, suggesting that either the length of this spacer domain or the amino acids around the cleavage site are important for proper anchor attachment (Lowe, 1992). Interfering with the synthesis of dolichol-P-mannose, and thus of the GPI glycan structure, leads to aberrant processing of PLAP by causing a prolonged accumulation of the proform, which is then gradually processed into secretory forms by proteolytic removal of the C-terminal hydrophobic peptide (Takami et al., 1992). In a review, Amthauer et al. (1992) describe their work using an engineered "mini-PLAP" polypeptide of 28 kDa in which the N- and C-termini are retained but most of the interior of the PLAP molecule is deleted. Naturally, this "mini-PLAP" was devoid of catalytic activity, but this construct was useful in helping sort out the residues needed for proper processing of the nascent proteins into mature GPI-anchored proteins. They found that processing of nascent mutant proteins occurs only when a small amino acid is located at the site of cleavage and PGI attachment (omega site, ω). Mutations adjacent to and C-terminal to the omega site revealed that the $\omega + 1$ site is promiscuous in its requirements but that only glycine and alanine are effective at the $\omega + 2$ site. Translocation of the proprotein precedes C-terminal processing. *In vitro* analyses of the microsomal processing of mini-PLAP demonstrated that the anchor donor initially transferred to promini-PLAP is acylated and then progressively deacylated and that this process is cell specific but also can be protein dependent and that deacylation occurs in the endoplasmic reticulum immediately after GPI transfer (Chen et al., 1998). PLAP and other APs can be released from the membrane by GPI-specific phospholipases. However, the sensitivity of APs to phospholipases is variable in magnitude (~20–90%). Wong and Low (1992) found that phospholipase resistance is the result of acylation of the inositol ring in the GPI anchor. They also found evidence for a phospholipase C-sensitive precursor and its post-attachment conversion into a phospholipase C-resistant form. Two phospholipases are of particular interest in the context of our discussion of PLAP and mammalian APs in general. The bacteria or Trypanosoma-derived phosphatidylinositol-specific phospholipase C (PI-PLC or GPI-PLC) is able to release PLAP from the membrane by cleaving be-

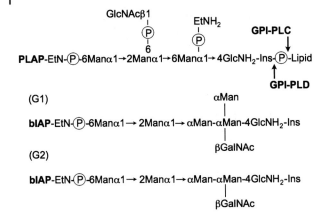

Fig. 17 Structure of GPI-anchor structure in PLAP and bIAP. The location of cleavage by GPI-PLC and GPI-PLD is shown for the PLAP anchor.

tween the phosphate and the diacylglycerol group, releasing this lipid anchor, which is left behind on the membrane (indicated in Fig. 17 in the anchor structure of PLAP). It is the activity of PI-PLC that is inhibited by inositol acylation. Serum-derived GPI-specific phospholipase D (GPI-PLD), in turn, cleaves between the inositol ring and the phosphatidic acid moiety and this activity is not affected by inositol acylation.

GPI anchors share a common core structure consisting of an ethanolamine phosphate, three mannose residues and a nonacetylated glucosamine linked to phosphatidylinositol. GPI-anchored proteins have the general structure protein–(ethanolamine-PO$_4$–6Manα1–2Manα1–6Manα1–4 GlcN1–6myo-inositol-1-P-lipid), where the structure within parentheses represents a minimal GPI moiety and the C-terminus of the protein is amide-linked to the GPI ethanolamine residue. The structure of the GPI-anchored glycan of PLAP was investigated by Redman et al. (1994), who reported it to be Thr–Asp–EtN-PO$_4$–6Manα1–2Manα1–6Manα1–4GlcN-(sn-1-O-alkyl-2-O-acylglycerol-3-PO$_4$-1-myo-D-inositol), with an additional ethanolamine phosphate group at an undetermined position. More recently, Fukushima et al. (2003) found that not only ethanolamine phosphate, but also α-N-acetylglucosaminyl phosphate diester (GlcNAc-P) residues are present in PLAP and are positioned as side-chains of its GPI-anchored glycan (Fig. 17). Of interest is the finding by these authors that the β-N-acetylglucosaminyl phosphate diester residue is important for GPI anchor recognition of aerolysin, a channel-forming toxin derived from the enteropathogenic bacterium *Aeromonas hydrophila*. In that context, it is of interest that both PLAP and IAP were also found to serve as receptors for the related *Aeromonas sobria* hemolysin (Wada et al., 2005). Since both PLAP and IAP are co-expressed in the neonatal intestine (Behrens et al., 1983), these AP isozymes are very likely involved in the pathogenesis of infantile diarrhea caused by these bacterial toxins.

Given that the detailed structure of the PLAP GPI anchor and also of its N-linked sugars have been elucidated, we are in a position to model these structures in the context of the 3D crystallographic structure of PLAP to provide a graphic view of how these post-translational modifications affect the overall surface of PLAP. Figure 18, kindly produced by Dr Mark R. Wormald (Oxford Glycobiology Institute, Oxford, UK), displays the 3D crystallographic structure of PLAP on which were attached the modeled structure of the GPI-anchor (prepared to scale) at Asp484 and also the N-linked carbohydrate chains (also prepared to scale) to both Asn122 and Asn249. Since Asp484 is perfectly conserved in GCAP and IAP, we can predict the same attachment site for these human isozymes. Experimental evidence also indicates that the bovine IAPs may be attached via residue 480 (Manes et al., 1998). However, the corresponding anchor residue for TNAP has not yet been identified. Pertinent to our discussion in Section 7.1 of skeletal mineralization, the bone isoform of TNAP functions as an ectoenzyme attached to the osteoblast cell membrane via its GPI anchor (Fedde et al., 1988; Hawrylak and Stinson, 1988; Hamilton et al., 1990; Hooper 1997). *In vitro* studies have demonstrated that bone TNAP is released from human osteoblast-line cells in an anchor-intact (insoluble) form attached to matrix vesicles (MVs) (Fedde, 1992; Anh et al., 1998), where it participates in the initiation of bone matrix mineralization (Tenenbaum, 1987; Bellows et al., 1991; Wennberg et al., 2000). *In vivo*, bone TNAP circulates as an anchor-depleted (soluble) homodimeric form indicating the conversion of the insoluble to the soluble form found in serum by the actions of two endogenous circulating phospholipases, GPI-PLC and GPI-PLD. Both of these cleavage enzymes are present in humans; but GPI-PLC is found at lower levels in serum compared with GPI-PLD, which is abundant in the circulation (Low et al.,

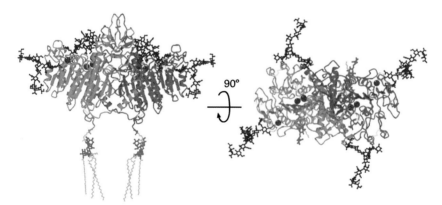

Fig. 18 Modeled structure of the N-linked oligosaccharides and GPI anchored attached to the 3D structure of PLAP. Five C-terminal residues, not present in the crystal structure, were added as per the sequence from SWIS-SPROT up to residue 484 (mature protein numbering). N-Linked complex glycan added at N122 and at N249. Light blue, peptide. red, crystallographic metal ions; yellow, disulfide bonds; orange, N122; purple, N-linked glycan; Green, GPI anchor; Dark green, glycan component of the GPI anchor. This figure was kindly produced and contributed by Dr Mark R. Wormald, Oxford Glycobiology Institute, University of Oxford, Oxford, UK.

1986; Davitz et al., 1987; Low and Prasad, 1988; Anh et al., 2001). Also of interest is a short discussion of the most common method for solubilizing membrane-bound AP activity, which involves extraction of tissues or cells with n-butanol to disrupt the cell membrane. Butanol extraction at pH 5.5 activates GPI-PLC, which then removes the 1,2-diacylglycerol moiety responsible for membrane anchoring, leaving a soluble, nonaggregated, hydrophilic, dimeric form of the enzyme (Malik and Low, 1986; Hawrylak and Stinson, 1988). In contrast, n-butanol extraction at neutral or alkaline pH results in the extraction of hydrophobic aggregated AP. These latter hydrophobic molecules, however, can be converted to hydrophilic dimeric enzymes either by GPI-PLC or protease treatment (Section 4.3.4).

Bublitz et al. (1993) found considerable heterogeneity of GPI-anchored IAP in the calf intestine. They purified the GPI-anchored IAP from the intestinal mucosa and chyme to homogeneity and found that both preparations contained ~2 mol fatty acid mol^{-1} of subunit and exhibited a very similar fatty acid composition with octadecanoate and hexadecanoate as prevalent components. No significant differences between the native GPI-anchored and hydrophilic APs from both sources were found regarding K_m, V_{max}, the type of inhibition and inhibition constants with the inhibitors L-Leu, L-Phe and L-Trp. The purified enzymes of both sources yielded diacylglycerol and phosphatidic acid, after treatment with GPI-PLC and GPI-PLD. Enzyme preparations of both sources appeared as heterogeneous mixtures of five fractions separable by octyl-Sepharose chromatography. Fraction I corresponded to the anchorless enzyme whereas fractions II–V differed in their susceptibility to phospholipases. The similarity of all measured parameters of both enzymes suggested that the GPI-anchored IAP of the mucosa is released into the chyme without changing the anchor molecule constituents (Bublitz et al., 1993). They also found that about 80% of IAP in the calf intestine is located intraluminally within chyme and only 20% within mucosa. The intraluminal AP is bound to vesicles, termed chymosomes, with diameters from 100 to 500 nm (Bublitz et al., 2001). Digestion of calf IAP with pronase and subsequent dephosphorylation of the released peptidyl-(Etn-P)$_2$-glycosylphosphatidylinositol with hydrogen fluoride generated eight glycosyl-inositol species, the largest of which had the structures G1 and G2 (Fig. 17, lower panel) as determined by Armesto et al. (1996). The other structures, G3 and G5, were homologues of G1 and G2, respectively, but with one l–2-linked mannopyranosyl residue shorter that G1 and G2. G4 was analogous to G2, lacking the N-acetylgalactosaminyl residue, and G6 was the next lower homologue of G4. Most of G4 and G6 occur substituted with a palmitoyl (G4, G6) or a myristoyl residue (G6,) probably attached to the inositol moiety. Hence the basic Man–Glc–inositol species are either substituted with an N-acetylgalactosaminyl residue or a fatty acid ester.

Given the fact that mammalian APs are membrane-anchored proteins, it is possible that the lipid environment of the cell membrane may affect the catalytic properties of the enzyme. In fact, Chakrabartty and Stinson (1985a) reported that the membrane-bound liver TNAP had the lowest affinity for eight different substrates and the K_i for the competitive inhibitors phenylphosphonate and P$_i$ was lower than for the purified soluble enzyme. Chang and Shiao (1994) studied the properties of PLAP embedded in a reverse micellar system prepared by dissolving the surfactant sodium bis(2-ethylhexyl)sulfosuccinate in isooctane as this microemulsion system provided a convenient instrumental tool to study the possible kinetic properties of the membranous enzyme in an immobilized form. Linear Lineweaver–Burk and Eadie–Hofstee plots for the substrate and linear Arrhenius plot for the temperature-dependent enzyme reaction were obtained in reverse micelles, suggesting that substrate diffusion was not a rate-limiting step in the system and exchange of materials between reverse micelles was very rapid. The catalytic constant (k_{cat}) of the enzyme was decreased and the K_m for the substrate and the K_i for phosphate were increased in reverse micelles. The enzyme was more stable in reverse micelles than in aqueous solution at 30 °C but, was unstable at higher temperature (65 °C). Further studies by this group (Huang et al., 1998) examined the pL (pH/p^{2}H) dependence of hydrolysis of pNPP over a pL range of 8.5–12.5 in both aqueous and reverse micellar systems. They proposed that the rate-limiting step of the hydrolytic reaction changes from phosphate release in aqueous solution to a covalent phosphorylation or dephosphorylation step in reverse micelles. Ierardi et al. (2002) incorporated polidocanol-solubilized TNAP, free of detergent, into resealed ghost cells. The enzyme could be completely released from the resealed ghost cell system using only GPI-PLC. Importantly, the enzyme bound to resealed ghost cell did not lose the ability to hydrolyze ATP, pyrophosphate and pNPP, but the presence of a ghost membrane, as a support of the enzyme, affected its kinetic properties specifically for pyrophosphate. Similar results were obtained when the enzyme was inserted into liposomes constituted from dimyristoylphosphatidylcholine, dilaurylphosphatidylcholine or dipalmitoylphosphatidylcholine (Camolezi et al., 2002). Milhiet et al. (2002) studied phospholipid bilayers made of a mixture of sphingomyelin–dioleoylphosphatidylcholine containing cholesterol or not in order to mimic the fluid-ordered lipid-phase separation in biological membranes. Spontaneous insertion of AP through its GPI anchor was observed inside both lipid bilayers with or without cholesterol, but AP insertion was markedly increased by the presence of cholesterol. Of interest is the fact that by separating human AP isoforms with and without intact GPI anchors in aqueous polymer phase systems, Raymond et al. (1994) demonstrated that the AP present in human serum is predominantly anchor degraded, whereas that in bile retains its anchor intact.

4.3.3
Nonenzymatic Glycation of APs

Pollak et al. (1983) reported studies on the nonenzymatic glycation of serum AP comprising short- and long-term exposure of the enzyme to glucose and other carbohydrates. Glucose and amino sugars clearly inhibited enzyme activity, which was in contrast to reducing and nonreducing disaccharides, which had an enhancing effect. After serum APs had been incubated with 18 nmol L^{-1} glucose for 180 min (short-term incubation), a subsequent extensive dialysis revealed full recovery of the enzymatic activity. This data, plus the demonstration of a sodium

[³H]borohydride-reducible glucose–protein adduct, indicated that initially a labile aldimine (Schiff base) had been formed. Binding experiments with [¹⁴C]glucose and failure of dialysis to achieve a recovery of enzymatic activity after long-term incubation suggested that subsequently a stable ketoamine product had been formed. Preliminary results further suggested that nonenzymatic glycation of AP also occurs *in vivo*. Streptozotocin diabetic rats had significantly lower serum AP activities than did nondiabetic controls. Subsequently, these authors determined blood glucose concentrations and serum AP in 32 healthy children during an oral glucose tolerance test and in ten type I diabetic children at 30-min intervals for 2.5 h (Pollak et al., 1984). A significant negative correlation was noted for blood glucose and AP during the oral glucose load. In diabetics, however, only children with marked blood glucose fluctuations showed this inverse relationship. McCarthy et al. (1998) confirmed that hyperglycemia in poorly controlled diabetic patients induces nonenzymatic glycation of human IAPs, altering the structure and activity.

4.3.4
Quaternary Structure of APs

Although, clearly, APs are dimeric enzymes, some reports have shown the existence of tetrameric forms of the enzyme and have even suggested that the tetrameric state is the normal state *in vivo*. Early reports had indicated that purified liver TNAP (Gerbitz et al., 1977), purified rat hepatoma TNAP (Yokota, 1978) and purified human first trimester placental AP (which at that stage is due to expression of TNAP) (Sakiyama et al., 1979) were tetrameric in structure. TNAP solubilized from human liver plasma membranes by nonionic detergents such as Triton X-100 or Nonidet P-40 was shown to have a tetrameric molecular weight when resolved on nondenaturing polyacrylamide gradient gels (Chakrabartty and Stinson, 1985b), and when this detergent-solubilized enzyme was treated with GPI-PLC, the tetrameric enzyme was converted to the dimeric, hydrophilic form (Hawrylak and Stinson, 1987, 1988; Nakamura et al., 1988). Protease treatment was also able to release hydrophilic dimeric APs. Subtilisin was shown to cleave a small (2 kDa) portion of the C-terminus from human PLAP (Abu-Hasan and Sutcliffe, 1984) and bromelain was also shown to cleave a similar-sized segment from PLAP (Jemmerson et al., 1984b), both treatments resulting in the release of hydrophilic forms of the enzyme. Hawrylak et al. (1990) showed that both purified TNAP and PLAP could be converted from hydrophobic tetrameric forms to hydrophilic dimeric forms by treatment with GPI-PLC. The authors proposed that interactions between the fatty acid chains of the GPI anchors and also ionic interactions between the protein subunits might be necessary to maintain a tetrameric structure and that this structure may have functional significance *in vivo*. The authors also pointed out that a tetrameric conformation has also been reported for boar seminal plasma AP (Iyer et al., 1988) and for bovine IAP (Krull et al., 1988). In fact, Bublitz et al. (1993), analyzing the GPI-anchored forms of bIAP, reported that enzyme preparations appeared as heterogeneous mixtures of five fractions separable by octyl-Sepharose chromatography. Fraction I corresponded to the anchorless enzyme whereas fractions II–V differed in their susceptibility to phospholipases. Fractions II and IV were completely split by GPI-PLC or GPI-PLD action and almost 50% of fraction III was split by GPI-PLC, whereas fraction V was resistant, and the susceptibility of these two fractions towards the action of GPI-PLD was considerably higher. Fatty acid analysis yielded molar ratios of fatty acids/IAP subunit of 1.78, 2.58, 2.24 and 3.37 for fractions II, III, IV and V, respectively. They concluded that fractions II and IV consisted of tetramers and octamers with two molecules of fatty acid per subunit. Fraction III was a tetramer, bearing one additional fatty acid molecule, localized on the dimer. Fraction V was an octamer, containing GPI-anchor molecules with three molecules of fatty acid per anchor molecule. Fedde et al. (1996b) examined the biochemical properties of TNAP in fibroblasts cultured from 16 patients with severe autosomal recessive hypophosphatasia. Gel filtration analysis indicated the presence of a mixture of dimeric and tetrameric TNAP in both patient and control samples.

So far, the significance of these observations remains unclear and the determination of the 3D structure of PLAP has not pointed to any obvious surface domain on the PLAP dimer that could be involved in tetramer formation. However, in reviewing the literature, it is of interest that Takeya et al. (1984) determined the molecular dimensions of the type 1 and 2 allozymes of PLAP using rotary shadowing and negative staining electron microscopic techniques. In the rotary shadowing technique, the molecules of the two phenotypes appeared to be approximately elliptical with slit-like structures in the center of the molecules (a 'donut' structure), suggestive of the groove between two subunits. The dimensions of the rotary-shadowed molecules were calculated as 10.1×5.7 nm for type 1 PLAP and 10.1×5.6 nm for type 2 PLAP allozymes. The predominant shape of the molecules in this method appeared to be rectangular, with a longitudinal stain-filled groove and with each of the half molecules (presumably 65 000 M_r subunit) very often appearing bi-lobed, accounting for molecules that appeared to have four pronounced electron-transparent regions. The dimensions of the negatively stained rectangular-shaped molecules were measured as 7.5×5.5 nm for type 1 PLAP and 7.6×5.4 nm for type 2 PLAP phenotypes (Takeya et al., 1984). Given the fact that the 3D structure reported by Le Du et al. (2001) does not account for a groove or slit separating the subunits of the dimer and also the fact that Takeya et al. used PLAP isolated at pH 7.4 and purified on an octyl-Sepharose column, it is very likely that what they observed in rotary shadowing and negative staining electron microscopy were indeed tetrameric, octameric or even more complex molecular forms of PLAP.

4.3.5
Subcellular Localization of APs

There is extensive information in the literature (McComb et al., 1979; and also references in this book) to support the statement that mammalian APs are ectoenzymes, anchored to the outer surface of the cytoplasmic membrane via a GPI anchor. Tokumitsu et al. (1981a) used an immunocytochemical approach to demonstrate that in the HeLa subline TCRC-1, in addition to its cell surface location, PLAP was also observed in the perinuclear space, endoplasmic reticulum and Golgi apparatus, consistent with the usual transport mechanism of membrane glycoprotein biosynthesis and transport. Clearly, the number of AP molecules present on the cell membrane at any one time on a particular cell type is bound to be highly variable and to depend on the stage of the cell cycle, state of differentiation, etc. However, some values have been obtained for some cell lines. Jemmerson et al. (1985d) used a radioimmunoassay to estimate the number of GCAP molecules in the A431 cell line to be $\sim 7.5 \times 10^5$ per cell, a value significantly higher than that observed for HeLa TCRC-1 cells (5×10^4) that express the type 1 PLAP allozyme. The number of the AP molecules could be significantly upregulated (up to 10-fold) by treating the cells with modulating agents, including sodium butyrate, prednisolone and hyperosmolar sodium chloride.

Although it remains obscure why proteins are endowed with a GPI anchor, its presence appears to confer some general functional characteristics to proteins, as summarized by Brown and Wanweck (1992). First, the GPI anchor is a strong apical targeting signal in polarized epithelial cells; second, GPI-anchored proteins do not cluster into clathrin-coated pits but instead are concentrated into specialized lipid domains in the membrane, including so-called smooth pinocytotic vesicles or caveoli; third, GPI-anchored proteins can act as activation antigens in the immune system; fourth, when the GPI anchor is cleaved by GPI-PLC or GPI-PLD, second messengers for signal transduction may be generated; and last, the GPI anchor can modulate antigen presentation by major histocompatibility complex molecules.

Some of these characteristics may apply to APs. It has been reported that the presence of the GPI anchor endows APs with high lateral mobility on the cell membrane (Noda et al., 1987a). Also, it seems that APs are not always distributed uniformly on the surface of the cells and the GPI modification also appears to be responsible for the presence of APs in rafts, a type of functionally important membrane microdomain enriched in sphingolipids and cholesterol. Makiya et al. (1992) found PLAP concentrated in clathrin-coated pits, where it was neither adsorbed on the surface of the vesicles nor appeared to be due to plasma membrane contaminants, but was located in the lumen of the vesicles instead, where it was postulated to serve a physiological role in the transcytosis of IgG molecules in the placenta (Section 7.3). Another important example of vesicle compartmentalization is the presence of TNAP on the surface of osteoblast-derived MVs. In skeletal tissue, TNAP is confined to the cell surface of osteoblasts and chondrocytes, including the membranes of their shed MVs (Ali et al., 1970; Bernard, 1978). In fact, by an unknown mechanism, MVs are markedly enriched in TNAP compared with both whole cells and the plasma membrane (Morris et al., 1992). Leach et al. (1995) reported that regulation of plasma membrane TNAP and that of MV TNAP are independent events and suggested that MV biogenesis is independent of and distinct from plasma membrane biogenesis. Furthermore, they showed that the alternative promoter usage of the TNAP gene was not responsible for the differential localization of this enzyme in MVs and suggested that MVs and plasma membrane TNAP are regulated differently at a post-transcriptional level.

Another example relates to leukocytes, particularly neutrophils. Smith et al. (1985) homogenized human polymorphonuclear leucocytes in isotonic sucrose and subjected them to analytical subcellular fractionation by centrifugation on a discontinuous sucrose density gradient to give a preparation of a highly purified phosphasome fraction, free of plasma membrane components. Electron microscope cytochemistry of the purified fraction identified the phosphasomes as regular- and irregular-shaped spheres and rods and TNAP was found to be associated with the inner surface of the vesicle membrane. Borregard et al. (1990) identified a highly mobilizable subset of human neutrophil intracellular vesicles that contain tetranectin and TNAP that are functionally distinguishable from other types of granules. These organelles may play an important role as stores of membrane proteins that are mobilized to the cell surface during stimulation by inflammatory mediators (Borregaard et al., 1990; Sengelov et al., 1992).

Yet another example relates to the compartmentalization of IAP on surfactant-like particles (SLPs) produced by enterocytes. IAP is found associated with the brush border of the intestinal epithelium and enriched in SLPs (Eliakim et al., 1989, 1997; Alpers et al., 1994, 1995; Goetz et al., 1997). Fat feeding increases SLP secretion (Eliakim et al., 1991; Tietze et al., 1992) and an IAP-containing membrane surrounds the lipid droplet in the enterocyte (Yamagishi et al., 1994b; Zhang et al., 1996b). These findings suggest a role for SLP in fat absorption involving the induction of IAP by fat, which in turn results in IAP-induced SLP secretion.

A more enigmatic localization of APs has also been reported. Tokumitsu et al. (1981b) used human gastric carcinoma KMK-2 cells that display plasma cell polarity and express TNAP, to demonstrate intense AP staining in the perimitochondrial membrane. They found no differences in intensity between the mitochondria near the plasma membrane or those close to the nucleus. Risco and Traba (1994) reported the presence of an AP in mitochondrial membrane preparations from pig renal epithelial LLC-PK1 cells. This AP activity was Mg^{2+} dependent and inhibited by levamisole, indicating that the enzyme was most likely TNAP. Preparations of mitochondrial membrane from LLC-PK1 cells also showed 25-$(HO)_2$D3-1-hydroxylase (1-hydroxylase) and 25-$(HO)_2$D3-24R-hydroxylase (24-hydroxylase) activities being both enzymes responsive to 8Br-cAMP-mediated regulation (Risco and Traba, 1994). It has been suggested that mitochondrial AP activity could be involved as an intracellular signal in the regulation of 25(OH)D3 metabolism for the synthesis of 1,25$(OH)_2$D3 and 24,25$(OH)_2$D3 in renal LLC-PK1 cells through the cAMP protein kinase system (Municio and Traba, 2003). Nevertheless, it is still unclear what the significance of this mitochondrial AP staining may be. Some reports have even linked high AP expression to apoptotic events (Hui et al., 1997b; Souvannavong et al., 1997) and cytochrome c, a mediator of the intrinsic pathway of apoptosis, has been shown to bind to AP. Dadak et al. (1999) reported that IAP affects ferricytochrome c (cyt c-Fe^{III}) by changing its optical properties, redox state and conformation. The proportion of products formed in the mixtures depends on pH, ionic strength, temperature and the buffer composition but the reaction reaches equilibrium between cyt c-Fe^{II} and other cyt c conformers (Dadak et al., 2002). We will need to wait for future work to shed light on these seemingly disparate observations.

Recently, a nuclear localization of AP has been observed in cultured human cancer cells by electron microscopic cytochemistry (Yamamoto et al., 2003). In a study using Hep-G2 hepatocellular carcinoma, A-375 malignant melanoma and BxP-3 pancreatic carcinoma cells, these authors found high AP activity in the nucleolus of these cells and the localization of AP changed with the stages of the cell cycle. The patterns of AP localization during the interphase were found to be either cytoplasmic, nuclear or a mixture of both types. Moreover, at the mitotic phase, the reaction products were observed on the chromosome. In the cultured malignant melanoma cells, the appearance ratio of the AP reaction products on the nucleolus (33.9%) showed a higher ratio compared with normal cultured fibroblasts (6.3%). These data suggest that a high level of AP may be related to a high level of proliferation in cancer cells (Yamamoto et al., 2003).

5
Enzymatic Properties

5.1
Catalytic Inhibition

Mammalian APs display the unique kinetic property, not shared by their bacterial ancestors, of being inhibited stereospecifically by L-amino acids and peptides through an uncompetitive mechanism (McComb et al., 1979; Fishman, 1990). The discovery of AP inhibitors stems from early work aimed at answering a clinical question: does an increased serum AP reflect a liver or a skeletal pathology? This challenge prompted Fishman and collaborators to evaluate 139 chemicals, including bile acids, metals, sulfhydryl compounds and amino acids for their ability to inhibit selectively different APs (Fishman and Sie, 1971). This effort did not succeed in differentiating liver from bone AP, but established the means to discriminate those two isoforms from the APs expressed in the intestine and placenta through the use of L-amino acid inhibition. L-Phenylalanine was found to inhibit the activity of IAP and PLAP isozymes at concentrations 30 times lower than the isoforms from liver, bone or kidney. Conversely, the liver/bone/kidney AP isoforms were inhibited by L-hArg at concentrations 10–20 times lower than those needed to inhibit IAP or PLAP. Several other isozyme-specific inhibitors of APs have since been reported, including L-Trp, L-Leu (Fishman and Sie, 1971; Doellgast and Fishman, 1977) and L-homoarginine (Lin and Fishman, 1972), and also some unrelated compounds, such as levamisole, the L-stereoisomer of tetramisole (Van Belle, 1976), and theophylline, (a 1,3-dimethyl derivative of xanthine) (Farley et al., 1980). The inhibition is of a rare uncompetitive type (Cornish-Bowden, 2004) and although the biological implications of this inhibition are not known, the inhibitors have proven to be useful in the differential determination of AP isozymes in clinical chemistry since their discovery (Fishman, 1974; Mulivor et al., 1978b).

There are several types of inhibition (Cornish-Bowden, 2004). Broadly, there are reversible and irreversible inhibition. Irreversible inhibition occurs when the inhibitor interaction with the enzyme inactivates the active site permanently. The catalytic poisons are included in this class. The reversible types of inhibition can be classified according to the apparent Michaelis–Menten parameters that are affected and these are termed competitive (the most common type), uncompetitive, mixed inhibition and noncompetitive. "Competitive" inhibition can be thought of mechanistically as competition between two molecules, substrate and inhibitor, for binding at the active site. Kinetically, competitive inhibition manifests itself as an apparent increase in the K_m of the enzyme; the V_{max} is unaffected, but the apparent V_{max}/K_m decreases. This can be thought of as the rate of reaction being lowered by an inhibitor, decreasing the number of active sites available for substrate. This effect can be overcome by adding more substrate; at high enough substrate concentrations, the substrate will compete out the inhibitor from all the active sites and the maximum reaction rate will once again be attained. "Uncompetitive" inhibition, on the other hand, occurs when an inhibitor binds to the active site only after the enzyme–substrate complex has formed. In this instance, the apparent V_{max} is affected and the apparent K_m is decreased to the same extent, i.e. apparent V_{max} is decreased and V_{max}/K_m is unchanged.

The following schemes and equations are presented in this section to help the reader understand better some of the conclusions derived from mutagenesis and enzyme kinetic experiments presented in Sections 4.1.5, 5.1.2 and 5.5. The general reaction scheme of APs can be depicted with the following equation (Scheme 1):

$$E + S \underset{k_{-1}}{\overset{k_1}{\rightleftharpoons}} E \cdot S \underset{k_{-2}}{\overset{k_2}{\rightleftharpoons}} E\text{-}P \underset{k_{-3}}{\overset{k_3}{\rightleftharpoons}} E + P$$

where E represents the enzyme, S the substrate and P the product and k_1, k_{-1}, k_2 and k_{-2} are the rate constants describing formation and disappearance of the noncovalent complex [E•S], and k_2, k_{-2}, k_3 and k_{-3} describe the formation and disappearance of the covalent catalytic intermediate [E-P]. From this general scheme, the competitive and uncompetitive inhibitors can be described in terms of inhibition constants, with reference to Scheme 2:

$$\begin{array}{ccc} E \cdot I & & E \cdot S \cdot I \\ \Updownarrow K_{ic} & & \Updownarrow K_{iu} \\ I & & I \\ + & & + \\ E + S \underset{k_{-1}}{\overset{k_1}{\rightleftharpoons}} & E \cdot S \underset{k_{-2}}{\overset{k_2}{\rightleftharpoons}} & E\text{-}P \underset{k_{-3}}{\overset{k_3}{\rightleftharpoons}} E + P \end{array}$$

where K_{ic} and K_{iu} represent the inhibition constants for competitive and uncompetitive inhibition, respectively. "Mixed" inhibition has both competitive and uncompetitive components. "Noncompetitive" inhibition refers to that rare case of mixed inhibition when $K_{ic} = K_{iu}$. This happens when the inhibitor binds to a secondary site (both in free and substrate-occupied enzyme), leading to enzyme inhibition. In the particular case of the uncompetitive inhibition of APs by amino acids, the reaction can be represented by reaction Scheme 3, in which the inhibitor prevents dephosphorylation of the phosphoenzyme intermediate:

$$\text{E-P} \cdot \text{I}$$
$$\updownarrow K_{EPI}$$

$$\text{E} + \text{S} \underset{k_{-1}}{\overset{k_1}{\rightleftharpoons}} \text{E} \cdot \text{S} \underset{k_{-2}}{\overset{k_2}{\rightleftharpoons}} \text{E-P} \underset{+\ \text{I}}{\overset{k_3}{\underset{k_{-3}}{\rightleftharpoons}}} \text{E} + \text{P}$$

Hence, the different kinetic constants, i.e. k_{cat}, K_m, k_{cat}/K_m, K_{EPI} and K_i, depend on the following parameters:

$$k_{cat} = \frac{k_{+2}}{(1 + k_{+2}/k_{+3})}$$

$$K_m = \frac{k_{-1} + k_{+2}}{k_{+1}(1 + k_{+2}/k_{+3})}$$

with

$$k_{cat}/K_m = \frac{k_1}{(1 + k_{-1}/k_{+2})}$$

and

$$K_{EPI} = \frac{[\text{E-P}][\text{I}]^o}{[\text{E-P} \cdot \text{I}]}$$

with

$$K_i = K_{EPI}(1 + k_{+3}/k_{+2}).$$

The Michaelis constant K_m and the catalytic rate constant k_{cat} (determined in the absence of inhibitor) are complex functions of k_1, k_2 and k_3. The dissociation constant K_{EPI} describes the affinity of reversible inhibitor binding.

5.1.1
Competitive and Noncompetitive Inhibitors of APs

Clearly, any chelating agent that removes Zn^{2+} and Mg^{2+} from the catalytic site results in a partially or completely demetalated apoAP preparation that will have lost part or all of its catalytic activity. As discussed by McComb et al. (1979), compounds such as EDTA, cyanide, 1,10-phenanthroline, 8-hydroxyquinoline-5-sulfonic acid and Chelex all fall in this category. Interestingly, also bisphosphonates, such as alendronate, pamidronate and zoledronate, which are pyrophosphate analogs used for the treatment of osteoporosis, appear to inhibit TNAP activity via chelation of the active site Zn^{2+} and Mg^{2+} (Vaisman et al., 2005). Understandably, given the reaction mechanism, APs are susceptible to product inhibition by phosphate (P_i) itself through a competitive mechanism (McComb et al., 1979) and therefore the concentrations of P_i in biological fluids will impact on the ability of APs to hydrolyze natural substrates. Indeed, Coburn et al. (1998) examined the kinetics of serum TNAP from 49 patients with endogenous phosphate concentrations ranging from 0.5 to 2.1 mmol L^{-1} and AP activity ranging from 41 to 165 nmol min^{-1} mL^{-1} using pyridoxine 5′-phosphate (PLP) as a substrate at pH 7.4. These authors found that under physiological conditions, serum TNAP activity toward PLP was reduced by ~50% by the normal physiological P_i concentration. Interestingly, however, the presence of P_i increases the half-life of skeletal TNAP when subjected to inactivation by EDTA treatment by preventing the dissociation of active-center Zn^{2+} (Hall et al., 1999).

Arsanilic acid was found to be a competitive inhibitor of calf IAP (Brenna et al., 1975) and this property was used for the chromatographic purification of this isozyme through binding to a Sepharose column containing a diazonium salt derived from 4-(p-aminophenylazo)phenylarsonic acid and elution by competition with P_i. Orthovanadate is another well-known and potent competitive inhibitor of all AP isozymes ($K_i < 1 \mu M$) as it competes with P_i for the same binding site on the enzyme (Seargeant and Stinson, 1979; Register and Wuthier, 1984). In fact, this property has been used to develop an assay to quantify trace levels of free vanadium(IV) and (V) (Crans et al., 1990). Bismuth was also shown to be a competitive inhibitor of AP, particularly effective against the PLAP isozyme (Komoda et al., 1981). Phenylphosphonate, a nonhydrolyzable analog of pNPP, is also a competitive inhibitor of APs (Chakrabartty and Stinson, 1985a).

Myo-inositol hexaphosphate (phytic acid), a naturally occurring plant constituent capable of forming coordinate complexes with polyvalent cations, reversibly inhibits the hydrolysis of pNPP by IAP (Martin and Evans, 1989). Inhibition is strictly competitive with $K_i = 260 \mu M$ at pH 8.0 and 25 °C. A similar result is also obtained in the presence of an inositol hexaphosphate–cupric ion coordinate complex. This inhibition of substrate hydrolysis by either inositol hexaphosphate or its Cu(II) complex should be differentiated from the effect of the latter on IAP in the absence of substrate. In this case, total but irreversible inhibition of the enzyme is obtained in a time-dependent process that apparently involves a metal ion-exchange reaction, i.e. substitution of the enzyme's zinc atoms for copper atoms (Martin and Evans, 1991a,b).

Isatin (indole-2,3-dione), a potent inhibitor of monoamine oxidase in human urine and rat tissues, was found to inhibit rat testicular AP through a noncompetitive mechanism (Kumar et al., 1978). Chang et al. (1981) showed that periodate-oxidized AMP could act as a substrate, in addition to a noncompetitive inhibitor and an affinity label of human PLAP. Two dyes, Procion Red HE-3B and Cibacron Blue F3G-A, were also found to bind to and inhibit calf IAP via a noncompetitive mechanism with an inhibition constant $K_i = 0.03$ mM. This property was used to develop a chromatographic purification procedure for calf IAP using the dyes immobilized onto a solid phase (Kirchberger et al., 1987). Orellanine (3,3′,4,4′-tetrahydroxy-2,2′-dipyridyl-1,1′-dioxide), a fungal toxin, specifically inhibits TNAP activity through a noncompetitive mechanism, whereas the IAP and PLAP isozymes are inhibited competitively (Ruedl et al., 1989). Okadaic acid, a toxin metabolite responsible for shellfish poisoning, is a known inhibitor of protein phosphatase 1 and 2A. However, okadaic acid is also a noncompetitive inhibitor of human PLAP and calf IAP with a K_i of 2.05 and 3.15 µM, respectively (Mestrovic and Pavela-Vrancic, 2003).

Acetazolamide, furosemide, ethacrynic acid and chlorothiazide, diuretics of considerable structural diversity, all inhibit rat and human kidney TNAP (Price, 1980). The inhibition is reversible and the mechanism is of a mixed type, having both competitive and noncompetitive characteristics. The K_i was calculated as 8.4, 7.0, 2.8 and 0.1 mM for acetazolamide, furosemide, ethacrynic acid and chlorothiazide, respectively. Chlorothiazide is a much more potent inhibitor of TNAP than the other three diuretics. Phenylene-1,3-diphosphonate, 2,6-dinitrophenylphosphonate and phosphonoacetaldehyde were found to be competitive inhibitors of rat IAP, with K_i values in the range 16–80 µM (Shirazi et al., 1981a,b). Adenosine 5'-(β-thio)diphosphate and adenosine 5'-(γ-thio)triphosphate are also very potent inhibitors, with K_i values of around 10 µM. The inhibition produced by these thiophosphates is mainly competitive but with a slight noncompetitive element. Adenosine 5'-(β,γ-imido)triphosphate is also a competitive inhibitor of APs, but oxidation of the ribose moiety of this compound with $NaIO_4$ results in an active site-directed irreversible inhibitor that could be of general use in studies of the mechanism of action of these isozymes. The dye Reactive Yellow 13, an affinity reagent for IAP, inhibits all human APs in solution. The inhibition depends markedly on the presence of a phosphate acceptor such as diethanolamine. In the case of IAP, the inhibition is noncompetitive with respect to the substrate and competitive with respect to the phosphate acceptor. However, the dye is an uncompetitive inhibitor with respect to both substrate and phosphate acceptor in the case of nonintestinal APs (Williams et al., 1985). Vovk et al. (2004) found that calix[4]arenes bearing one or two methylenebisphosphonic acid fragments displayed stronger inhibition of calf IAP than simple methylenebisphosphonic or 4-hydroxyphenyl methylenebisphosphonic acids according to a partial mixed-type inhibition with inhibition constants of 0.38 and 2.8 µM, respectively, in Tris–HCl buffer at pH 9.

5.1.2
Uncompetitive Inhibition

5.1.2.1 Mechanism of Inhibition in PLAP/GCAP

The mechanism of uncompetitive inhibition of mammalian APs is best understood for the human PLAP and GCAP isozymes. The most invariant feature shared by the GCAP phenotypes that distinguishes them from PLAP is the inhibition by L-Leu. Two independent studies, one starting with PLAP and mutagenizing residues to the GCAP homologues (Hummer and Millán, 1991) and the other starting with GCAP and mutagenizing residues to PLAP homologues (Watanabe et al., 1991), showed that position 429 was responsible for the differential inhibition by L-Leu. A single E429G substitution in GCAP accounts for the sensitivity of

GCAP for L-Leu and the reduced heat stability and decreased K_m of this AP isozyme (Hoylaerts and Millán, 1991; Hummer and Millán, 1991; Hoylaerts et al., 1992a).

The inhibition of wt PLAP and wt GCAP by L-Phe and L-Leu follows an uncompetitive mechanism, i.e. double reciprocal plots of the rate of product formation versus substrate concentration, $1/v$ versus $1/[S]$, yield parallel lines. Furthermore, L-Leu inhibition is ~17-fold more effective for wt GCAP ($K_i = 0.54$ mM) than for wt PLAP ($K_i = 9.2$ mM) and this difference turned out to be key in elucidating the pivotal role of residue 429 in this mechanism of inhibition (Hummer and Millán, 1991). Both L-Phe and L-Leu inhibit [G429]PLAP to comparable extents (Table 4), but to a higher degree than wt PLAP. Figure 19 shows the uncompetitive nature of the inhibition of [G429]PLAP by L-Leu. This analysis surprisingly indicated that, although with considerably reduced affinities, D-Leu and D-Phe are also uncompetitive inhibitors of [G429]PLAP and to a somewhat lower extent also of the wt PLAP and wt GCAP. Mutagenesis of Glu429 in wt PLAP for Ser429 (present in IAP) or His429 (found in TNAP) produces mutants with K_i values for L-Leu inhibition comparable to those of [G429]PLAP, i.e. for [S429]PLAP $K_i = 0.25$ mM and for [H429]PLAP $K_i = 0.3$ mM.

Table 4 Inhibition properties of some human enzymes and mutants. Data taken from Hoylaerts et al. (1992) and Kozlenkov et al. (2004).

Isozyme	K_i				
	L-Phe (mM)	L-Leu (mM)	L-hArg (mM)	Levamisole (µM)	Theophylline (µM)
hPLAP	0.61	9.2	59	563	947
hGCAP	0.6	0.54	n.d.	n.d.	n.d.
[G429]PLAP	0.15	0.2	12	474	200
hTNAP	19	n.d.	1.4	16	25

Clearly, mutations at position 429 in wt PLAP have very pronounced effects on the inhibition efficiency of amino acids. While these mutations were found to affect greatly the conformation of the surface loop of PLAP (Hoylaerts and Millán, 1991), they do not introduce any major structural changes in the active site pocket, as was revealed experimentally by performing inhibition experiments using free P_i as a competitive inhibitor. P_i displays very similar affinities for the different isozymes, i.e. K_i (wt PLAP) = 40 µM, K_i ([G429]PLAP) = 40 µM and K_i (wt GCAP) = 25 µM. Also, the E429G substitution has no major effects on k_{cat} (Table 2), indicating that the rates of phosphorylation (k_2) and dephosphorylation (k_3) (see schemes and equations in Section 5.1) are not affected in [G429]PLAP. However, [G429]PLAP displays 3–5 times lower K_m and K_i values (measured with D-Leu, D-Phe and L-Phe) than wt PLAP, indicating an increased accessibility of substrate

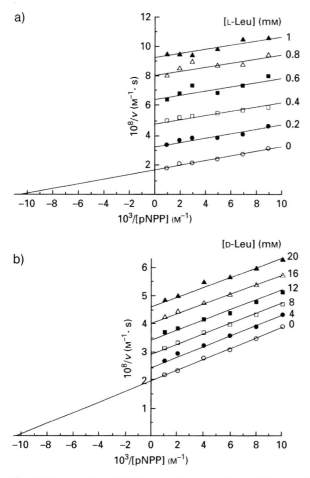

Fig. 19 Uncompetitive inhibition. Double reciprocal plot of the rate of p-nitrophenol formation (v) versus substrate concentration ([pNPP]) during inhibition of [G429]PLAP (0.5 U L^{-1}) by increasing [L-Leu] (0–1 mM) (a) and by increasing [D-Leu] (0–20 mM) (b). Taken from Hoylaerts et al. (1992) and reproduced with permission from the Biochemical Journal.

(higher k_1 and/or lower k_{-1}) and inhibitors (lower K_{EPI}) in the active site of the mutant enzyme. Curiously, one rare variant of PLAP, the D allozyme (Section 4.2), displays enzymatic properties at variance with those of the common PLAP allozymes including a slow migration in starch gels and inhibition not only by L-Phe, but also by L-Leu (Doellgast and Fishman, 1976). This allelic variant was found to display two substitutions, [RArg209; G429]PLAP (Wennberg et al., 2002), which explain the differential behavior of this allozyme due to the presence of a E429G mutation.

Substitution of the active site residue Asp91 (equivalent to Asp101 in ECAP) for Ala91 or Asn91 generates PLAP mutants that are still active but display considerably reduced turnover numbers (Table 2). Whereas the K_ms in [A91]PLAP and

[N91]PLAP rose by only 2–3-fold compared with wt PLAP, the catalytic rate constants for [A91]PLAP and [N91]PLAP dropped by a factor of 13–16 (Table 2). Both [A91]PLAP and [N91]PLAP, are inhibited by L-Leu according to an uncompetitive mechanism and their K_i values are comparable to that of wt PLAP, indicating that Asp91 is not involved in stabilizing L-Leu during enzyme inhibition. Very different effects were observed when mutagenizing Arg166. In the ECAP active site, residue Arg166 has been attributed a stabilizing role during the positioning of the phosphorylated substrate in the active site of the enzyme (Fig. 7) (Stec et al., 2000). Introduction of an R166A substitution in wt PLAP resulted in a 12-fold drop in K_m and a 24- fold drop in k_{cat} (Table 2) indicating that this mutation causes the catalytic efficiency (k_{cat}/K_m) to drop by a factor of almost 300. More importantly, the replacement of the positively charged Arg166 residue by a neutral Ala166 residue had a major effect on the mechanism of inhibition by L-Leu and by L-Phe, i.e. inhibition was observed only at higher concentrations and in the case of L-Leu, the inhibition mechanism changed from uncompetitive to a more complex mixed-type inhibition, with double reciprocal plots that intersect below the x-axis (Hoylaerts et al., 1992a).

Upon further characterization of the kinetic behavior of single and double PLAP mutants in the presence of the D- and L-enantiomers of L-Phe and L-Leu and also L-leucinamide and leucinol, Hoylaerts et al. (1992a) proposed the following molecular mechanism for the uncompetitive inhibition of PLAP and GCAP: The inhibition occurs through three interaction points of the inhibitor, i.e. the carboxylic group of L-Phe or L-Leu attacks the active site Arg166 during catalysis, the amino group of the inhibitor interacts with Zn1 in the active site, while the side-chain of the inhibitor is stabilized by the loop containing residue 429. This mechanism has now been largely confirmed, but also refined, by new structural information obtained recently after the phosphorylated intermediate of PLAP with L-Phe in the active site was resolved at 1.6 Å resolution (Llinas et al., 2005). In that structure, the amine group of L-Phe interacts with the phosphate moiety of the phosphoseryl while Glu429 stabilizes its carboxylic group through two water molecules and its phenyl ring stacks against Tyr367 and Phe108 (Fig. 20A). Therefore, owing to the positioning of L-Phe, the water molecule involved in the nucleophilic attack on the phosphoseryl can no longer reach the phosphoseryl intermediate and therefore the hydrolysis of the phosphoserine cannot occur and the reaction is blocked. Given that L-Phe makes a direct interaction with the phosphate, the inhibitor interacts better with the phosphoseryl intermediate than with the free serine, which would be too far to stabilize the ligand properly. This explains the uncompetitive character of this inhibition and why Glu429 is vital to this network and crucial for inhibitor selectivity. In PLAP, two water molecules bridge the gap from Glu429 to the phosphate moiety of the phosphoserine, one of these being the highly conserved water molecule involved in the nucleophilic attack on the phosphoseryl (Fig. 20B). Since the step of hydrolysis of the P-Ser92 is the limiting step in the catalytic process of APs, the environment around this water molecule will affect its nucleophilic strength and therefore the catalytic rate constant of the reaction. Thus, Glu429 interacts with the nucleophilic water molecule through an

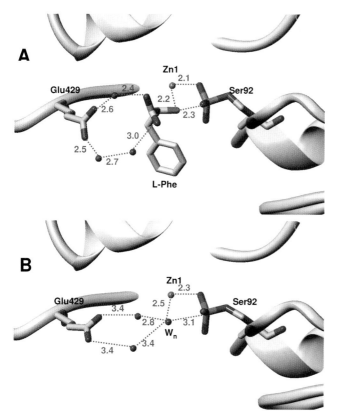

Fig. 20 Positioning of the uncompetitive inhibitor L-Phe in the active site of PLAP. (A) Interaction of L-Phe with Glu429 and PO$_3$-Ser92, Phe108 and Tyr367 in the PLAP active site. (B) Network of water molecules between Glu429 and PO$_3$-Ser92 in the active site of the native PLAP. Wn, nucleophilic water molecule. Taken from Llinas et al. (2005) and reproduced with permission from the *Journal of Molecular Biology*.

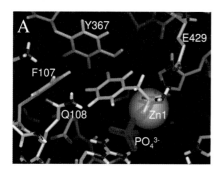 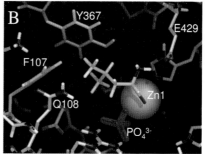

Fig. 21 Computer modeling of L-Phe (A) and L-Leu (B) in the active site of PLAP. The program FlexX was used for the docking predictions. The conformations of the ligands with the best energy score are shown. The program Pymol (http://pymol.sourceforge.net) was used to generate the figures. Taken from Kozlenkov et al. (2002) and reproduced with permission from the *Journal of Biological Chemistry*.

additional water molecule (Fig. 20B) and the nucleophilic character of this additional water molecule will vary depending of the identity of the residue at position 429.

Other residues located near Glu429 include His317 (Section 4.1.2) and Tyr367 (Section 4.1.6). Kozlenkov et al. (2002) examined the inhibition properties of the H317A, Y367A and Y367F PLAP mutants towards L-Phe, L-Leu and also L-(2-phenyl)glycine. Since all these enzyme variants had comparable kinetic parameters (Table 2), any observed changes in inhibition properties should mainly reflect changes in the true binding constants for the inhibitors. The Y367A mutation in PLAP was found to have a very profound destabilizing effect on the inhibition by L-Phe and L-Leu, increasing their K_is 17.5- and 12.7-fold respectively. The Y367F substitution had similar, although less pronounced, consequences, increasing K_is by 3.2- and 4-fold, respectively. These results show that the side-chain of Tyr367 is crucial for the binding of inhibitors to PLAP, most likely by providing a local binding area for the hydrophobic side-chains of Leu and Phe (Fig. 21). In the double mutants, [A367, G429]PLAP and [F367, G429]PLAP, the pattern of inhibition by L-Leu and L-Phe was intermediate. The effect of the Y367A substitution overweighed that of E429G, while the effect of Y367F was completely rescued by the E429G mutation. Inhibition of [F367]PLAP and [A367]PLAP by D-Phe was even less pronounced than the inhibition by L-Phe and so low that accurate determination of K_i was difficult ($K_i > 90$ mM). While the effect of the Y367A mutation was the most significant, the H317A substitution also had a mild influence on the inhibition by both L-Phe and L-Leu, decreasing K_i 2.6- and 2.5- fold respectively. The [A317; G429]PLAP double mutant displayed a similar decrease in K_i compared with [G429]PLAP. They also studied the inhibition of PLAP mutants by L-(2-phenyl)glycine, which from a structural viewpoint can be considered as a truncated form of L-Phe. The results showed that L-(2-phenyl)glycine is as potent an inhibitor of PLAP and PLAP mutants as L-Phe and L-Leu. However, L-(2-phenyl)glycine does not discriminate between the Y367A and Y367F mutations, suggesting that its shorter hydrophobic side-chain make less contact with Tyr367. These results indicate that substitutions at positions 367 and 429 can act independently in determining inhibition properties of PLAP. It is clear that removal of Glu429 facilitates access of the inhibitor to the active site, whereas removal of Tyr367 eliminates stabilization of the inhibitor's positioning at the active site.

5.1.2.2 Inhibitor Binding in TNAP

Recently, Kozlenkov et al. (2004) clarified which amino acid residues are responsible for the marked differences in inhibition selectivity between TNAP and PLAP. In that study, the modeled structure of TNAP was superimposed onto the PLAP structure and the amino acid differences in the active site area (the 12 Å region around the catalytic Zn1 ion) were pinpointed. In total, six TNAP residues were identified that cluster into two groups; the first group includes residues 433 and 434 and the second includes residues 108, 109, 120 and 166 (using TNAP numbers). Using a large number of PLAP and TNAP site-directed mutants, the authors established the likely location of the binding of L-hArg, levamisole and theophylline on the TNAP structure. The first conclusion relates to the crucial role of TNAP residue 108 in determining the selectivity of inhibition by L-amino acids. A single substitution, i.e. E108F in TNAP, completely reversed the pattern of TNAP inhibition, rendering it >10 times more sensitive to L-Phe than to L-hArg inhibition. The reciprocal F107E mutation in PLAP made the enzyme nearly equally sensitive to L-Phe and L-hArg. However, the double [E108; G109]PLAP mutant displayed a 2-fold selectivity towards L-hArg compared with L-Phe. The importance of residue 108 in TNAP is further supported by analysis of the [A108]TNAP mutant, which displayed a 10-fold reduction in inhibition by L-hArg and a slight reduction in L-Phe inhibition. These data suggested that residues 108 and 109 largely determined the selectivity towards L- amino acid inhibition in both TNAP and PLAP. Residue 108 lies within the same pocket as Tyr371 (Tyr367 in PLAP), which, as was discussed in Section 5.1.2.1, is important for the efficient inhibition of PLAP by L-Phe. If the suggested mode of binding of amino acid sidechains in the active site of TNAP or PLAP is correct, one should expect reduced inhibition in the case of Ala371 substitution in TNAP. The data obtained for Ala371 proves that this is true. When tested for either L-Phe or L-hArg inhibition, [A371]TNAP was one of the least efficiently inhibited mutants. In addition to residues 108, 109 and 371, another position which has a large effect on the inhibition is His434 (Glu429 in PLAP). However, changes at His434 in TNAP do not significantly affect the inhibition by either L-Phe or L-hArg (Kozlenkov et al., 2004). Summarizing the data for the L-Phe and L-hArg inhibition, the binding of amino acid inhibitors occurs in the area E108–G109–Y371–H434 in TNAP (equivalent to F107–Q108–Y367–E429 in PLAP). The specificity of L-hArg inhibition in TNAP is most likely determined by electrostatic interaction of the positively charged side-chain of the inhibitor with residue Glu108 in the active site (Fig. 22A). The conserved Tyr371 is also necessary for the binding, most likely by providing the hydrophobic area for the nonpolar alkyl part of the inhibitor sidechain. In PLAP, the residues Phe107 and Gln108, together with Tyr367, provide a hydrophobic pocket that accommodates the phenyl ring of L-Phe. Finally, residue 434 in TNAP (429 in PLAP) modulates the inhibition. In this case an H434E replacement is clearly unfavorable for the inhibition by either L-hArg or L-Phe. However, other substitutions at 434 show nearly the same level of inhibition as in wt TNAP.

Levamisole, the L-stereoisomer of tetramisole, is a well-known potent uncompetitive inhibitor of TNAP (Van Belle, 1976) and it had been reported that chicken TNAP is about eight times less sensitive to levamisole than human TNAP (Reynolds and Dew, 1977). Chicken TNAP has only a few substitutions in the active site area compared with human TNAP, one of them being Gln434. Kozlenkov et al. (2004) showed that chicken TNAP is inhibited comparably to, although slightly better than, the [Q434]TNAP variant of TNAP. The fact that such different amino acids as Gly or Glu at this position all showed a marked decrease in inhibition suggests that the effect is not due to steric hindrance, but rather depends on

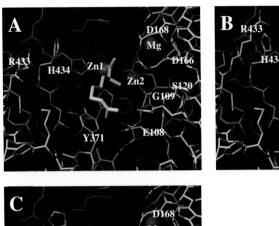
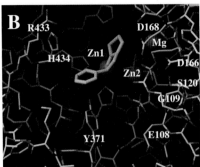
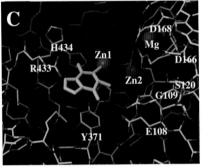

Fig. 22 Calculated optimal docking of the inhibitors L-homoarginine (A), levamisole (B) and theophylline (C) into the modeled active site of TNAP. The figures were produced with the use of TURBO-FRODO. Taken from Kozlenkov et al. (2004) and reproduced with permission from the *Journal of Bone and Mineral Research*.

the particular presence of the amino acid His at this position. A possible explanation for the role of His434 is a stacking interaction between the flat hydrophobic groups of the protein and the ligand. The reciprocal E429H mutation in PLAP improved the inhibition by levamisole about 10-fold, further confirming the crucial role of this residue. Kozlenkov et al. (2004) concluded that the binding area for levamisole partially overlaps the binding pocket for amino acid inhibitors, but is more spatially restricted than the latter. However, in contrast with the inhibition by amino acids, the selectivity of levamisole towards the TNAP isozyme is nearly fully explained by the substitution at residue 434 (Fig. 22B). Of interest here is the data of Butterworth (1994), who showed that the combination of adenosine and nicotinamide was a more effective uncompetitive inhibitor than either of them alone, and suggested that a dinucleotide structure was required for inhibition of TNAP. He also reported that oxidized adenosine could irreversibly inactivate the enzyme and concluded that His434 was the likely residue involved in this inactivation. His434 also appears to be the main determinant of theophylline binding to TNAP, according to Kozlenkov et al. (2004), although their data did not establish the exact orientation of the theophylline molecule in the active site. Data in the literature have documented the effect of several substitutions in the xanthine structure (Croce et al., 1979). For example, substitutions at N7, C8 and N9 in theophylline lead to compounds with no inhibitory activity. One can conclude that N7 or N9 is involved in important interactions with the enzyme, probably via Zn1 ion in the active site. In contrast, several substitutions at N1 and N3 in theophylline gave active compounds with altered selectivities towards bovine TNAP or intestinal isozymes. For example, a negatively charged 1-carboxymethyl derivative of theophylline has a greatly decreased activity against the TNAP isozyme but enhanced activity against the intestinal isozyme, thus being a selective inhibitor of the latter. Interestingly, bovine IAP isozymes have a positively charged residue (Arg or Lys) at the position homologous to Gly109 in TNAP and a noncharged tyrosine at the position homologous to negatively charged Glu108 in TNAP. These data, together with the positive effect of the Phe108 substitution on theophylline inhibition in the study by Kozlenkov et al. (2004), suggest that the binding area for the large ring of theophylline might lie near TNAP residues 108/109 and that the nature of the substitutions at N1 (or N3) in theophylline should be complementary to the nature of residues 108/109 for effective inhibition to take place. Mutations at residue 434 lead to a wide range of theophylline inhibition levels, suggestive of close interactions between residue 434 and the inhibitor (Fig. 22C). Thus, these mutagenesis studies revealed that residue 108 in TNAP largely determines the specificity of inhibition by L-hArg, while the conserved Tyr371 (equivalent to Y367 in PLAP) also contributes to L-hArg positioning. In contrast, the binding of levamisole is mostly dependent on His434 and Tyr371, but not on residues 108 or 109, while the main determinant of sensitivity to theophylline is His434 (Kozlenkov et al., 2004). These studies open the door to drug design efforts aimed at developing more specific inhibitors of TNAP that could be used therapeutically to treat conditions of ectopic calcification, where TNAP is know to play a role (Section 13.4).

Other uncompetitive inhibitors of APs have also been described (McComb et al., 1979). In particular, bromotetramisole is a strong uncompetitive inhibitor of TNAP. Initially described by Van Belle et al. (1977), bromotetramisole was used to develop methods to quantitate IAP and PLAP by inhibiting serum TNAP activity (Kuwana and Rosalki, 1991). This inhibitor has also been used to examine the role of TNAP in renal phosphate transport (Skillen and Harrison, 1980; Brunette et al., 1982; Onsgard-Meyer et al., 1996) and in mineralizing tooth germs (Lyaruu et al., 1983, 1984, 1987; Woltgens et al., 1985). Insulin also inhibits TNAP through an uncompetitive mechanism (Gazzarrini et al., 1989). The lowest active concentration of insulin was 10^{-6} M and maximum inhibition was obtained at about 10^{-4} M. Full recovery of the hormone-inhibited enzyme could be obtained with 10^{-4} M 2-mercaptoethanol, suggesting a stable interaction between TNAP and insulin molecules, involving either disulfide cross-linkages or the metal chelating activity of insulin.

5.2
Allosteric Behavior

The analysis of subunit interactions invokes the concepts of cooperativity and allosterism. Cooperativity implies the interaction of subunits based on their interaction with ligand. Positive cooperativity refers to that situation in which binding of the ligand in one subunit increases the affinity of binding to the same ligand in another subunit. The classical model for this is hemoglobin, in which binding of molecular oxygen in one of the tetramers increases the affinity for oxygen in the others through conformational changes translated through the protein to the binding sites. Negative cooperativity refers to the same dynamic, but with a decrease in affinity for the ligand. Allosterism implies interaction of subunits based on allosteric effectors. Allosteric effectors are those molecules that affect the enzyme by binding specifically and reversibly to a site other than the active site and alter the kinetics of reaction. These sites that the effector binds to were termed allosteric sites by Monod et al. (1963). The word "allosteric" is derived from the Greek words *allo*, meaning other or different, and *stere*, meaning structure or solid. This was meant to emphasize the difference in structure between substrate and effector. Formation of the enzyme–allosteric effector complex is assumed to bring about a discrete reversible alteration of the enzyme's structure, an allosteric transition, which modifies the properties of the active site (Monod et al., 1963).

Early studies done with the ECAP had indicated that upon substrate binding only a single active site per dimer was phosphorylated. These results changed once sample preparation techniques were altered to ensure that the enzyme was fully metalated. With fully metalated enzyme, both active sites of the dimer were occupied and molecular symmetry was re-established (for a review, see Coleman, 1992). However, these studies did suggest that there was negative cooperativity between subunits that was dependent on metalation. Other approaches taken to study subunit interactions in ECAP involved the creation of heterodimers either by reconstitution of denatured native and partially modified enzyme or by controlled proteolysis (Olafsdottir and Chlebowski, 1989). These studies revealed a structural and functional asymmetry in the hybrids, indicating catalytically relevant subunit communication.

Hoylaerts et al. (1997) investigated whether the monomers in a given PLAP or GCAP dimer are subject to cooperativity during catalysis following an allosteric model or act via a half-of-sites model, in which at any time only a single monomer is operative. The authors used wt, single- and double-mutant PLAP homodimers and heterodimers for this analysis. The active site Zn^{2+} stability was probed in two ways: by competitive inhibition by EDTA and by measuring the spontaneous dissociation of Zn^{2+} from the active site. EDTA had a 30–40-fold higher affinity for Zn^{2+} metal ions located in the GCAP and [G429]PLAP active site than in the wt PLAP and [S84]PLAP active site. In fact, [G429]PLAP could be completely demetalated by treatment with a chelating agent, but the activity could be restored by Zn^{2+} in a dose-dependent manner. By comparing the activity of partially and fully reconstituted enzymes as a function of substrate concentrations, it could be seen that the partially metalated enzyme displayed negative cooperativity, but when fully metalated the AP dimers function noncooperatively (Hoylaerts et al., 1997). To facilitate the further study of subunit interactions in mammalian APs, the authors produced heterodimers consisting of distinguishable subunits so that the resulting hybrids could be analyzed with respect to that property. Exploiting the 100-fold difference in L-Leu inhibition displayed by [S84]PLAP and [G429]PLAP, heterodimers consisting of these subunits were produced and evaluated with regard to L-Leu inhibition. Rate equations based on an allosteric model or a half-of-sites model provided predictions regarding the inhibition curves (Fig. 23). The resultant inhibition curves only matched the model predicted by the allosteric model.

To confirm that mammalian APs are classical allosteric but noncooperative enzymes, heterodimers in which one of the subunits, i.e. [A92]PLAP (Table 2), was completely inactive were produced. Analysis of these heterodimers revealed

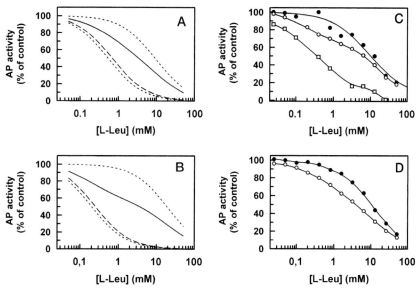

Fig. 23 Allosteric behavior of PLAP. Prediction of AP heterodimer inhibition by L-Leu (left panels A and B). A, residual enzyme activity for PLAP-GCAP heterodimers according to the allosteric (*solid line*) or the half-of-sites (*dashed line*) enzyme model, in comparison with the predicted inhibition curves by L-Leu for the parent homodimers PLAP (*right dotted line*) and GCAP (*left dotted line*). B, a similar analysis for the inhibition curves of [S]PLAP-[G]PLAP heterodimers during inhibition by L-Leu, as predicted by the allosteric (*solid line*) or the half-of-sites (*dashed line*) model, in comparison with the predicted inhibition curves of [S]PLAP (*right dotted line*) and [G]PLAP (*left dotted line*). L-Leu inhibition of [S]PLAP-[RGR]PLAP heterodimers (Right panels C and D). C, inhibition by L-Leu of chromatographically isolated [RGR]PLAP homodimers (□), [S]PLAP-t-[RGR]PLAP heterodimers (○), and [S]PLAP-t homodimers (●). D, inhibition by L-Leu of chromatographically isolated [S]PLAP-[ARGR]PLAP heterodimers (○) and [S]PLAP homodimers (●). Taken from Hoylaerts et al. (1997) and reproduced with permission from the *Journal of Biological Chemistry*.

that one active subunit could function independently of the inactive one. L-Leu inhibition analysis of this heterodimer revealed that the structural asymmetry affected the accessibility of the single active site by L-Leu. A second line of evidence for structural cross-talk between monomers in an asymmetric heterodimer was provided by the use of starch gel electrophoresis. PLAP and GCAP differ in conformation as a result of the identity of residue 429 and this causes the retarded migration of GCAP on starch gels compared with PLAP (Hoylaerts et al., 1992b). The allelic variants of PLAP (F, S and I) are so named after their relative migrations in starch gel electrophoresis (fast, slow and intermediate) and the heterodimers of these variants migrate equidistant between the respective homodimers (see Figs 15 and 25). However, PLAP/GCAP heterodimers migrate closer to GCAP homodimers than PLAP homodimers, indicating that the overall conformation of the PLAP/GCAP heterodimers resembles GCAP more than PLAP. Third, differences in heat stability provided another means to investigate the structural effect that one subunit has on the other. Heterodimers of the heat-stable [S84]PLAP and less heat-stable [G429]PLAP mutants behave more like the [G429]PLAP homodimer than predicted by the weighted average. That study clearly showed that mammalian APs are allosteric enzymes in which both monomers act independently, at least when both AP subunits are completely metalated. It is clear that for each AP isozymes subtle amino acid substitutions in positions close to the active site may dramatically affect the affinity for Zn^{2+} binding in the active site pocket. Therefore, in different tissues the mechanism of AP catalysis may be modulated by the nature of the isozyme and the local concentrations of zinc ions. It is also evident that heterodimers can form between structurally related mammalian APs. These heterodimers are not the weighted average of the parent homodimers. Hence mammalian APs are noncooperative allosteric enzymes, but the stability and catalytic properties of each monomer are controlled by the conformation of the second AP subunit (Hoylaerts et al., 1997).

The 3D structure of PLAP determined by Le Du et al. (2001) contributed the visualization of the residues postulated to be involved in conferring the allosteric character to PLAP. First, Asn84 is located at the dimer interface with its δ oxygen within hydrogen-bonding distance of Asp91-N (3.2 Å) and probably intervenes through this residue in the stabilization of the catalytic Ser92. On the same loop, Val85 and Asp86 interact with Ile1 and Ile2 from the other monomer and Lys87 interacts with Leu369 and Tyr367, also from the other monomer. Therefore, Asn84 may affect the enzyme activity through its interaction with Asp91 and may intervene in the allosteric process because of its involvement in dimerization (Fig. 24). The allosteric behavior is probably further favored by the quality of the dimer interface, by the long N-terminal α-helix from one monomer that embraces the other subunit and by the protrusion of Tyr367 from one monomer into the active site of the other.

Fig. 24 Residues involved in the allosteric behavior of PLAP. Ribbon representation of the active site is shown, with monomer 1 in green and monomer 2 in magenta. The network Glu429–His320–Zn1 is circled in blue, the network Tyr367(#2)–His432(#1)–Zn1 in yellow and the are circled in brown. Taken from Le Du et al. (2001) and reproduced with permission from the *Journal of Biological Chemistry*.

5.3
Catalytic Efficiency of Mammalian APs

A major property that remains to be explained in terms of structure is the large variability in catalytic activity displayed by mammalian APs, which can range from 10- to 100-fold higher k_{cat}s than ECAP (Table 5). The k_{cat} reflects the turnover of an enzyme, i.e. the number of conversions that it can carry out per second on condition it is kept saturated with substrate. What determines how efficiently an enzyme can act, however, is its degree of saturation as reflected by the K_m. So, for enzymes that function in an environment where the substrate is in large molar excess, k_{cat} is an accurate descriptor of enzyme efficiency. In general, however, such is not the case and the enzyme will have a certain degree of occupation, depending on the actual substrate concentration. Thus, the ratio k_{cat}/K_m, which has the dimensions of a second-order rate reaction ($s^{-1} M^{-1}$) and couples the speed of the reaction with the concentration of its substrate, takes both parameters into account and therefore better describes the properties of the enzyme.

Table 5 Kinetic properties of some representative mammalian AP isozymes in comparison with ECAP.

Isozyme	k_{cat} (s^{-1})	K_m (mM)	k_{cat}/K_m (s^{-1} M^{-1}) × (10^{-6})	Reference
ECAP	79.8	0.021	3.8	Stec et al. (1998)[a]
hPLAP	460	0.35	1.31	Hoylaerts et al. (1992)
hGCAP	277	0.11	2.52	Hoylaerts et al. (1992)
[G429]PLAP	344	0.09	3.82	Hoylaerts et al. (1992)
mEAP	8.4	0.14	0.06	Narisawa et al.[b]
mAKP6	59	0.79	0.07	Narisawa et al.[b]
mIAP	339	1.10	0.30	Narisawa et al.[b]
bIAPI	1800	1.30	1.38	Manes et al. (1998)
bIAPII	5900	3.40	1.76	Manes et al. (1998)
bIAPIII	4200	2.40	1.75	Manes et al. (1998)
bIAPIV	6100	3.90	1.56	Manes et al. (1998)
hTNAP	2100	0.36	5.83	Kozlenkov et al. (2004)

a Values determined at pH 8.0 in the presence of a phosphate acceptor.
b Manuscript in preparation.

Murphy et al. (1995) attributed part of this increase in k_{cat} to the nature of residues 153 and 317, as discussed in Section 4.1.2. The substitution of D153H and K328H in ECAP produced enzymes with kinetic properties similar to those of mammalian APs. For example, the D153H/K328H double ECAP mutant displayed a 5.6-fold higher k_{cat} and a 30-fold lower affinity both for the substrate and the product P$_i$, a decrease in heat stability and a shift in pH optimum to alkaline pH values (Murphy et al., 1995). They argued that since the release of P$_i$ from the noncovalent E•Pi complex is the slow step in the mechanism, the enhanced catalytic activity is due to the reduced affinity of the mutant enzyme for P$_i$. When Kozlenkov et al. (2002) tested the reverse mutation in PLAP, i.e. H153D and H328K, however, the K_ms consistently remained increased for the mutants and there were no significant changes in the pH dependence or heat stability of the mutants compared with the wt PLAP.

Llinas et al. (2005) argue that the absence of a side-chain at position 429, such as for Gly429 in GCAP, results in a low k_{cat} of 277 s^{-1} (Table 5), the Glu429 in PLAP yields a k_{cat} of 460 s^{-1} and His429 in TNAP or Ser429 in IAP accelerate the reaction rate to 2100 s^{-1}, or higher. Therefore, the presence of an acidic side-chain such as Glu may favor the formation of a hydrogen-bonding network around the nucleophilic water molecule that modulates the catalytic efficiency of the enzyme, but the acidic character of Glu is less favorable than that of Ser or His to reinforce the nucleophilic tendency of the water molecule. In fact, the substitution E429H doubles the PLAP rate constant to a k_{cat} of $950\,s^{-1}$ (Kozlenkov et al., 2004). This simple mutation does not explain entirely the differences between the catalytic constants of TNAP and PLAP, although the improved activity clearly shows that position 429 is an important residue.

In one mutagenesis study, Manes et al. (1998) compared two closely related bovine IAPs that differed significantly in their k_{cat} values to identify one residue as responsible for their increased turnover number. A G322D mutation was single-handedly able to convert the kinetic properties of bIAP II into those of bIAP I. The changes included a 3-fold decrease in k_{cat} and K_m to values comparable to those of bIAP I (Table 5). The converse mutation gave entirely consistent results, since by introducing a D322G mutation into bIAP I, the k_{cat} and K_m were increased in the resulting mutant to values comparable to that of bIAP II itself. Similarly, the introduction of an S322G substitution in bIAP III increased its k_{cat} value to $5900\,s^{-1}$ whereas the S322D mutation reduced its k_{cat} value to $1200\,s^{-1}$, comparable to those of [Asp322]bIAP II and bIAP I. Despite the fact that no 3D structure is yet available for any of the IAPs, sequence comparisons indicate that residue 322 is located two amino acids away from a sequence absolutely conserved in APs throughout evolution, i.e. ^{311}EGGRIDHGHH^{320}, which contains three crucial ligands (E311, D316 and H320) coordinating to the active site Zn^{2+} and Mg^{2+} ions as detailed in Section 4.1.2. Although further experimentation will be necessary to understand the detailed mechanistic effect of the D322G substitution, we can conclude from these data that the additional Asp at position 322 in bIAP I is impairing the hydrolysis of the phosphoenzyme complex during catalysis. Based on the general reaction scheme of APs (Scheme 1 and equations in Section 5.1), the expression for k_{cat} is $k_{cat} = 1/(1/k_2 + 1/k_3)$. Hence a rise in k_{cat} can only occur as a consequence of an improved enzyme phosphorylation (k_2) or an improved hydrolysis of the phosphoenzyme complex (k_3). As these kinetic data reveal, a rise in k_{cat} in [G322]bIAP I is paralleled by a proportionally identical rise in K_m, resulting in a constant k_{cat}/K_m ratio. Since this ratio equals $k_1/(1 + k_{-1}/k_2)$, it follows that k_2 is not affected. Hence the enhanced catalytic activity in bIAP II and other Gly322 containing bIAP mutants in comparison with bIAP I results from an increase in k_3. This is to say that Asp322 does not restrict phosphate positioning in the bIAP I active site pocket and does not impair the covalent phosphoenzyme complex formation, but it impairs the subsequent changes in coordination of the phosphate group during its hydrolysis from the active site Ser. The D322G substitution present in bIAP II relieves this interference.

5.4
Substrate Specificities

Mammalian APs have broad substrate specificity and are able to hydrolyze or transphosphorylate a wide variety of phosphated compounds *in vitro*. However, only a select few of those compounds have to date been confirmed to serve as natural substrates for some of the AP isozymes. The inorganic pyrophosphohydrolase activity associated with osseous plate TNAP has been documented by several investigators (McComb et al., 1979; Whyte et al., 1995). Rezende et al. (1998) reported that purified membrane-bound TNAP from rat osseous plate hydrolyzed pyrophosphate in the presence of Mg^{2+} ions with an optimal apparent pH of 8.0. The pyrophosphatase activity was rapidly destroyed at temperatures above 40 °C, but Mg^{2+} ions apparently protected the enzyme against denaturation. As will be discussed in Sections 7.1 and 8.1, abnormalities in the metabolism of PP_i largely explain the skeletal abnormalities in hypophosphatasia.

Pyridoxal-5′-phosphate (PLP) has also been shown to be a physiological substrate for the TNAP present in leukocytes (Smith and Peters, 1981; Gainer and Stinson, 1982; Wilson et al., 1983). TNAP isolated from human SAOS-2 osteosarcoma cells hydrolyses phosphoethanolamine (PEA) and PLP at physiological pH (Fedde et al., 1988). Clearly, abnormalities in PLP metabolism explain the epileptic seizures experienced by patients with hypophosphatasia (Sections 7.1 and 8.1). However, the contention that PEA is also a substrate for TNAP *in vivo* remains to be confirmed. In a clinical study, Whyte et al. (1995) documented that, during pregnancy, carriers of hypophosphatasia corrected their hypophosphatasemia during the third trimester because of PLAP in maternal blood. Blood or urine concentrations of PEA, PP_i and PLP diminished substantially during that time. After childbirth, maternal circulating levels of PLAP decreased and PEA, PP_i and PLP levels abruptly increased. In serum, unremarkable concentrations of IAP and the low residual levels of TNAP did not change. This study also pointed to the ability of PLAP to metabolize the same substrates as TNAP (Whyte et al., 1995). Chakrabartty and Stinson (1985a) determined the kinetic properties of membrane-anchored and solubilized forms of liver TNAP against eight putative substrates, i.e. pNPP, α-naphthylphosphate, β-glycerophosphate, phosphoserine, PLP, phosphothreonine, PP_i and phosphotyrosine, and showed that, even though binding of TNAP to the plasma membrane is not essential for catalytic function, the properties of the enzyme in the membrane are different from those of the soluble form (see also Section 4.3.2).

Other described substrates include monofluorophosphate (MFP), which can also be hydrolyzed by TNAP (Farley et al., 1987). In fact, a kinetic potentiometric assay for APs has been devised based on the hydrolytic cleavage of the P–F bond (Venetz et al., 1990). Human AP isozymes are able to hydrolyze phosphatidates with various fatty acyl chains (egg phosphatidate and dioleoyl, distearoyl, dipalmitoyl, dimyristoyl and dilauroyl phosphatidates). PLAP and IAP were capable of hydrolyzing all the phosphatidates examined with maximal activity in the presence of 10 g L^{-1} sodium deoxycholate and dilauroyl phosphatidate being the best substrate. The phosphatidate hydrolytic activity of PLAP was 2–3 times higher than that of the IAP enzyme, whereas TNAP did not hydrolyze phosphatidates with long fatty acyl chains (C_{16-18}) even in the presence of sodium deoxycholate (Sumikawa et al., 1990). Inorganic polyphosphates (polyP), being energy-rich linear polymers of orthophosphate residues known from bacteria and yeast, also exist in higher eukaryotes. Calf IAP is able to cleave polyP molecules up to a chain length of about 800, acting as an exopolyphosphatase progressively degrading polyP. The pH optimum is in the alkaline range. TNAP was not able to hydrolyze polyP under the conditions tested whereas PLAP and ECAP displayed polyP-degrading activity (Lorenz and Schroder, 2001).

APs appear also to be involved in the metabolism of nucleotides. Say et al. (1991) reported that purified osseous plate TNAP displayed broad substrate specificity and was able to hydrolyze ATP, ADP, AMP, pyrophosphate, glucose-1-phosphate, glucose-6-phosphate, fructose-6-phosphate, β-glycerophosphate and bis(p-nitrophenyl)phosphate in addition to pNPP. However, ATP, bis(p-nitrophenyl)-phosphate and PP_i were among the less hydrolyzed substrates assayed. Nevertheless, TNAP appears to be able to hydrolyze ATP at both pH 7.5 and 9.4 (Demenis and Leone, 2000). However, Pizauro et al. (1998) concluded that the membrane-specific ATPase activity present in osseous plate membranes and TNAP are different proteins. Several reports have indicated that TNAP is involved in AMP hydrolysis (Ohkubo et al., 2000; Picher et al., 2003). Extracellular adenine nucleotides induce cyclic AMP elevation through local adenosine production at the membrane surface and subsequent activation of adenosine A (2A) receptors in NG108-15 neuronal cells. NG108-15 cells hydrolyzed AMP to adenosine and this activity was suppressed at pH 6.5, but markedly increased at pH 8.5. The AMP hydrolysis was also blocked by levamisole and the cells expressed TNAP mRNA. These results indicated that AMP phosphohydrolase activity in NG108-15 cells is due to TNAP and suggest that this enzyme plays an essential role for the P1 antagonist-sensitive ATP-induced cyclic AMP accumulation in NG108-15 cells (Ohkubo et al., 2000). Similarly, Picher et al. (2003) found that two ectonucleotidases mediated the conversion of AMP to adenosine on the mucosal surface of human airway epithelia, i.e. ecto 5'-nucleotidase (CD73) and TNAP. Whereas both mucosal and serosal epithelial surfaces displayed ecto 5'-nucleotidase activity, TNAP activity was restricted to the mucosal surface. These experiments support a major role for extracellular nucleotide catalysis and for the involvement of TNAP in the regulation of adenosine concentrations on airway surfaces (Picher et al., 2003).

Other catalytic properties of AP are also worth mentioning. AP isolated from the mouse intestine is able to catalyze the synthesis of PP_i from P_i during hydrolysis of glucose 6-phosphate, ATP, ADP, PP_i or pNPP (Nayudu and de Meis, 1989). Whereas the rate of PP_i synthesis is 1000-fold lower than the rate of substrate hydrolysis, PP_i synthesis was increased by the addition of Mg^{2+} and also by decreasing the pH from 8.5 to 6.0. The data indicated that at the catalytic site of APs, the energies of hydrolysis of the phosphoserine residue and of PP_i are different from those measured in aqueous solutions. In another study, calf IAP was found to be able to transphosphorylate effectively thiamine to thiamine monophosphate using β-glycerophosphate or creatine phosphate as phosphate donors at pH 8.5. TMP production in the brush border membrane, however, was very small and corresponded to 0.001–0.01% of the total P_i simultaneously released by hydrolytic activity (Rindi et al., 1995).

Rezende et al. (1994) reported phosphodiesterase activity as a novel property of osseous plate TNAP. Bis(p-nitrophenyl)phosphate was hydrolyzed at both pH 7.5

and 9.4 with an apparent dissociation constant of 1.9 and 3.9 mM, respectively. The hydrolysis of *p*-nitrophenyl-5'-thymidinephosphate followed hyberbolic kinetics with a $K_{0.5}$ of 500 µM. The hydrolysis of cyclic AMP by the enzyme followed more complex kinetics, showing site–site interactions ($h = 1.7$) and $K_{0.5} = 300$ µM for high-affinity sites. ATP and cyclic AMP were competitive inhibitors of bis(*p*-nitrophenyl)phosphatase activity of the enzyme and K_i values (25 and 0.6 mM for cyclic AMP and ATP, respectively) very close to those of the $K_{0.5}$ (22 and 0.7 mM for cyclic AMP and ATP, respectively), determined by direct assay, indicated that a single catalytic site was responsible for the hydrolysis of both substrates. The alkaline apparent pH optima, the requirement for bivalent metal ions and the inhibition by methylxanthines, amrinone and amiloride demonstrated that rat osseous plate TNAP was a type I phosphodiesterase (Rezende et al., 1994).

Various *in vitro* studies have led to the suggestion that APs may function as a plasma membrane phosphoprotein phosphatase. Sarrouilhe et al. (1992) used purified rat liver plasma membranes to study endogenous phosphorylation and dephosphorylation events and detected an 18 kDa phosphoprotein as a potential substrate for TNAP. Fedde et al. (1993), however, compared the phosphorylation of plasma membrane proteins from control fibroblasts with those from profoundly TNAP-deficient fibroblasts of hypophosphatasia patients. Autoradiography of ^{32}P-labeled proteins resolved on 2D gels demonstrated 63 plasma membrane phosphoproteins with molecular weights ranging from 15 to 152 kDa and predominantly acidic pIs. They found no consistent different among all identifiable plasma membrane phosphoproteins in the control and hypophosphatasia samples and concluded that TNAP does not modulate the phosphorylation of plasma membrane proteins.

5.5
APs as Members of a Superfamily of Enzymes

As was discussed in the previous section, APs appear to be able to act on a wide variety of phosphorylated substrates. Of interest is the fact that APs appear to have structural similarity to a large number of other enzymes. Cofactor-independent phosphoglycerate mutase (iPGM) (EC 5.4.2.1) has been identified previously as a member of the AP superfamily of enzymes, based on the conservation of the predicted metal-binding residues (Galperin et al., 1998). Iterative searches using the PSI-BLAST program resulted in the identification of similarly conserved regions in phosphopentomutases (EC 5.4.2.7), APs (EC 3.1.3.1) and several previously uncharacterized proteins. All the amino acid residues that interact with Zn1 (Asp327, His331 and His412) and Zn2 (Asp51, Asp369 and His370) in ECAP (Kim and Wyckoff, 1991) are absolutely conserved in phosphocarbohydrate-binding proteins of the AP superfamily (Galperin et al., 1998). On the other hand, the ligands to the Mg site of ECAP are much less conserved, since an E322N substitution is found in phosphopentomutases and iPGMs, while Asp153 and Thr155 do not seem to be conserved at all. The strong conservation of the metal-binding residues in both phosphopentomutase and iPGM indicates that both these enzymes are metal dependent. Structural alignment of iPGM with ECAP and cerebroside sulfatase confirmed that all these enzymes have a common core structure and revealed similarly located conserved Ser (in iPGM, ECAP and mammalian APs) or Cys (in sulfatases) residues in their active sites. In ECAP and mammalian APs, this Ser residue is phosphorylated during catalysis, whereas in sulfatases the active site Cys residue is modified to formylglycine and sulfated. The similarly located Thr residue forms a phosphoenzyme intermediate in phosphodiesterase/nucleotide pyrophosphatase-1 (NPP1). In fact, APs are known to have phosphotransferase (Coleman, 1992) and also phosphodiesterase (Rezende et al., 1994) activities, and iPGM and NPP1 can also function as phosphatases (Breathnach and Knowles, 1977; Gijsbers et al., 2001). Using structure-based sequence alignment, Galperin and Jedrezjas (2001) identified homologous Ser, Thr or Cys residues in other enzymes of the AP superfamily, such as phosphopentomutase, phosphoglycerol transferase, phosphonoacetate hydrolase and GPI-anchoring enzymes (GPI-phosphoethanolamine transferases). Hence this AP superfamily includes enzymes with substantially different activities (isomerases, hydrolases and a putative lyase), which, however, all act on similar phosphocarbohydrate (or sulfocarbohydrate) substrates.

Although these bioinformatics studies were done using the ECAP structure as paradigm, the same structural similarity is bound to exist with the mammalian APs since, as discussed, the active site of mammalian APs is well conserved with that of ECAP. The authors predict that the catalytic cycles of all the members of this superfamily involve phosphorylation or sulfation or phosphonation of these conserved Ser/Thr/Cys residues. This would imply that all enzymes of the AP superfamily would have the same reaction scheme as was originally proposed for AP (Scheme 4; see also Scheme 1 in Section 5.1) where phosphate (sulfate or phosphonate) acceptor can be either water (R_2 = H, in phosphatases, sulfatases and phosphonate hydrolases) or a second substrate (in phosphoglycerol and phosphoethanolamine transferases) or just a different hydroxyl group of the same original substrate R_1 (in iPGM and phosphopentomutase).

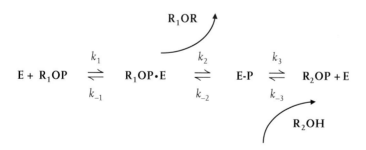

6
Epitope Maps

6.1
Epitopes in PLAP and GCAP

It is interesting that the initial publications concerning the development of monoclonal antibodies (MAbs) to human PLAP were mainly concerned with further understanding of the allelic polymorphism of this isozyme and in clarifying whether the tumor-derived enzymes were identical with the pregnancy-associated enzyme. The use of polyclonal anti-PLAP antisera, rendered allotype-specific by extensive cross-absorption, enabled Wei and Doellgast (1980) to detect epitope differences between the common S, F and I allozymes and the D allozyme of PLAP. These allozyme-specific polyclonal antisera and the MAbs described below also provided the first clues that some tumor-derived PLAP-like enzymes might be derived from a distinct locus (Wei and Doellgast, 1981; Millán et al., 1982b). This locus was subsequently cloned and shown to correspond to the GCAP gene (*ALPP2*) often expressed in choriocarcinomas and in testicular cancer (Millán and Manes, 1988).

The first paper on the development of MAbs to PLAP was published by Harris's group (Slaughter et al., 1981), documenting the reactivity of six out of a panel of 18 MAbs against PLAP, i.e. *ALPP*/P3/1, *ALPP*/Sp2/2, *ALPP*/Sp2/3, *ALPP*/Sp2/4, *ALPP*/Sp2/5, *ALPP*/Sp2/11, for reactivity against a series of 295 placental extracts that had been typed electrophoretically (Table 6). Enzyme–MAb complexes, formed between six different MAbs and the six phenotypes of PLAP representing the homozygous and heterozygous combinations of the three common alleles, were examined by electrophoresis in starch, acrylamide and agarose gels (Gogolin et al., 1981). The products of the three common PLAP alleles and also some rare alleles could be discriminated in this manner by the six MAbs and evidence for allelic differences not detectable electrophoretically was obtained (Slaughter et al., 1981). In a follow-up report, the remaining 12 MAbs were also studied (Slaughter et al., 1983). Among the total of 18 MAbs, six of them discriminated in some degree among the products of the common PLAP alleles as defined electrophoretically. Four of the MAbs displayed reduced reactivity with the products of the Pl^2 allele and two with the product of the Pl^3 allele.

Table 6 Monoclonal antibodies to human PLAP and IAP; their specifity and epitopes, if known.

Mab	Subtype	Immunogen	Antigen specificity	Epitope	Reference
ALPp/P3/1		PLAP			Slaughter et al. (1981)
ALPp/Sp2/2		PLAP			Slaughter et al. (1981)
ALPp/Sp2/4		PLAP			Slaughter et al. (1981)
ALPp/Sp2/4		PLAP			Slaughter et al. (1981)
ALPp/Sp2/5		PLAP			Slaughter et al. (1981)
ALPp/Sp2/11		PLAP			Slaughter et al. (1981)
NDOG2		Sincitio-trophoblast	PLAP		Sunderland et al. (1981)
F11	IgG1	PLAP1-2	PLAP1-1>2-2/GCAP	R209	Millán et al. (1982)
H317		Sincitio-trophoblast	PLAP		McLaughlin et al. (1982)
D10	IgG1	PLAP1-2	PLAP1-1>2-2/GCAP		Millán and Stigbrand (1983)
C2	IgG1	PLAP1-2	PLAP	133/361?	Millán and Stigbrand (1983)
H7	IgG2a	PLAP1-2	PLAP/GCAP		Millán and Stigbrand (1983)
B10	IgG2a	PLAP1-2	PLAP/GCAP/IAP		Millán and Stigbrand (1983)
E6	IgG2b		PLAP	R209	De Groote et al. (1983)
H17E2			PLAP/GCAP		Epenetos et al. (1984)
D20L					Epenetos et al. (1984)
G10	IgG2a	TCRC-1 cells		R209	Jemmerson et al. (1985)
B2	IgG2a			R209	Jemmerson et al. (1985)
H5	IgG1				Jemmerson et al. (1985)
A3	IgG1				Jemmerson et al. (1985)
F6	IgG2a			K62-L63?	Jemmerson et al. (1985)
E5	IgG2a			K62-L63?	Jemmerson et al. (1985)
C4	IgG2b			K62-L63?	Jemmerson et al. (1985)
HPMS-1	IgG1	PLAP	PLAP		Hirano et al. (1986)
11-D-10					Mano et al. (1986)

Table 6 Continued.

Mab	Subtype	Immunogen	Antigen specificity	Epitope	Reference
8B6					Durbin et al. 1988
3F6					Durbin et al. 1988
		PLAP1-2		M254?	De Broe and Pollet (1988)
17E3		PLAP1-2		R241	Hendrix et al. (1990)
7E8	IgG1	PLAP1-2			Hendrix et al. (1990)
ATC2		TGCTs	PLAP/GCAP		Nouri et al. (2000)
AAP1	IgG2a	D98/AH-2 cells	IAP		Arklie et al. (1981)
2HIMS-1	IgG2b	IAP			Hirano et al. (1987)
2HIMS-3	IgG2b				Hirano et al. (1987)
IAP151	IgA	IAP	IAP/PLAP		Verpooten et al. (1989)
IAP130		IAP			Verpooten et al. (1989)
IAP250	IgG1	IAP			Verpooten et al. (1989)

Working independently with MAbs to PLAP, Millán and colleagues (1982a,c) were also able to raise an allotype-specific antibody (F11), which, as in the work of Slaughter et al., selectively reacted with the products of the Pl^1 (S) and Pl^3 (I) alleles, but not of the Pl^2 (F) allele (Table 6). This antibody was useful in electrophoretic gel retardation experiments to demonstrate that the antibody could selectively retard the bands corresponding to the SS or FS dimers but not the FF dimers (Fig. 25). The F11 antibody proved to bind also to the D variant of PLAP, shown in Fig. 25 as a heterodimer with the type 2 (F) allozyme, and two clear examples of nonelectrophoretic allozymes were also detected (Millán et al., 1982a). These observations, by two independent groups, firmly established the utility of MAbs as an alternative and complementary approach to the electrophoretic system of Donald and Robson (1974) in the study of PLAP polymorphism. In a subsequent study, Millán and Stigbrand (1983) developed four additional MAbs, D10, C2, H7 and B10. The D10 antibody, similar to the F11 MAb, reacted less well with the Pl^2 allozyme, whereas the B10 antibody was found to cross-react with the IAP isozyme. This panel of five MAbs was used to examine the reactivity of the PLAP-like enzyme produced by normal human testis (now called GCAP). Most testicular samples (26/32) were found to be nonreactive with the F11 and C2 antibodies and a total of four different patterns of reactivity were observed with the testicular GCAP samples (types I–IV) (Millán and Stigbrand, 1983). Subsequently, using this panel of five MAbs to investigate the sera of patients with testicular seminoma, evidence for up to six different variants of the tumor-derived GCAP isozyme was revealed (Wahren et al., 1986). From these early studies, two antibodies from this panel were selected for future diagnostic and immunotherapeutic use, i.e. the H7 antibody reacting with most of the known PLAP and GCAP allozymes and the C2 antibody reacting preferentially with PLAP but not with GCAP allozymes (Sections 10 and 16.6). De Broe's group extended the degree of enzyme polymorphism of GCAP to nine allotypic variants using a panel of four MAbs raised against PLAP, i.e. 7E8, 17E3 and 327 and E6 (Hendrix et al., 1990). They found that the E6 MAb also reacted poorly with the Pl^2 allelic product.

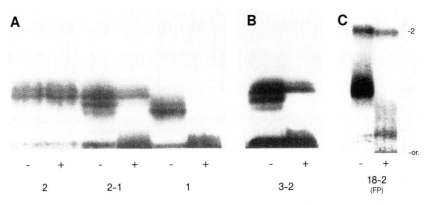

Fig. 25 Starch gel electrophoresis of PLAP variants and MAb retardation of allozymes. Electrophoresis on starch gels under acidic conditions (pH 5.6) of samples of the 1, 2-1 and 2 phenotypes (a) and the 3-2 phenotype (b) or of the 18-2 (FD) phenotype of PLAP (c) under alkaline conditions (pH 8.6) after incubation with normal mouse IgG (– lane) or the F11 MAb (+ lane). Taken from Millán et al. (1982) and reproduced with permission from *Human Genetics*.

Other groups developed MAbs with a focus on monitoring trophoblastic markers during pregnancy and also to detect PLAP/GCAP as cancer markers. McLaughlin et al. (1982) established six MAbs reacting with isolated human placental syncytiotrophoblast microvillous plasma membranes. One MAb, H317, detected a fetal differentiation antigen expressed only on the membranes of normal placental trophoblast and of certain tumor cell lines and was found to react with heat-stable L-phenylalanine-inhibitable PLAP isozyme. Similarly, Sunderland et al. (1984) produced two antibodies directed towards human syncytiotrophoblast, one of which, NDOG2, was shown to react with PLAP and was used in the immunohistochemical characterization of ovarian cystadenocarcinoma. Wray and Harris (1984) used the panel of MAbs to type PLAP in cell lines established from malignant human tumors. They found that HeLa-derived cell lines (Hep2 and WISH) had the type 1 PLAP phenotype whereas a non-HeLa cell line (HT-3) had the type 2 PLAP allozyme. The authors indicated that this approach could prove of value for the phenotyping of enzymes and proteins with poorly resolved or altered electrophoretic patterns (Wray and Harris, 1984). In turn, Jemmerson et al.

(1985c) developed a panel of seven MAbs to the AP expressed by the HeLa TCRC-1 human adenocarcinoma cell line. These antibodies, G10, B2, A3, E5, H5, C4 and F6, demonstrated identity between the normal placenta-derived type 1 allozyme and the tumor-derived counterpart. In another report, specific MAbs were produced against a single cell suspension from tissue fragments of a patient with seminoma of the testis. Out of 17 characterized MAbs. ATC2 had high specificity for human germ cell tumors and the target antigen was recognized as PLAP/GCAP (Nouri et al., 2000). Hence MAbs to PLAP proved to be very useful structural probes that enabled researchers to characterize further the genetic polymorphism of the PLAP locus, to define a fairly extensive genetic variability in the GCAP locus and to develop assays for the tumor-derived APs and even diagnostic tests (Section 16.6).

Using site-directed mutagenesis on the PLAP cDNA and antibody competition experiments, Hoylaerts and Millán (1991) were able to map three main antigenic domains (I, II and III) on the molecule by the use of a combined panel of 18 MAbs contributed by several investigators (Fig. 26). Domain I was defined by MAbs D10, F11, B2, G10, E6 and H5, which showed various degrees of competition and overlapping specificities. The first two antibodies (D10 and F11) are competing antibodies although they react with different epitopes (Millán and Stigbrand, 1983). F11 defines a subgroup that differentiates between the F and S PLAP-phenotypes, a property that F11 shares with B2, G10, E6 and partially H5 but not with D10. The authors further determined that a point mutation in PLAP, a P209R substitution, could explain the reactivity difference between the F/S discriminating antibodies. The replacement of a neutral by a positively charged amino acid is in line with the "slow" electrophoretic mobility of the S PLAP-phenotype in starch gel electrophoresis and polyacrylamide gel electrophoresis (Figs 15 and 25). The location of this residue in an exposed loop on the AP molecule may explain the pronounced effect that the Pro to Arg substitution has on *in vivo* antigen recognition, since five of the 18 antibodies were able to recognize this

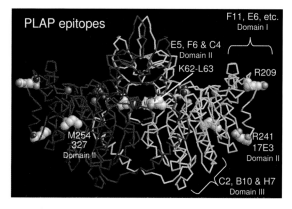

Fig. 26 Epitope map of PLAP. Approximate placement of antigenic domains on the 3D structure of PLAP as defined by a panel of 18 MAbs.

mutation and also three independent groups had obtained MAbs with similar specificity. These findings also imply that the previously reported immunological discrimination of GCAP phenotypes by the MAb F11 (Millán and Stigbrand, 1983; Wahren et al., 1986) is due to the presence of Pro209 in the most common GCAP type I allotype. Finally, the anti-intestinal AP antibodies 151 and 130 (cross-reacting with PLAP) were also found to compete slightly with F11 and map in the area recognized by antibodies D10 and H5, respectively, so they were both included in domain I.

Domain II included the reactivity of MAbs A3, E5, F6, C4, 17E3, 327 and 7E8 that appeared to bind at the center of the AP molecule. Antibodies A3, E5, F6 and C4 comprise a subgroup of antibodies that inhibit each other and are only partially inhibited by the domain I antibodies B2, G10 and H5 (Stigbrand et al., 1987). Antibodies E5, F6 and C4, furthermore, protect the only trypsin cleavage site, the Lys62–Leu63 bond, on the native PLAP polypeptide chain (Jemmerson and Stigbrand, 1984). Antibodies 17E3, 327 and 7E8 compete or interfere sterically with antibodies A3, E5, F6 and C4. Experiments in which antibody E6, 7E8 or 327 were allowed to compete with 12 of these anti-PLAP antibodies for binding to an S PLAP phenotype resulted in comparable competition patterns for antibodies 7E8 and 327, but different from the pattern found with domain I antibody E6. While 7E8 and 327 compete with each other, they bind PLAP independently from 17E3 (Hendrix et al., 1990). Arg241 was found to constitute the epitope for the antibody 17E3 that discriminates between PLAP and GCAP, whereas Met254 is more crucial for MAb 327 recognition, providing a basis for the simultaneous antibody binding to the antigen. Domain III was defined by the antibodies C2, B10 and H7, which have been described as competing antibodies, able to bind PLAP in two-site assays with antibodies in domain I (Millán and Stigbrand, 1983) and that recognize a C-terminal region in PLAP (Jemmerson et al., 1985b). Antibodies in this domain do not compete or interfere with domain I MAb E6 or domain II MAbs 7E8 or 327 (Fig. 26).

Introduction in PLAP of point mutations at residue 429, such as E429G (as in GCAP), E429S (as in IAP) or E429H (as in TNAP), induced a generalized decrease in binding affinities detected by 16 of the 18 antibodies (Hoylaerts and Millán, 1991) indicating the crucial role played by residue 429 in determining the conformation and stability of PLAP (Section 4.1.6). It was evident from these studies that MAb reactivity could also be used to probe for conformational changes when introducing point mutations in AP isozymes. However, the finding that all of the antibodies tested detected GCAP with considerably reduced affinity as compared with wt PLAP was surprising and indicative of a general conformational difference between PLAP and GCAP, despite their 98% sequence identity. Therefore, studies were undertaken to elucidate the extent to which various allelic GCAP positions were critical in determining the enzymatic, structural and immunological properties of GCAP phenotypes. Three homozygous GCAP phenotypes, i.e. JEG3 (Watanabe et al., 1989), BeWo (Lowe and Strauss, 1990) and wt GCAP (Millán and Manes, 1988), were analyzed and compared with a "core" GCAP mutant which contains the seven amino acid substitutions consistently different between

PLAP and GCAP, i.e. E15Q, I67T, P68F, N84S, R241H, M254L and E429G. These seven exchanges are found in all known GCAP allotypes. Although some substitutions could influence the electrophoretic behavior of the allotypes, the allelic differences had no effect on the kinetic properties of GCAP. However, the allelic amino acid differences affected the immunoreactivity and conformation of the variants as detected with the panel of 18 epitope-mapped anti-PLAP monoclonal antibodies (Fig. 27). The selective immunoreactivity of the PLAP/GCAP discrimi-

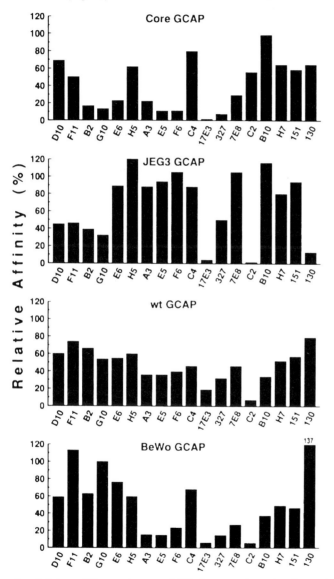

Fig. 27 Epitope differences between GCAP allozymes. Reactivity of a panel of 18 epitope-mapped MAbs to PLAP against four different allelic variants of GCAP.

nating monoclonal antibody C2 was found to be critically dependent on the nature of the allelic residues 133 and 361 in GCAP. Residue 133 was also found to be important for the general stability of the molecule, since BeWo GCAP and wt GCAP, which have Asn133 and Val133, respectively, instead of Met133, display a consistently reduced heat stability compared with core GCAP and JEG3 GCAP.

The immune patterns of the three GCAP phenotypes investigated were far from identical and provided a rational explanation for the differences in immunoreactivity that had been observed previously during the screening of GCAP-positive tumor extracts and serum samples (Millán and Stigbrand, 1983; Jeppsson et al., 1984a; Hendrix et al., 1990). GCAP, like PLAP, is a dimeric enzyme that results from the random association of two either identical or allelic GCAP monomers. Hence these three investigated genotypes could give rise to three heterozygous GCAP phenotypes, i.e. wt GCAP-JEG3 GCAP, wt GCAP-BeWo GCAP and JEG3-BeWo GCAP phenotypes. These heterozygous GCAP phenotypes would have immunoreactivity intermediate between those depicted for the homozygote phenotypes. These genotypes can, therefore, account for the existence of six of the nine allotypes identified by MAbs. Therefore, as reported above, the epitopes for the monoclonal antibodies F11 and 17E3 were almost entirely constituted by Arg209 and Arg241, respectively (Hoylaerts and Millán, 1991). Previously, a type II GCAP phenotype (Millán and Stigbrand, 1983) was characterized as being F11 fully reactive compared with PLAP (S phenotype). This indicates that the type II GCAP is likely to harbor a P209R substitution. Similarly, the type VII and VIII GCAP allozymes were defined as highly reactive with antibody 17E3 (Hendrix et al., 1990). This points to the existence of GCAP phenotypes that have an H241R substitution and identifies residue 241 as another allelic position. The core GCAP mutant constructed in that study contained the minimal number of substitutions that confer a GCAP character to the molecule. The monoclonal antibody reactivity of core GCAP, i.e. low reactivity with antibodies 7E8, 17E3, 327 and E6, is compatible with the definition of type IX GCAP (Hendrix et al., 1990) and hence may well represent a naturally existing genotype. Finally, the reactivity pattern of the same antibodies indicates that JEG3 GCAP is compatible with a type V GCAP phenotype, whereas the BeWo and wt GCAP cannot at this point be correlated with any of the previously defined allotypes. These studies illustrate the inherent genetic variability of the PLAP and GCAP isozyme loci and how critical the selection of the right MAb would be to detect PLAP and GCAP as tumor markers in immunodiagnostic and immunotherapy applications (Section 10).

6.2
Epitopes in IAP

The first reported MAb to human IAP, AAP1, was produced by immunizing a mouse with D98/AH-2 (HeLa) cells that produce the enzyme ectopically (Arklie et al., 1981) (Table 6). The antibody, which did not inhibit enzyme activity using pNPP as the substrate, was of the IgG2a class and did not show complement-dependent cytotoxicity. In nondenaturing conditions, the AAP1 MAb prevented the migration of IAP activity into the gel when cell-free extracts were made from human adult or fetal intestine or D98/AH-2 cells. Similarly, MAb AAP1 could be used to precipitate IAP activity from these extracts but not from extracts of human liver, kidney or placenta. Vockley et al. (1984b) produced six MAb to human fetal IAP, designated ALPFI/Sp2/36, ALPFI/Sp2/37, ALPFI/Sp2/38, ALPFI/Sp2/39, ALPFI/Sp2/41 and ALP-D98/Sp2/22. All the antibodies cross-reacted with the adult IAP and also with the IAP-like enzyme present in D98/AH-2 cells, but four MAb were able to distinguish between fetal and adult IAP in titration-binding studies. Subsequently, Ray et al. (1984) were able to map the binding of these MAbs to three distinct epitopes on the IAP molecule. However, these epitopes were present in both adult and fetal IAP. MAb binding to epitopes 1 and 3 also reacted with PLAP, whereas some antibodies to epitope 2 did not. Hirano et al. (1987b) reported the development of MAbs to human adult IAP and also meconial IAP. MAb 2HIMS-1 detected IAP and meconial IAP with an affinity of 5.8×10^9 and $2.5 \times 10^9\,M^{-1}$, respectively, whereas 2HIMS-3 bound to IAP with an affinity of $1.2 \times 10^{10}\,M^{-1}$. MAb 2HIMS-1 was used to develop a monoclonal antibody immunocatalytic assay to detect IAP in human sera (Hirano et al., 1987b). MAbs specific to human IAP, i.e. IAP250 (IgG1) and IAP151 (IgA), were also developed by Verpooten et al. (1989). Antibody IAP250 was specific for IAP, but antibody IAP151 cross-reacted with PLAP.

6.3
Discrimination Between Bone and Liver TNAP

Whereas the motivation to produce and characterize MAbs to PLAP/GCAP was driven by the desire to study the genetic polymorphism of PLAP/GCAP and to understand and monitor their re-expression in tumors, the impetus to develop MAbs to TNAP was mandated by the need to discriminate between changes in the bone versus liver isoforms of TNAP in a variety of clinical conditions. In healthy adults, the bone and liver isoforms of TNAP contribute approximately equally to about 95% of the total serum AP activity. Changes in the serum levels of the bone isoform can provide an index of the rate of osteoblastic bone formation such as in growing children or during the development of bone metastasis (Farley and Baylink, 1986; Van Straalen et al., 1991). In turn, changes in the serum levels of the liver isoform can reveal hepatic afflictions such as cholestasis or hepatoma.

A MAb to human liver TNAP was produced by Meyer et al. (1982). The antibody cross-reacted strongly with human kidney and bone TNAP but not with PLAP or IAP and it also cross-reacted with liver and kidney TNAPs from gorilla, chimpanzee and orang-utan. This MAb was used in an affinity column for the final stages of purification of liver TNAP. Lawson et al. (1985) prepared MAbs against isolated human bone TNAP. Hybridoma supernatants were separately screened for reactivity against both the liver and the bone isoforms. Although most antibody-positive hybrids showed similar reactivity against both isozymes, one hybridoma produced an antibody that interacted preferentially with liver TNAP. This antibody was purified and used to establish an immunoassay to differentiate liver from bone TNAP. When equal activities of the two isozymes were measured by the immunoassay, a 5-fold greater response was obtained with the liver than with the bone isoform. The cross-reactivity with human PLAP and IAP was <3% relative to liver TNAP. Bailyes et al. (1987) also purified bone and liver TNAPs and used them to raise antibodies to both isoforms. Altogether, 27 antibody-producing hybridomas were cloned and several of these MAbs showed a >2-fold preference for the bone isoform in the binding assay used for screening. No MAbs showing a preference for liver TNAP were cloned. None of the antibodies showed any significant cross-reactivity with PLAP or IAP. The epitope analysis gave rise to six groupings, with four antibodies unclassified. Hirano et al. (1987b) produced a MAb HLMS-1 (IgG1) against human liver TNAP, which reacted with an affinity of 4.07 $\times 10^{10} M^{-1}$ with all isoforms of TNAP. This antibody was used in an immunocapture assay to detect the presence of TNAP in a variety of normal tissues (Hirano et al., 1987b). Subsequently, Seabrook et al. (1988) indicated that a MAb BAP 1/9 showed a preference for binding AP from serum samples containing predominantly bone TNAP compared with those containing predominantly liver TNAP.

Finally, Hill and Wolfert (1990) reported the preparation of MAbs with high specificity for the bone isoform of TNAP. They immunized mice with the human SAOS-2 osteosarcoma cell line that expresses high levels of bone TNAP and obtained five MAbs. One of them, BA1G 121, displayed <5% cross-reactivity with liver TNAP (Hill and Wolfert, 1990). Masuhara et al. (1991) reported the generation of a MAb that reacted 3.5-fold better with human bone-type TNAP also using human osteosarcoma bone AP as an immunogen. The antibody characterized by Hill and Wolfert allowed the development of the first commercially available IRMA, Tandem-R Ostase™ (Beckman Coulter, formerly Hybritech, San Diego, CA, USA) to measure specifically bone TNAP. In 1998, a single MAb microplate-based enzyme immunoassay, Tandem-MP Ostase™, was introduced by the same company (Broyles et al., 1998). The MAb used in this new assay was reported to be identical with one of the MAbs used in the Tandem-R Ostase IRMA (Hill and Wolfert, 1990; Broyles et al., 1998). The cross-reactivity with liver TNAP was reported to be in the range 7–18% for the Tandem-R and Tandem-MP Ostase assays (Deftos et al., 1991; Garnero and Delmas, 1993; Farley et al., 1994; Panigrahi et al., 1994; Price et al., 1997; Broyles et al., 1998). In 1995, Gomez et al. reported an enzyme immunoassay specific for BALP named Alkphase-B™ (Quidel, formerly Metra Biosystems, Palo Alto, CA, USA). The cross-reactivity with liver TNAP was reported to be in the range 3–15% for the Alkphase-B assay (Gomez et al., 1995; Hata et al., 1996; Milligan et al., 1997; Price et al., 1997; Broyles et al., 1998). An ELISA for the determination of bone TNAP with a cross-reactivity with the liver isoform of <1% has also been developed by Syva (Palo Alto, CA, USA) (Kurn et al., 1994; Demers et al., 1995). However, this assay is not available for commercial use, only for "in-house" research purposes. Another company developing and manufacturing assays of markers of bone turnover, Osteometer BioTech (Herlev, Denmark), reported a purification process for the bone isoform and might therefore have an interest in developing a bone TNAP immunoassay (Nakayama et al., 1998). The development of these commercial immunoassays for the clinical assessment of serum levels of bone TNAP has indeed increased the use of this clinical parameter.

In an effort to understand the antigenic domains of TNAP, an international collaborative study was initiated to collect and characterized a panel of 19 MAbs against TNAP from various investigators and companies (Magnusson et al., 2002a). These studies became the Tissue Differentiation-9 (TD-9) Workshop, part of a series of international collaborative studies under the auspices of the International Society for Oncodevelopmental Biology and Medicine (ISOBM) that deal with different aspects of detection methods for TD antigens. The ISOBM TD-9 MAbs (Table 7) were generated with antigens obtained from human bone tissue, human osteosarcoma cell lines (SaOS-2 and TPX) and human liver tissue. The evaluation included the following antigen forms: (a) commercially available preparations of human bone and liver isoforms of TNAP (Calzyme Laboratories Inc., San Luis Obispo, California); (b) human bone TNAP isoforms, B/I, B1 and B2 (Section 9.1.1.); and (c) soluble-secreted epitope-tagged recombinant human TNAP (*set*TNAP) expressed in COS-1, human osteosarcoma (SaOS-2) and human hepatoma (Hhu2) cell lines. In addition, 16 TNAP mutant cDNAs corresponding to a wide range of reported hypophosphatasia mutations (Section 7.1) were used in an attempt to map specific immunoreactive epitopes on the surface of the TNAP molecule. The TD-9 MAbs were evaluated by immunoradiometric (IRMA) assays, cross-inhibition and different enzyme immunoassay designs. No indications of explicit tissue discriminatory immunoreactivity of the investigated MAbs against TNAP were found. However, certain IRMA combinations of MAbs increased the specificity of bone TNAP isoform measurements. All MAbs bound to the three bone isoforms B/I, B1 and B2, but none of the investigated MAbs were specific for any of the isoforms. Significant differences were found, however, in immunoreactivity between these isoforms with cross-reactivities ranging from 21 to 109% between the two major bone isoforms B1 and B2. Desialylation with neuraminidase significantly increased the MAb affinity for the bone TNAP isoforms B/I, B1 and B2 and also decreased the observed differences in cross-reactivity between these isoforms. These data suggested that the MAb affinity is dependent on the amount/number of terminal sialic acid residues located at the five putative *N*-glycosylation sites.

Table 7 Monoclonal antibodies to human TNAP.

ISOBM No.	MAb name	Isotype	Source of antigen	Reference
314	TP-1	IgG2a	Human osteosarcoma TPX	Bruland et al. (1986)
315	TP-3	IgG2b	Human osteosarcoma TPX	Bruland et al. (1986)
316	BA1G 017.107	IgG2a	Human osteosarcoma SaOS-2	Hill and Wolfert (1990)
317	BA1F 419.2.3	IgG2b	Human osteosarcoma SaOS-2	Hill and Wolfert (1990)
318	LAP 1/5	IgG1	Human liver tissue	Bailyes et al. (1987)
319	LAP 1/12	IgG1	Human liver tissue	Bailyes et al. (1987)
320	BAP 1/8	IgG1	Human bone tissue	Bailyes et al. (1987)
321	BAP 1/9	IgG1	Human bone tissue	Bailyes et al. (1987)
322	BAP 1/10	IgG1	Human bone tissue	Bailyes et al. (1987)
323	BAP 1/15	IgG1	Human bone tissue	Bailyes et al. (1987)
324	BAP 4A5	IgG1	Human bone tissue	Bailyes et al. (1987)
325	BAP 5A5	IgG1	Human bone tissue	Bailyes et al. (1987)
326	BAP 5D4	IgG1	Human bone tissue	Bailyes et al. (1987)
327	BAP 7B3	IgG1	Human bone tissue	Bailyes et al. (1987)
328	BAP A	IgG1	Human bone tissue	Bailyes et al. (1987)
332	BAPIII-4B6	IgG2b	Human osteosarcoma SaOS-2	Gomez et al. (1995)
333	BAPI-1F4	IgG1	Human osteosarcoma SaOS-2	Gomez et al. (1995)
334	BAPIII-2B4	IgG1	Human osteosarcoma SaOS-2	Gomez et al. (1995)
335	HLMS-1	IgG1	Human liver tissue	Hirano et al. (1987)

Based on those overall results, Magnusson et al. (2002a) developed a tentative three-dimensional model of the TNAP molecule with positioning of four major antigenic domains (designated A–D) as defined by the investigated MAbs (Fig. 28). Domain A was defined by MAbs 315, 316, 317, 325 and 332, all of which affected the catalytic activity of TNAP, indicating that they bind close to the active site. Interestingly, both MAbs included in the commercial bone TNAP kits (from Beckman Coulter and Quidel) grouped with domain A MAbs. In addition, desialylation appeared to improve the binding of 317 and 325, which suggested that these MAb epitopes are located close to a sugar chain. As the authors had no way of placing O-linked oligosaccharides (since their location is unknown), they only used the N-linked sugar sites for support in the placement. Domain B included MAbs 314 and 334. MAb 314 appeared to cross-inhibit domain C MAbs and MAb

334 appeared to interfere slightly with domain A MAbs. Domain B was, therefore, placed on the crown domain (Section 4.1.6) of TNAP since this region can allow interference with MAbs from both domains A and C. Domain C, including MAbs 320, 321, 322, 323, 324, 326, 327 and 335, is well defined by the IRMA assay combinations. In addition, the mutagenesis studies demonstrated a 70% drop in reactivity by the E281K mutation with MAb 320. Therefore, domain C antibodies were placed just around residue E281, which is located close to the fourth metal binding site likely to be occupied by a calcium ion (Mornet et al., 2001). Domain D included MAbs 318, 319, 328 and 333. According to the mutagenesis studies, MAb 318 binds close to residues A115, A162 and E174 and the binding of MAb 333 was affected by the attached FLAG tag sequence in *set*TNAP, which suggests that this antibody binds close to the GPI anchor site. Consequently, this domain comprises the flank of the TNAP molecule and extends to the membrane-anchoring area. MAbs 318 and 333 also cross-inhibit binding of some domain C MAbs, hence these domains partially overlap. Therefore, none of the 19 investigated TD-9 MAbs were entirely specific for the bone or liver TNAP isoforms, indicating that all MAbs bind mainly to epitopes on the common protein core of TNAP and/or common glycosylated epitopes. However, some MAbs (either single or in combination with other MAbs) work sufficiently well for measurement of the bone isoform if the assayed samples do not contain very highly levels of the liver isoform.

More recently, also as part of the ISOBM TD-9 workshop, the same panel of MAbs to TNAP was investigated for possible cross-reactivity to the mouse isozyme. Surprisingly, antibodies 314 and 315 reacted strongly with mouse TNAP in Western blots, whereas all other antibodies were negative. By immunohistochemistry, antibodies 314, 315 and 333 produced strong positive staining using frozen sections, whereas antibody 334 was moderately positive. Enzyme immunoassays indicated that MAb 333 was also able to bind to serum TNAP (Narisawa et al., 2005). These antibodies represent very useful reagents to study the pathophysiological expression of TNAP in mouse tissues and in mouse serum.

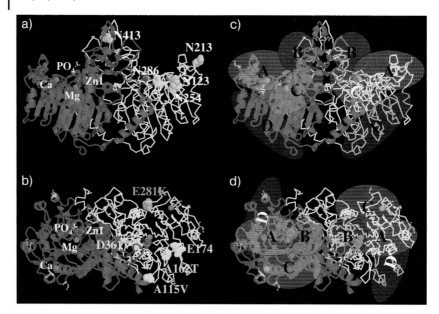

Fig. 28 Epitope map of TNAP. Overall structure of the TNAP molecule with suggested placements of the antigenic group domains designated A–D. The TNAP molecule is illustrated as a homodimer. Subunit A is shown in white with backbone representation and subunit B is shown in magenta with ribbon representation. Most (but not necessarily all) MAb epitopes are present twice in the TNAP homodimer molecule although each group domain is only exemplified once. (a) Side view of the entire TNAP homodimer. The active site phosphate (PO_4^{3-}), Mg and Zn1 (Zn2 is not visible) ions are indicated in subunit B for reference in addition to the fourth metal binding site likely to be occupied by Ca. Each of the five putative N-glycosylation sites (i.e. Asn residues) with terminal sialic acid residues are highlighted in space-filling models in cyan with the residue numbers indicated in subunit A. Our experiments with desialylation by neuraminidase suggests that the MAb affinity and bone AP isoform specificity are dependent on the amount/number of sialic acid residues. (b) Top view of the TNAP homodimer [i.e. rotating the top of the molecule, where N413 is located in (a), "out from the paper" using the paper plane as an axis]. The five Asn residues that serve as N-glycosylation sites are still shown in cyan but now in a less conspicuous manner using a ball-and-stick conformation on both subunits. The active site phosphate (PO_4^{3-}), Mg, Zn1 (Zn2 is not visible) and Ca ions are also indicated in subunit B for reference. Hypophosphatasia mutations that were informative in the experiments are illustrated in subunit A. Mutations A115V, A162T, E174K and D361V are shown in yellow and had an effect on the binding of MAb 318. Mutation E281K is shown in orange and had an effect on the binding of MAb 320. (c) This recapitulates the TNAP molecule shown in (a) apart from illustrating the antigenic group domains. Based on the overall results, suggestive placements are shown for the antigenic domains A–D. Most (but not necessarily all) epitopes are present twice in the TNAP homodimer molecule so each domain is shown in each subunit. Domain C in subunit A is really not visible from this angle, so it has been labeled with the letter C in white to indicate this fact. (d) Top view of the TNAP homodimer with suggestive placements for the antigenic domains A–D. This top view of the TNAP molecule leaves domain D hidden from view so the domain is shown with the letter D in white to indicate this fact. Taken from Magnusson et al. (2002) and reproduced with permission from *Tumor Biology*.

Part II
In Vivo Functions

7
The *In Vivo* Role of TNAP

7.1
Function of TNAP in Bone

7.1.1
Hypophosphatasia

The clearest evidence that APs are important *in vivo* has been provided by studies of human hypophosphatasia, where a deficiency in the TNAP isozyme is associated with a rare form of rickets and osteomalacia, i.e. poorly mineralized cartilage or bone respectively (Frazer, 1957; Whyte, 1994, 1995). Unlike most types of rickets/osteomalacia, neither calcium nor inorganic phosphate levels in serum are subnormal in hypophosphatasia. In fact hypercalcemia and hyperphosphatemia may exist and hypercalciuria is common in infantile hypophosphatasia. The clinical severity in hypophosphatasia patients varies widely. The different syndromes, listed from the most severe to the mildest forms, are perinatal hypophosphatasia, infantile hypophosphatasia, childhood hypophosphatasia, adult hypophosphatasia, odontohypophosphatasia and pseudohypophosphatasia (Whyte, 1995, and references cited therein). These phenotypes range from complete absence of bone mineralization and stillbirth to spontaneous fractures and loss of decidual teeth in adult life. Perinatal (lethal) hypophosphatasia is expressed *in utero* and can cause stillbirth (Fallon et al., 1984). Some neonates may survive for several days but suffer increased respiratory compromise due to the hypoplastic and rachitic disease of the chest. Common features are a failure to gain weight, irritability, high-pitched cry, periodic apnea, myelophthisic anemia, intracranial hemorrhage and idiopathic seizures. Infantile hypophosphatasia presents itself before 6 months of age. Postnatal development often appears normal until the onset of poor feeding, inadequate weight gain and rickets. The marked radiological features are characteristic and resemble those found in the perinatal form although they are less severe. Serial radiological studies may reveal persistence of impaired skeletal mineralization and gradual demineralization of osseous tissue. Childhood hypophosphatasia also has highly variable clinical expression. Premature loss of deciduous teeth results from aplasia, hypoplasia or dysplasia of dental cementum that connects the tooth root with the periodontal ligament. Rickets causes short stature and the skeletal deformities may include bowed legs and enlargement of the wrists, knees and ankles as a result of flared metaphysis. Adult hypophosphatasia usually presents during middle age, although frequently there is a history of rickets and/or early loss of teeth followed by good health during adolescence and young adult life (Whyte et al., 1979). Recurrent metatarsal stress fractures are common and calcium pyrophosphate dihydrate (CPPD) deposition causes attacks of arthritis and pyrophosphate arthropathy (Whyte et al., 1982a). Odontohypophosphatasia is diagnosed when the only clinical abnormality is dental disease and radiological studies and even bone biopsies reveal no signs of rickets or osteomalacia. Pseudohypophosphatasia has only been documented convincingly in two infants. Clinical, radiological and biochemical findings are typical of patients with infantile hypophosphatasia with the important exception that serum AP activity is normal or even increased.

The primary biochemical defect in this condition is a deficiency of the TNAP isozyme, which leads to greatly elevated levels of extracellular pyrophosphate (PP_i) and increased urinary excretion of pyridoxal-5'-phosphate (PLP), PP_i and phosphoethanolamine (PEA). In bone, TNAP is confined to the cell surface of osteoblasts and chondrocytes, including the membranes of their shed matrix vesicles (MVs) (Ali et al., 1970; Bernard, 1978) where the enzyme is particularly enriched compared with the plasma membrane (Morris et al., 1992). It has been proposed that the role of TNAP in the bone matrix is to generate the P_i needed for hydroxyapatite crystallization (Robison, 1923; Majeska and Wuthier, 1975; Fallon et al., 1980). However, TNAP has also been hypothesized to hydrolyze the mineralization inhibitor PP_i (Meyer, 1984) to facilitate mineral precipitation and growth (Moss et al., 1967; Felix and Fleisch, 1974; McComb et al., 1979; Rezende et al., 1998). PP_i inhibits mineralization by preventing initial crystal formation and also crystal growth by physically coating the nascent hydroxyapatite crystals. Electron microscopy revealed that TNAP-deficient MVs, in both humans and mice, contain apatite crystals, but that extravesicular crystal propagation is retarded (Anderson et al., 1997, 2004, 2005a). This growth retardation could be due to either the lack of TNAP's pyrophosphatase function or the lack of P_i generation. Recent studies on TNAP-deficient mice (Section 8.1) have provided compelling proof that a major function of TNAP in bone tissue consists in hydrolyzing PP_i to maintain a proper concentration of this mineralization inhibitor to ensure normal bone mineralization. Of interest is the fact that given the elevated levels of PP_i in these patients, abnormal calcium pyrophosphate dihydrate (CPPD) precipitation has been observed in association with TNAP deficiency (Whyte et al., 1982a, 1995). Since fibroblasts from normal and hypophosphatasia patients appear to have similar levels of nuclotidetriphosphate pyrophosphohydrolase (NPP) activity (Caswell et al., 1986), this accumulation of PP_i is probably not due to elevated synthesis of PP_i but rather to unopposed accumulation of extracellular PP_i levels that result from the missing pyrophosphatase activity of TNAP.

7.1.2
Hypophosphatasia Mutations

The first identified hypophosphatasia mutation was a missense mutation in exon 6, A162T, of the *ALPL* gene (Weiss et al., 1988). Subsequently, compound heterozygosity and new mutations (i.e. R54C, R54P, E174K, Q190P, Y246H, D277A, D361V and Y419H) were reported (Henthorn and Whyte, 1992; Henthorn et al., 1992). Other mutations were subsequently described such as G317D (Greenberg et al., 1993), E281K, A160T, F310L and G439R (Ozono et al., 1996) and a frame shift mutation at position 328 and another at position 503 (Orimo et al., 1994). Mornet et al. (1998) reported 16 new missense mutations in European patients (i.e. S-1F, A23V, R58S, G103R, G112R, N153D, R167W, R206W, W253X, E274K, S428P, R433C, G456S, G474R) and two splice mutations in intron 6 and 9, respectively. Interestingly, in a recent historical vignette, the mutations present in the original 3-week-old male infant with hypophosphatasia studied by Rathbun in 1948 were identified (Mumm et al., 2001). That patient was a compound heterozygote for the A97T and D277A mutations.

Mornet and colleagues (2001) used the modeled structure of TNAP to identify the location of 73 missense mutations associated with hypophosphatasia. Their placements provide further supporting evidence for the structural/functional domains of the molecules described in Section 4. The first group of mutations were located in the active site where 20 substitutions cluster within a 15 Å sphere centered on the P-Ser92 group (Table 8). Except for M45L, H154R and R433H, all the mutations found in this sphere were classified as severe alleles as previously reported (Zurutuza et al., 1999). The second group was located in the active site valley, which extends on both sides of the active site. Here, six mutations affecting four residues were observed: T117N, R119H, E174K, E174G, R433H and R433C. The location of Arg119 and Arg433 in this region, both unique to TNAP, suggests that these two highly basic residues may control access to the active site by the incoming substrate and/or its stabilization. Amino acid substitutions seem to be better tolerated in this area, since at least four of the mutations, R119H, E174K, E174G and R433H, have a moderate phenotype. The third important group of substitutions is found on the homodimer interface. Since dimerization is such a fundamental aspect of APs, it is not surprising to find that of the 20 mutations in the homodimer interface, at least 15 of them are severe alleles. These residues may be directly involved in homodimer interactions or play a role in maintaining the correct fold to allow these interactions to form.

Table 8 Location and severity of some hypophosphatasia mutations. For mutations followed by "?", the severity remains unclear as these mutations were found in patients with moderate hypophosphatasia and compound heterozygote for two missense mutations for which the respective effects were not determined by mutagenesis experiments. A few mutations appear more than once as they belong to overlapping domains. Reproduced from Mornet et al. (2001) with the authors' and publisher's permission.

Severe	Moderate
Active site or active site vicinity	
G46V, T83M, A94T, A99T, G112R, N153D, H154Y, A159T, R167W, D277A, D277Y, G317D, A331T, A360V, D361V, H364R, R433C	M45L, H154R, R433H
Active site valley	
R433C, T117N	R119H, E174K, E174G, R433H
Homodimer interface	
A23V, R54C, R54P, R54H, G58S, G103R, R374C, N400S, S428P, V442M, G456S, E459K, E459G, N461I, G474R	A16V?, V365I?, D389G, G439R?, I473F
Crown domain	
R374C, N400S, V406A, G409C, S428P, R433C	R433H, A382S, D389G, Y419H?
Calcium site or calcium site vicinity	
R206W, K207E, E218G, E274K, D277A, D277Y, E281K, D289V	G203V?, L272F, M278V?
Others	
A34V, R135H, G145V, A162T, C184Y, Q190P, N194D, R229S, G232V, F310C, F310L, C472S	Q59R, A160T, S164L?, F310G?

The fourth cluster of mutations was located on the loop 405–435 within the crown domain, composed of a total of 65 residues. Among the 10 mutations associated with hypophosphatasia identified in this domain, V406A, G409C, Y419H, S428P, R433C and R433H are located within this loop. The others, R374C, A382S, D389G and N400S, in the portion of the crown domain involved in monomer–monomer interactions, are disruptive of the interface, as is the case also for V406A. The substitution G409C rigidifies the loop backbone at the tip of the crown domain. Y419H belongs to an α-helix of the crown domain and S428P occurs within a β-strand. Arg433 is found at the entrance of the active site pocket and has a role in substrate positioning. Its mutation to His433 is probably less severe than its mutation to Cys433, probably because it is more conservative. This is corroborated by the lethal effect associated with the homozygous genotype

R433C and the moderate phenotype of the R433H exchange in childhood hypophosphatasia with just dental abnormalities. Most of the mutations affecting the crown domain are severe, reinforcing the idea that collagen binding (Section 4.1.6) may be important for TNAP function.

The fifth group of substitutions surrounds the calcium-binding site (M4). Eleven distinct mutations were found in this region. The calcium atom is coordinated by the carboxylates of Glu218, Glu274 and Asp289, by the carboxyl of Phe273 and by a water molecule. Mutations of any of the three ligands are related to a severe form of hypophosphatasia, probably because this abrogates calcium binding. E218G was found in a patient with the adult form of the disease and the genotype E218G/A382S (Taillandier et al., 2001). Thus, the second mutation could mask the effect of the mutation E218G, which could be either severe or moderate. To ascertain the role of the calcium atom at M4, Mornet et al. (2001) performed site-directed mutagenesis experiments at this position. They found that COS-1 cells transfected with E218G had no residual enzymatic activity, attesting to the severity of this mutation. The moderate phenotype of the patient was due to a compensatory effect of A382S. Mutation E274K was detected in a patient with childhood hypophosphatasia, carrying the genotype E274K/E174K (Mornet et al., 1998). Previous site-directed mutagenesis experiments showed that this mutation allowed 8% of wt activity and was therefore classified as moderate, although close to the limit for severe alleles (Zurutuza et al., 1999). Mutation D289V was found in a homozygous patient affected with lethal hypophosphatasia (Taillandier et al., 1999) and therefore had to be classified as severe. The other mutations in the vicinity of M4 are severe except for L272F, which has 50% of wild-type activity in site-directed mutagenesis experiments (Sugimoto et al., 1998) and G203V and M278V, which are relatively conservative. The severe character of the mutations surrounding the calcium site indicates that this metal is fundamental to TNAP activity. However, as discussed by Mornet et al. (2001), site-directed mutagenesis experiments do not differentiate between structural and functional mutations. It remains possible, therefore, that some mutations may have a structural defect that results in protein misfolding and degradation, an effect that would not directly involve the metal-binding site function. However, the finding of a cluster of mutated residues in this particular region strongly suggests that at least some of these mutations are related to the function of the calcium binding site. Recently, four out of 11 newly described TNAP mutations, i.e. R255H, P275T, M278T and Y280D, where shown to cluster around the calcium site, providing additional support to the functional/structural significance of this domain (Brun-Heath et al., 2005)

In addition to these five critical functional domains of TNAP, severe hypophosphatasia mutations were also found in other structural regions. Cysteine mutations C184Y and C472S affect one or the other disulfide bridge and other mutations (R135H, G145V, Q190P, N194D, F310C, F310L) are buried and important for secondary structure. Finally, mutation F310C results in the introduction of an extra cysteine that could affect the proper folding of the protein. The severity of these mutations is linked to their location within the core of the protein, likely to lead to structural destabilization of TNAP. Mumm et al. (2002) performed a comprehensive mutational analysis of TNAP using denaturing gradient gel electrophoresis (DGGE). DGGE analysis was 100% efficient in detecting mutations in the coding exons and adjacent splice sites of TNAP in this group of severely affected patients but, as expected, failed to detect a large deletion. In a large subset of severely affected patients, they identified eight novel TNAP mutations (A34S, V111M, ΔG392, T117H, R206Q, G322R, L397M and G409D) and one new TNAP polymorphism (R135H), expanding the considerable genotypic variability of hypophosphatasia. Recently, Komaru et al. (2005) investigated the 1559delT frame-shift mutation that had been reported only in Japanese patients with high allele frequency (Orimo et al., 1994). The authors found that the mutant protein was larger than wt TNAP by ~12 kDa, reflecting an 80 amino acid-long extension at its C-terminus and also lacked a GPI anchor. Only a limited amount of the newly synthesized protein was released into the medium and the rest was polyubiquitinated, followed by degradation in the proteasome, although the aggregated form was still enzymatically active. SDS–PAGE and analysis by sucrose density gradient analysis indicated that the 1559delT mutant TNAP forms a disulfide-bonded high molecular weight aggregate via three cysteines at positions of 506, 521 and 577 of the mutant enzyme and that the aggregation disappeared when these Cys residues where mutagenized to Ser, indicating that these cysteine residues in the C-terminal region are solely responsible for aggregate formation by cross-linking the catalytically active dimers. Mornet has compiled a downloadable database of all known TNAP mutations and polymorphism in the TNAP gene (http://www.sesep.uvsq.fr/Database.html). As of July 2005, a total of 167 distinct mutations had been listed in that database, which is reproduced with permission in Table 9.

7.1 Function of TNAP in Bone | 113

Table 9 Hypophosphatasia mutations identified as of July 2005, taken from the site developed and maintained by Etienne Mornet (http://www.sesep.uvsq.fr/Database.html). Nucleotide numbering is given according to Weiss et al. (1988) and the Nomenclature Working Group (Antonarakis et al., 1988): the first nucleotide (+1) corresponds to the A of the ATG initiation codon. Amino acid numbering is given according to both the nonstandardized and standardized nomenclatures. Nonstandardized nomenclature is according to Weiss et al. (1988) and takes into account the 17-residue signal peptide, i.e. the ATG initiation codon is numbered as residue −17. The standardized nomenclature follows the recommendations of the Human Genome Variation Society (http://www.hgvs.org/), i.e. the first codon is the ATG initiation codon. p, Protein numbering; c, nucleotide numbering of the coding sequence. Reproduced with permission from Etienne Mornet.

Exon	Base change	Amino acid change		Reference	Clinical form	Genotype of patient
		Nonstandardized	Standardized			
1	c.−195C>T			Taillandier et al. (2000)	Perinatal	c.−195C>T/C184Y
2	c.27T>A	L−12X	p.L5X	Taillandier et al. (2000)	Childhood	L−12X/?
2	c.50C>T	S−1F	p.S16F	Mornet et al. (1998)	Infantile	S−1F/G58S
3	c.83A>G	Y11C	p.Y28C	Taillandier et al. (2001)	Infantile	Y11C/R119H
3	c.98C>T	A16V	p.A33V	Henthorn et al. (1992)	Childhood	A16V/Y419H
3	c.110T>C	L20P	p.L37P	Versailles Lab. (Oct. 2003)	Perinatal	L20P/L20P
3	c.119C>T	A23V	p.A40V	Mornet et al. (1998)	Perinatal	A23V/G456S
3	c.132C>T	Q27X	p.Q44X	Mornet, E., unpublished	Perinatal	Q27X/c.662insG
3	c.151G>T	A34S	p.A51S	Mumm et al. (2002)	Infantile	A34S/T117H
3	c.152G>T	A34V	p.A51V	Taillandier et al. (2001)	Infantile	A34V/V442M
4	c.184A>T	M45L	p.M62L	Taillandier et al. (1999)	Infantile	M45L/c.1172delC
4	c.184A>G	M45V	p.M62V	Spentchian et al. (2003)	Infantile	M45V/M45V

Table 9 Continued.

Exon	Base change	Amino acid change		Reference	Clinical form	Genotype of patient
		Nonstandardized	Standardized			
4	c.186G>C	M45I	p.M62I	Taillandier et al.(2005)	Childhood	M45I/E174K
4	c.187G>C	G46R	p.G63R	Spentchian et al. (2003)	Infantile	G46R/G46R
4	c.188G>T	G46V	p.G63V	Lia-Baldini et al. (2001)	Infantile	G46V/N
4	c.203C>T	T51M	p.T68M	Orimo et al. (2002)	Childhood	T51M/A160T
4	c.211C>T	R54C	p.R71C	Henthorn et al. (1992)	Infantile	R54C/D277A
4	c.211C>A	R54S	p.R71S	Orimo et al. (2002)	Childhood	R54S/?
4	c.212G>C	R54P	p.R71P	Henthorn et al. (1992)	Perinatal	R54P/Q190P
4	c.212G>A	R54H	p.R71H	Taillandier et al. (2001)	Perinatal	A23V/R54H
4	c.219T>C	I55T	p.I72T	Versailles Lab. (Oct. 2004)	Odonto	I55T/N
4	c.223G>A	G58S	p.G75S	Mornet et al. (1998)	Infantile	S-1F/G58S
4	c.227A>G	Q59R	p.Q76R	Mornet et al. (2001)	Infantile	Q59R/T117N
IVS4	c.298−2A>G			Taillandier et al. (2000)	Perinatal	c.298−2A>G/c.997+3A>C
5	c.299C>T	T83M	p.T100M	Mornet et al. (2001)	Infantile	T83M/E174K
5	c.323C>T	P91L	p.P108L	Herasse et al. (2003)	Odonto	P91L/N
5	c.331G>A	A94T	p.A111T	Goseki-Sone et al. (1998)	Odonto	A94T/?
5	c.334G>A	G95S	p.G112S	Witters et al. (2004)	Infantile	G95S/R374C
5	c.340G>A	A97T	p.A114T	Mumm et al. (2001)	Infantile	A97T/D277A

Exon	Base change	Amino acid change		Reference	Clinical form	Genotype of patient
		Nonstandardized	Standardized			
5	c.341C>G	A97G	p.A114G	Mornet, E., unpublished	Perinatal	A97G+c.348_349insACCGTC/G309R
5	c.348_349insACCGTC			Mornet, E., unpublished	Perinatal	A97G+c.348_349insACCGTC/G309R
5	c.346G>A	A99T	p.A116T	Hu et al. (2000)	Adult	A99T/N
5	c.358G>A	G103R	p.G120R	Mornet et al. (1998)	Perinatal	G103R/648+1G>A
5	c.382G>A	V111M	p.V128M	Mumm et al. (2002)	Perinatal	V111M/R206W
5	c.385G>A	G112R	p.G129R	Mornet et al. (1998)	Perinatal	G112R/G474R
5	c.388_391delGTAA			Spentchian et al. (2003)	Perinatal	E294K/388_391delGTAA
5	c.389delT			Spentchian et al. (2003)	Perinatal	c.389delT/c.389delT
5	c.392delG			Mumm et al. (2002)	Perinat/infant	c.392delG/A331T
5	c.395C>T	A115V	p.A132V	Watanabe et al. (2001)	Adult	A115V/?
5	c.400_401AC>CA	T117H	p.T134H	Mumm et al. (2002)	Perinatal	T117H/F310del
5	c.401C>A	T117N	p.T134N	Taillandier et al. (2000)	Perinatal	T117N/T117N
5	c.406C>T	R119C	p.R136C	Versailles Lab. (Oct. 2003)	Odonto	R119C/R119H
5	c.407G>A	R119H	p.R136H	Taillandier et al. (1999)	Infantile	R119H/G145V
5	c.442A>G	T131A	p.T148A	Michigami et al. (2005)	Perinatal	T131A/?
5	c.443C>T	T131I	p.T148I	Spentchian et al. (2003)	Infantile	T131I/G145S
6	c.484G>A	G145S	p.G162S	Spentchian et al. (2003)	Infantile	T131I/G145S

Table 9 Continued.

Exon	Base change	Amino acid change		Reference	Clinical form	Genotype of patient
		Nonstandardized	Standardized			
6	c.485G>T	G145V	p.G162V	Taillandier et al. (1999)	Infantile	R119H/G145V
6	c.500C>T	T150M	p.T167M	Versailles Lab. (Oct. 2003)	Infantile	T150M/E174K
6	c.508A>G	N153D	p.N170D	Mornet et al. (1998)	Perinatal	N153D/N153D
6	c.511C>T	H154Y	p.H171Y	Taillandier et al. (1999)	Infantile	H154Y/E174K
6	c.512A>G	H154R	p.H171R	Mornet, E., unpublished	Adult	H154R/E174K
6	c.526G>A	A159T	p.A176T	Taillandier et al. (2000)	Childhood	A159T/R229S
6	c.529G>A	A160T	p.A177T	Goseki-Sone et al. (1998)	Adult	A160T/F310L
6	c.535G>A	A162T	p.A179T	Weiss et al. (1988)	Perinatal	A162T/A162T
6	c.542C>T	S164L	p.S181L	Lia-Baldini et al. (2001)	Infantile	S164L/?
6	c.544delG			Taillandier et al. (1999)	Perinatal	G232V/544delG
6	c.550C>T	R167W	p.R184W	Mornet et al. (1998)	Perinatal	R167W/W253X
6	c.567C>A	D172E	p.D189E	Spentchian et al. (2003)	Perinatal	D172E/D172E
6	c.568_570delAAC	N173del	p.N190del	Michigami et al. (2005)	Perinatal	c.1559delT/N173del
6	c.571G>A	E174K	p.E191K	Henthorn et al. (1992)	Infantile	E174K/D361V
6	c.572A>G	E174G	p.E191G	Goseki-Sone et al. (1998)	Odonto	E174G/c.1559delT
6	c.577C>G	P176A	p.P193A	Mumm et al. (2002)	Adult	A97T/P176A
6	c.602G>A	C184Y	p.C201Y	Taillandier et al. (1999)	Perinatal	c.-195C>T/C184Y
6	c.609C>G	D186E	p.D203E	Versailles Lab. (Oct. 2004)	Perinatal	D186E/D186E

Exon	Base change	Amino acid change		Reference	Clinical form	Genotype of patient
		Nonstandardized	Standardized			
6	c.620A>C	Q190P	p.Q207P	Henthorn et al. (1992)	Perinatal	R54P/Q190P
6	c.631A>G	N194D	p.N211D	Taillandier et al. (2001)	Infantile	A99T/N194D
6	c.634A>T	I195F	p.I212F	Souka et al. (2002)	Perinatal	I195F/E337D
IVS6	c.648+1G>T			Brun-Heath et al. (2005)	Perinatal	c.648+1G>T/D277A
IVS6	c.648+1G>A			Mornet et al. (1998)	Perinatal	G103R/c.648+1G>A
7	c.653T>C	I201T	p.I218T	Utsch et al. (2005), contact	Perinatal	I201T/R374C
7	659G>T	G203V	p.G220V	Taillandier et al. (2001)	Odonto	E174K/G203V
7	659G>C	G203A	p.G220A	Spentchian et al. (2003)	Perinatal	G203A/G203A
7	662insG			Mornet, E., unpublished	Perinatal	Q27X/662insG
7	c.662delG			Spentchian et al. (2003)	Perinatal	R255L/c.662delG
7	c.662G>T	G204V	p.G221V	Versailles Lab. (Oct. 2004)	Perinatal	G204V/M338T
7	c.667C>T	R206W	p.R223W	Mornet et al. (1998)	Perinatal	R206W/?
7	c.668G>A	R206Q	p.R223Q	Mumm et al. (2002)	Perinatal	R206Q/deletion
7	c.670A>G	K207E	p.K224E	Mochizuki et al. (2000)	Infantile	K207E/G409C
7	c.704A>G	E218G	p.E235G	Taillandier et al. (2001)	Adult	E218G/A382S
7	c.738G>T	R229S	p.R246S	Taillandier et al. (2000)	Childhood	A159T/R229S
7	c.746G>T	G232V	p.G249V	Fedde et al. (1996)	Perinatal	G232V/N
8	c.809G>A	W253X	p.W270W	Mornet et al. (1998)	Perinatal	R167W/W253X

Table 9 Continued.

Exon	Base change	Amino acid change		Reference	Clinical form	Genotype of patient
		Nonstandardized	Standardized			
8	c.815G>T	R255L	p.R272L	Spentchian et al. (2003)	Perinatal	R255L/c.662delG
8	c.815G>A	R255H	p.R272H	Brun-Heath et al. (2005)	Infantile	R255H/R255H
8	c.824T>C	L258P	p.L275P	Orimo et al. (2002)	Childhood	L258P/A160T
8	c.853_854insGATC	Y268X	p.Y285X	Michigami et al. (2005)	Perinatal	c1559delT/Y268X
IVS8	c.862+5G>A			Taillandier et al. (1999)	Infantile	c.862+5G>A/c.862+5G>A
9	c.865C>T	L272F	p.L289F	Sugimoto et al. (1998)	Infantile	L272F/?
9	c.871G>A	E274K	p.E291K	Mornet et al. (1998)	Infantile	E174K/E274K
9	c.871G>T	E274X	p.E291X	Taillandier et al. (2000)	Perinatal	A94T/E274X
9	c.874C>A	P275T	p.P292T	Brun-Heath et al. (2005)	Infantile	P275T/A16V
9	c.876_881delAGGGGA	G276_D277del		Spentchian et al. (2003)	Perinatal	G276_D277del/c.962delG
9	c.880G>T	D277Y	p.D294Y	Taillandier et al. (2001)	Infantile	A159T/D277Y
9	c.881A>C	D277A	p.D294A	Henthorn et al. (1992)	Infantile	R54C/D277A
9	c.883A>G	M278V	p.M295V	Mornet et al. (2001)	Childhood	E174K/M278V
9	c.884T>C	M278T	p.M295T	Brun-Heath et al. (2005)	Perinatal	M278T/R206W
9	c.889T>G	Y280D	p.Y297D	Brun-Heath et al. (2005)	Childhood	R119H/Y280D
9	c.892G>A	E281K	p.E298K	Orimo et al. (1994)	Infantile	E281K/1559delT
9	c.896T>C	L282P	p.L299P	Versailles Lab. (Oct. 2003)	Infantile	L282P/L282P

Exon	Base change	Amino acid change		Reference	Clinical form	Genotype of patient
		Nonstandardized	Standardized			
9	c.917A>T	D289V	p.D306V	Taillandier et al. (1999)	Infantile	D289V/D289V
9	c.919C>T	P290S	p.307S	Versailles Lab. (Oct. 2004)	Infantile	P290S/M450T
9	c.928_929delTC			Brun-Heath et al. (2005)	Perinatal	T394A/c.928_929delTC
9	c.931G>A	E294K	p.E311K	Spentchian et al. (2003)	Perinatal	E294K/c.388_391delGTAA
9	c.962delG			Spentchian et al. (2003)	Perinatal	G276_D277del/c.962delG
9	c.976G>C	G309R	p.G326R	Litmanovitz et al. (2002)	Perinatal	G309R/E274K
9	c.981_983delCTT	F310del	p.F327del	Orimo et al. (1997)	Infantile	F310del/c.1559delT
9	c.979T>G	F310C	p.F327C	Mornet et al. (2001)	Perinatal	T117N/F310C
9	c.979_980T]>GG	F310G	p.F327G	Taillandier et al. (2001)	Adult	E174K/F310G
9	c.979T>C	F310L	p.F327L	Ozono et al. (1996)	Infantile	F310L/G439R
9	c.982T>A	F311L	p.F328L	Michigami et al. (2005)	Perinatal nonlethal	F311L/T83M
IVS9	c.997+2T>A			Taillandier et al. (2000)	Perinatal	c.997+2T>A/C472S
IVS9	c.997+2T>G			Brun-Heath et al. (2005)		c.997+2T>G/c.997+2T>G
IVS9	c.997+3A>C			Mornet et al. (1998)	Perinatal	c.997+3A>C/c.997+3A>C
IVS9	c.998–1G>T			Taillandier et al. (2001)	Perinatal	E174K/c.998–1G>T
10	c.1001G>A	G317D	p.G334D	Greenberg et al. (1993)	Perinatal	G317D/G317D
10	c.1015G>A	G322R	p.G339R	Mumm et al. (2002)	Perinatal	G322R/A159T
10	c.1016G>A	G322E	p.G339E	Versailles Lab. (Oct. 2004)	Infantile	G322E/V111M

Table 9 Continued.

Exon	Base change	Amino acid change		Reference	Clinical form	Genotype of patient
		Nonstandardized	Standardized			
10	c.1042G>A	A331T	p.A348T	Taillandier et al. (2000)	Infantile	E174K/A331T
10	c.1062G>C	E337D	p.E354D	Souka et al. (2002)	Perinatal	I195F/E337D
10	c.1064A>C	M338T	p.M355T	Versailles Lab. (Oct. 2004)	Perinatal	G204V/M338T
10	c.1101_1103delCTC	S351del	p.S368del	Versailles Lab. (Oct. 2004)	Perinatal	c.1101_1103delCTC/T372I
10	c.1120G>A	V357M	p.V374M	Versailles Lab. (Oct. 2004)	Adult	V357M/E281K
10	c.1130C>T	A360V	p.A377V	Mornet et al. (2001)	Perinatal	A360V/A360V
10	c.1133A>T	D361V	p.D378V	Henthorn et al. (1992)	Infantile	E174K/D361V
10	c.1142A>G	H364R	p.H381R	Taillandier et al. (2000)	Infantile	A23V/H364R
10	c.1144G>A	V365I	p.V382I	Goseki-Sone et al. (1998)	Childhood	F310L/V365I
10	c.1166C>T	T372I	p.T389I	Versailles Lab. (Oct. 2004)	Perinatal	T372I/S351del
10	c.1171C>T	R374C	p.R391C	Zurutuza et al. (1999)	Childhood	E174K/R374C
10	c.1172G>A	R374H	p.R3P1H	Orimo et al. (2002)	Childhood	R374H/?
10	c.1172delC			Taillandier et al. (1999)	Infantile	M45L/c.1172delC
11	c.1195G>T	A382S	p.A399S	Taillandier et al. (2001)	Adult	E218G/A382S
11	c.1216_1219delGACA			Brun-Heath et al. (2005)	Perinatal	c.1216_1219delGACA/?
11	c.1217A>G	D389G	p.D406G	Taillandier et al. (2000)	Odonto.	D389G/R433H
11	c.1228T>C	F393L	p.F410L	Versailles Lab. (Oct. 2004)	Infantile	F393L/E174K

Exon	Base change	Amino acid change		Reference	Clinical form	Genotype of patient
		Nonstandardized	Standardized			
11	c.1231C>G	T394A	p.T411A	Brun-Heath et al. (2005)	Perinatal	T394A/c.926_927delTC
11	c.1240C>A	L397M	p.L414M	Mumm et al. (2002)	Perinatal	L397M/D277A
11	c.1250A>G	N400S	p.N417S	Sergi et al. (2001)	Perinatal	N400S/c.648+1G>A
11	c.1256delC			Taillandier et al. (2000)	Perinatal	c.1256delC/?
11	c.1258G>A	G403S	p.G420S	Glaser et al. (2004)	Perinatal	G403S/G403S
11	c.1268T>C	V406A	p.V423A	Taillandier et al. (2001)	Perinatal	A99T/V406A
11	c.1276G>T	G409C	p.G426C	Mochizuki et al. (2000)	Infantile	K207A/G409C
11	c.1277G>A	G409D	p.G426D	Mumm et al. (2002)	Childhood	G409D/E174K
11	c.1282C>T	R411X	p.R428X	Taillandier et al. (1999)	Perinatal	R411X/R411X
11	c.1306T>C	Y419H	p.Y436H	Henthorn et al. (1992)	Childhood	A16V/Y419H
12	c.1333T>C	S428P	p.S445P	Mornet et al. (1998)	Infantile	S428P/?
12	c.1349G>A	R433H	p.R450H	Taillandier et al. (2000)	Odonto.	D389G/R433H
12	c.1348C>T	R433C	p.R450C	Mornet et al. (1998)	Infantile	R433C/R433C
12	c.1354G>A	E435K	p.E452K	Spentchian et al. (2003)	Perinatal	A94T/E435K
12	c.1361A>G	H437R	p.E454R	Versailles Lab. (Oct. 2003)	Childhood	E174K/H437R
12	c.1363G>A	G438S	p.G455S	Versailles Lab. (Oct. 2003)	Adult	G438S/G474R
12	c.1366G>T	G439W	p.G456W	Versailles Lab. (2 Oct. 003)	Childhood	G439W/?
12	c.1366G>A	G439R	p.G456R	Ozono et al. (1996)	Infantile	G439R/?

Table 9 Continued.

Exon	Base change	Amino acid change		Reference	Clinical form	Genotype of patient
		Nonstandardized	Standardized			
12	c.1375G>A	V442M	p.V459M	Taillandier et al. (2000)	Infantile	A34V/V442M
12	c.1375G>T	V442L	p.V459L	Versailles Lab. (Oct. 2004)	Perinatal	V442L/E435K
12	c.1396C>T	P449L	p.P466L	Versailles Lab. (2 Oct. 003)	Perinatal	P449L/?
12	c.1400T>C	M450T	p.M467T	Versailles Lab. (Oct. 2004)	Infantile	M450T/P290S
12	c.1402G>A	A451T	p.A468T	Spentchian et al. (2003)	Perinatal	A451T/A451T
12	c.1417G>A	G456S	p.G473S	Mornet et al. (1998)	Perinatal	A23V/G456S
12	c.1426G>A	E476K	p.E476K	Taillandier et al. (1999)	Perinatal	A94T/E459K
12	c.1427A>G	E459G	p.E476G	Mornet et al. (2001)	Perinatal	E459G/E459G
12	c.1433A>T	N461I	p.N478I	Taillandier et al. (2000)	Childhood	N461I/N
12	c.1444_1445insC			Brun-Heath et al. (2005)	Perinatal	c.1444_1445insC/G317D
12	c.1456G>C	C472S	p.C489S	Taillandier et al. (2000)	Perinatal	C472S/c.997+2T>A
12	c.1468A>T	I473F	p.I490F	Lia-Baldini et al. (2001)	Adult	I473F/?
12	c.1471G>A	G474R	p.G491R	Mornet et al. (1998)	Perinatal	G112R/G474R
12	c.1471delG			Brun-Heath et al. (2005)	Odonto	c.1471delG/R119H
12	c.1559delT			Orimo et al. (1994)	Infantile	E281K/c.1559delT

7.1.3
Variable Penetrance and Expressivity

The severe forms of hypophosphatasia are usually inherited as autosomal recessive traits with parents of such patients showing subnormal levels of serum AP activity. Variable expressivity and incomplete penetrance are common features (Whyte, 1995). For the milder forms of hypophosphatasia, i.e. adult and odonto-hypophosphatasia, an autosomal dominant pattern of inheritance has also been documented. The molecular bases for these complex patterns of inheritance are not completely understood, although some mechanisms are becoming apparent.

Defective trafficking of several hypophosphatasia mutations, i.e. R54C, A162T and G317D, through the Golgi apparatus has been established as one mechanism that contributes to the variable expressivity of this disease. For example, R54C not only abolishes completely the catalytic activity of TNAP (Di Mauro et al., 2002), it has also been shown to remain trapped inside transfected cells (Fukushi-Irie et al., 2000). In contrast, two mutations, A162T and D277A, were found to maintain high catalytic efficiency towards pNPP as substrate but not against PLP or PP_i (Di Mauro et al., 2002). When the R54C mutation was co-expressed with a milder active D277A mutation in COS-1 cells, the R54C mutant interfered with folding and assembly of the D277A mutant in trans (Fukushi-Irie et al., 2000), thus explaining why a compound heterozygote for these mutant alleles developed severe hypophosphatasia (Henthorn et al., 1992). The A162T substitution was found to form an interchain disulfide-bonded high molecular weight aggregate within the cells that resulted in impaired intracellular transport of the mutant molecule (Shibata et al., 1998). However, data using a soluble, epitope-tagged mutant enzyme indicated that the lack of activity of A162T mutation is not only due to inappropriate trafficking but also that this mutation abolishes catalytic activity (Di Mauro et al., 2002). The same is true for the G317D mutation (Greenberg et al., 1993). When expressed in COS-1 cells, the mutant does not exhibit any residual TNAP activity (Fukushi et al., 1998; Di Mauro et al., 2002). Pulse-chase experiments showed that the newly synthesized mutant failed to acquire endo H-resistance and to reach the cell surface. This mutant was also found to form a disulfide-bonded high molecular weight aggregate and was rapidly degraded within the cell, despite it being modified by a GPI anchor (Fukushi et al., 1998). Similarly, an N153D mutation causes misfolding and incorrect assembly of TNAP, which results in its retention at the cis-Golgi *en route* to the cell surface, followed by a delayed degradation, presumably as part of a quality control process (Ito et al., 2002). Also, the D289V mutant fails to reach the cell surface and undergoes proteasome-mediated degradation. Another naturally occurring TNAP mutation, i.e. E218G, was also found to be polyubiquitinated and degraded in the proteasome. Interestingly, since these acidic amino acids at positions 289 and 218 of TNAP are thought to be directly involved in Ca^{2+} coordination, these data suggest the critical importance of calcium binding in post-translational folding and assembly of the TNAP molecule (Ishida et al., 2003). Clearly, aberrant intracellular trafficking and protein folding are a major mechanism leading to TNAP deficiency, irrespective of whether the mutant enzyme may have retained partial catalytic activity.

Another mechanism that helps explain variable expressivity is the differential utilization of the two confirmed substrates of TNAP, i.e. pyridoxal-5-phosphate (PLP) and inorganic pyrophosphate (PP_i), by the mutant enzymes. Di Mauro et al. (2002) characterized the kinetic behavior, inhibition and heat stability properties of 16 hypophosphatasia mutations that span the severity of the disease including the ability of the mutants to metabolize PLP and PP_i at physiological pH. Six of the mutant enzymes were completely devoid of catalytic activity (R54C, R54P, A94T, R206W, G317D and V365I) and 10 others (A16V, A115V, A160T, A162T, E174K, E174G, D277A, E281K, D361V and G439R) showed various levels of residual activity. The A160T substitution was found to decrease the catalytic efficiency of the mutant enzyme towards pNPP, to retain normal activity towards PP_i and to display increased activity towards PLP. The A162T substitution caused a considerable reduction in the activity towards pNPP, PLP and PP_i. The D277A mutant was found to maintain high catalytic efficiency towards pNPP as substrate but not against PLP or PP_i. Three mutations (E174G, E174K and E281K) were found to retain normal or slightly subnormal catalytic efficiency towards pNPP and PP_i but not against PLP. Because abnormalities in PLP metabolism have been shown to cause epileptic seizures in mice null for the TNAP gene (Section 8.1), the kinetic data on these mutants help explain the variable expressivity of epileptic seizures in hypophosphatasia patients. Patients who inherit mutations such as E174G, D277A or E281K, which are very inefficient in hydrolyzing PLP, would be expected to manifest more severe seizures and apnea than patients with other mutations. Litmanovitz et al. (2002) described a patient diagnosed with lethal perinatal hypophosphatasia with a unique clinical presentation of convulsions that responded to vitamin B_6. The patient was a compound heterozygote for the E274K and G309R mutations. Since the E274K mutation has been characterized previously as moderate, phenotype/genotype correlation indicates that G309R is a deleterious mutation that can also lead to seizures and a lethal outcome (Litmanovitz et al., 2002). However, this substitution remains to be tested by mutagenesis *in vitro* and assayed for its ability to hydrolyze PLP and PP_i.

Given the extensive compound heterozygocity found in hypophosphatasia, preferential substrate utilization by different mutations, impaired intracellular trafficking and our current understanding that heterodimers have catalytic properties that are not the compounded average of those of the corresponding homodimers, we can begin to visualize the plethora of mechanisms that explain the remarkable variability in terms of expressivity and penetrance that characterizes this disease.

7.2
Role of TNAP in Nonskeletal Tissues

While the skeletal role of TNAP has been very well defined by studies of patients with hypophosphatasia and also by the detailed characterization of TNAP-deficient mouse models (Section 8.1), the role of TNAP in other organs, e.g. liver, kidney and skin, are much less understood. Little is known about the physiological role of TNAP in normal liver. Most of the studies on liver TNAP have revolved around the understanding of enzyme release in cholestasis and these studies are described in Section 9.1.3. Warnes et al. (1981) examined the AP isozyme composition in human bile obtained from patients with gallstones and found that bile AP was partly secreted by the liver cells and partly derived from the small intestine, suggesting the existence of an enterohepatic circulation. Some recent reports have suggested a role for TNAP in the regulation of intrahepatic biliary epithelium secretory activities (Alvaro et al., 2000). Given that one of the natural substrates of TNAP is PLP (a phosphated form of vitamin B_6), it is possible that a role for TNAP in liver is to participate in the metabolism of this B_6 vitamer. Early studies on rat hepatocytes by Li et al. (1974) had concluded that protein binding and hydrolysis of excess unbound PLP by TNAP controlled the hepatic concentrations of PLP. Whereas a study of three infants with hypophosphatasia found no changes in tissue concentrations of vitamin B_6 (Whyte et al., 1988), PLP concentrations in liver, brain, heart, kidney and skeletal muscle of TNAP-deficient mice (Section 8.1) were reduced by ~40% (Waymire et al., 1995). In fact, since the liver enzyme O-phosphorylethanolamine phospholyase (PEA-P-lyase) requires PLP as cofactor (Fleshood and Pitot, 1969, 1970) and since TNAP is crucial for the normal metabolism of PLP (Narisawa et al., 2001), it is possible that it is the insufficient amount of PLP inside the hepatocytes that leads to suboptimal activity of PEA-P-lyase, which in turn leads to increased excretion of PEA in hypophosphatasia. Although one study has addressed this issue (Grøn, 1978) and did not find support for this mechanism, further studies in this regard seem warranted.

Vitamin B_6 is also important for skin development and maintenance and pellagra-like dermatitis is one of the earliest recognized symptoms of vitamin B_6 deficiency. For unknown reasons, the pilosebaceous unit displays prominent TNAP activity and alterations in TNAP activity are seen in alopecia areata. Histochemical correlation studies performed by Handjiski et al. (1994) showed hair cycle and compartment-dependent changes in AP activity. Enzyme activity in outer root sheath keratinocytes was present only during late anagen and catagen, whereas in sebaceous glands the highest activity was found during catagen and telogen and in dermal papilla equivalent levels of activity were present throughout the entire hair cycle (Handjiski et al., 1994). Coburn et al. (2003) attempted to correlate the levels of TNAP expression in skin with the ability to metabolize PLP. By Northern blot analysis, the 2.4 kb mRNA transcript was detected in mouse anagen VI skin. Very low expression was found in telogen skin. Gene expression was absent or below detectability in human neonatal and HaCaT keratinocytes and human skin biopsy. Thus, the pattern of gene expression matched the changes in enzyme activity in mouse tissues but not in human skin and human keratinocytes. Therefore, PLPase activity in mouse skin would appear to correspond predominantly to TNAP, but unexpectedly TNAP knockout mice showed a surprising absence of pathological changes in the skin and adnexal structures (Coburn et al., 2003). Hence an important role for TNAP in vitamin B_6 metabolism in skin appears doubtful at this point.

A role of TNAP in phosphate resorption through the brush border of proximal convoluted kidney tubules has been proposed (McComb et al., 1979). However, Brunette and Dennis (1982) examined this putative role using bromotetramisole (Section 5.1.2.2) to inhibit TNAP activity using isolated rabbit tubules and rat brush border vesicles and could not find support for such a role. However, experiments *in vivo* showed that administration of bromotetramisole in fact increased the fractional excretion of P_i in rats (Onsgard-Meyer et al., 1996). TNAP activity on endothelial cell surfaces appears to be responsible, in part, for the conversion of adenosine nucleotides to adenosine, a potent vasodilator and anti-inflammatory mediator that can protect tissues from the ischemic damage that results from injury. This TNAP activity is induced by IL-6 released during wound repair (Gallo et al., 1997). Also, TNAP has been postulated to play a major role in extracellular nucleotide catalysis and in the regulation of adenosine concentrations on airway surfaces (Picher et al., 2003).

7.3
Proposed Biological Functions of TSAPs

7.3.1
Proposed Functions of IAP

As discussed by McComb et al. (1979), APs are often found in membranes where there is active transport. The presumption for years has been that APs are involved in P_i production, transport and/or absorption. However, some data indicate that this presumed role may be misinterpreted. Tenenhouse et al. (1980) concluded that AP activity does not mediate P_i transport in the renal–cortical brush-border membrane. They studied first the effect of primary modulators of P_i transport, namely the hypophosphatemic mouse mutant (*Hyp*) and low-P_i diet, on AP activity in mouse renal–cortex brush-border membrane vesicles and second the effect of several primary inhibitors of AP on phosphate transport. Brush-border membrane vesicles from the kidneys of *Hyp* mice had 50% loss of Na^+-dependent P_i transport, but only 18% decrease in AP activity. The low-P_i diet effectively stimulated Na^+/P_i co-transport in brush-border membrane vesicles, but increased AP activity only slightly. Levamisole (0.1 mM) and EDTA (1.0 mM) inhibited brush-border membrane-vesicle AP activity up to 82 and 93%, respectively, but had no significant effect on Na^+/P_i co-transport. The authors concluded that AP does not play a direct role in P_i transport across the brush-border membrane of mouse kidney. Similarly, Shirazi et al. (1981b) used potent inhibitors of AP to investigate the role of the rat enzyme in intestinal transport of P_i. When the inhibitor effect of potent inhibitors of AP, such as phosphonates and phosphate analogues, was tested, it was found that there was little, if any, inhibition of P_i transport under conditions in which the inhibition of AP activity was total. Incubation of the intestinal vesicles with oxidized adenosine 5'-(β,γ-imido)triphosphate followed by rapid gel filtration to remove the inhibitor resulted in an irreversible loss of AP activity, but left P_i transport unimpaired. Conversely, a similar prolonged incubation with adenosine 5'-(β-thio)diphosphate or adenosine 5'-(γ-thio)triphosphate had no effect on AP activity but resulted in a permanent partial loss of transport capability. The authors concluded that failure to demonstrate an inhibition of P_i transport resulting from inhibition of AP and the different responses of enzymatic activity and P_i transport to irreversible inhibition make it very unlikely that the enzyme is directly involved in the transport system (Shirazi et al., 1981b). However, Hirano et al. (1985) examined the effects of various inhibitors and antibody to human IAP on P_i uptake into brush-border membrane vesicles and suggested that human IAP may function as a P_i binding protein at low P_i concentrations under physiological conditions.

In birds and mammals, the tissue-specific IAP is localized to the apical brush-border membrane of the enterocyte. The amount of IAP activity increases progressively with phylogenetic development from birds to humans (Komoda et al., 1986). One of the most intriguing observations is that concerning the striking increase in lymph and serum levels of IAP after a fatty meal in rodents (McComb et al., 1979; Young et al., 1981; Alpers et al., 1990) and in humans, especially those of blood group A and/or positive blood group secretor status (Langman et al., 1966; McComb et al., 1979; Domar et al., 1991). The post-prandial rise in serum IAP activity is significantly greater following a long-chain fatty acid meal than following a medium-chain fatty acid meal (Day et al., 1992a). In turn, IAP levels are dramatically decreased upon starvation (Hodin et al., 1994). IAP is found associated with the brush-border of the intestinal epithelium and enriched in SLP (Section 4.3.5) (Eliakim et al., 1989; Alpers et al., 1994, 1995; Eliakim et al., 1997; Goetz et al., 1997). After fat feeding, an IAP-containing membrane surrounds the lipid droplet in the enterocyte (Yamagishi et al., 1994a,b; Zhang et al., 1996a). These findings suggest a role for SLP in fat absorption involving the induction of IAP by fat, which in turn results in IAP-induced SLP secretion. This proposed role of IAP in lipid absorption seems to be further supported by studies on the $Akp3^{-/-}$ mice (Section 8.2).

Colonic SLP also appear to play a role in binding pili of uropathogenic *E. coli*, so SLPs may serve as reservoirs for those organisms (Goetz et al., 1999). Interestingly, PLAP and IAP are able to bind *Aeromonas sobria* hemolysin and may act as receptors mediating gastroenteritis by pathogenic bacteria (Wada et al., 2005). Interestingly, in patients suffering from acute bacterial infections, specific IgG autoantibodies to IAP are transiently expressed in high titer (Kolbus et al., 1996). Of particular interest are the studies by Poelstra et al. (1997a,b) that demonstrate the ability of IAP to dephosphorylate lipopolysaccharide (LPS) at a physiological pH and the proposal that IAP may have an important physiological role in lipopolysaccharide detoxification as a host defense mechanism (see also Section 13.3). This anti-endotoxin principle may also be inducible in the glomerulus in the kidney (Kapojos et al., 2003).

7.3.2
Putative Functions of GCAP and PLAP

Whereas the IAP isozyme is likely to be involved in the intestinal absorption of lipids via its association with surfactant-like particles and in detoxification, currently there are no clues concerning the function of the GCAP isozyme. GCAP is expressed in primordial germ cells during their migration through the genital ridges (Heath, 1978), as well as in trace amounts during the first steps of germ cell maturation in normal adult testis (Hustin et al., 1987), in the thymus and in the type I pneumocytes in the normal lung (Nouwen et al., 1990), and this latter expression most likely accounts for the normal basal serum levels of GCAP detected in all individuals (Millán et al., 1985a). It is possible that GCAP plays a role in early embryonic development, as suggested by the subtle abnormalities observed after inactivating the mouse ortholog, *Akp5* (Section 8.3). The enzyme may also have a role to play in germ cell development, although to date there is no evidence in this regard.

However, more indications exist as to the possible functions of PLAP, which is 98% identical with GCAP. In addition to its catalytic properties, PLAP has the capacity to bind to the Fc portion of human IgG (Makiya and Stigbrand, 1992a,b). The dissociation constant for the interaction ($3.86\,\mu mol\,L^{-1}$) indicates that the PLAP–IgG complex probably occurs *in vivo*. PLAP also promotes the internalization of IgG in HEp2 cells in culture (Makiya and Stigbrand, 1992c) where PLAP appears clustered in clathrin-coated vesicles (Makiya et al., 1992). In a subsequent clinical study, venous blood samples from term pregnant women and cord samples from their fetuses were obtained, together with the corresponding placentas. Mean PLAP levels were determined as 23.7 and $1.2\,ng\,mL^{-1}$ in maternal and fetal blood, respectively. The mean IgG level of the fetal samples was significantly higher than that of the maternal samples. The S-type PLAP had a larger dissociation constant to IgG than did F-type PLAP and was found to have a mean fetal IgG level higher than those of the F type. The PLAP phenotype was found to be related to the levels of its ligand IgG in fetal blood, although the mechanism for this remains to be established (Friden et al., 1994). Stefaner et al. (1997) also analyzed the ability of PLAP to bind, internalize and transcytose IgG in BeWo choriocarcinoma cells endogenously expressing the protein and in Madin–Darby canine kidney (MDCK) cells transfected with the PLAP cDNA. Although PLAP expression in MDCK cells resulted in increased IgG binding to intact cells, binding was not correlated with the level of PLAP expressed in the different cell lines. The authors concluded that his findings did not support a role for PLAP in IgG endocytosis or transcytosis (Stefaner et al., 1997). Hence the hypothesized role of PLAP in mediating IgG transport through the placenta remains to be confirmed. Of related interest to those studies, Shinozaki et al. (1998) examined the effects of immunoglobulins on the catalytic activity of both bovine IAP and human PLAP *in vitro*. They found that bovine but not rabbit immunoglobulin enhanced the AP activity of bIAP. Similarly, human but not bovine IgG enhanced human PLAP activity. They concluded that in clinical conditions with high immunoglobulin concentrations, the serum AP determinations might give spuriously high values. However, the molecular basis for this effect is currently unknown.

Another postulated role for PLAP is in the regulation of cell division. Telfer and Green (1993b) found that PLAP inducers, sodium butyrate, dexamethasone, bromodeoxyuridine and dibutyryl-cAMP caused a dose-dependent reduction in growth rate in HeLaS3 cells. Mimosine, an agent that blocks the cell cycle in Gl, caused an increase in PLAP activity whereas the mitogen epidermal growth factor caused a corresponding decrease in PLAP activity. PLAP has also been shown to stimulate DNA synthesis and cell proliferation when added exogenously. She et al. (2000a) showed that in human fetus fibroblasts and also normal and H-Ras-transformed mouse embryo fibroblasts, PLAP stimulates DNA synthesis and cell proliferation in synergism with insulin, zinc and calcium. The mitogenic effects of PLAP were associated with the activation of c-Raf-1, p42/p44 mitogen-activated protein kinases, p70 S6 kinase, Akt/PKB kinase and phosphatidylinositol 3′-kinase. The results suggested that *in vivo* PLAP might promote fetalus development and also the growth of cancer cells that express oncogenic Ras. Subsequently, these authors also reported that in mouse embryo (NIH 3T3) and human fetalus (HTB-157) fibroblasts, PLAP (200 nM) alone provided full protection against serum starvation-induced cell death for 5 days. After 12 days, substantial effects of PLAP on cell survival required the presence of insulin (500 nM) and ATP or adenosine (100 μM). In serum-starved NIH 3T3 cells, PLAP induced activating phosphorylation of p42/p44 mitogen-activated protein (MAP) kinases; insulin, but not ATP, had small additional effects. PLAP also stimulated the expression of various cyclins; ATP both prolonged and enhanced PLAP-induced expression of cyclins A and E. Finally, ATP/adenosine enhanced activation of Akt kinase by insulin. The results suggested that PLAP might be a regulator of growth and remodeling of fetal tissues during the second and third trimesters of pregnancy (She et al., 2000b).

8
Knockout Mouse Models

The identification and study of inborn errors of metabolism often provide invaluable clues as to the *in vivo* function(s) of a molecule. In humans, only deficiencies in the *ALPL* gene have been observed. There are no reported cases of deficiencies in the *ALPI*, *ALPP* or *ALPP2* genes, so the *in vivo* functions of the IAP, PLAP and GCAP isozymes remain unclear. However, important information can be derived from gene targeting experiments in mice. Missense mutations in the human *ALPL* gene lead to the inborn error of metabolism known as hypophosphatasia (Sections 7.1.1. and 7.1.2) and studies of this disease both in humans and in *Akp2* null mice have provided compelling evidence for an important role for TNAP during the development and mineralization of the skeleton. Functional deletion of the *Akp3* gene leads to an accelerated transport of fat through the intestinal epithelium, suggesting that IAP may be involved in a rate-limiting step during fat absorption. Deletion of the *Akp5* gene leads to a 12 h delay in preimplantation development and increased embryo lethality *in utero*. Whether similar functions could be attributable to the human orthologs remains to be determined. A recapitulation of what has been learned from each of these mouse AP null models is given below.

8.1
Phenotypic Abnormalities in *Akp2*$^{-/-}$ mice

8.1.1
Developmental and Skeletal Defects

Two *Akp2* (TNAP) knockout mouse models have been developed independently (MacGregor et al., 1995; Waymire et al., 1995; Narisawa et al., 1997). To understand the significance of TNAP expression in primordial germ cells (PGCs), MacGregor et al. created a null allele of the *Akp2* gene by replacing the coding sequences with a *βgeo* (*lacZ/neo^r*) reporter gene. Whereas the authors named this allele TNAP^{βgeo} in their paper, it is referred to here as *Akp2*^{βgeo} to conform to the current gene nomenclature cited in Table 1. Since the TNAP isozyme is expressed at relatively low levels prior to gastrulation at stages where EAP expression is

Mammalian Alkaline Phosphatases: From Biology to Applications in Medicine and Biotechnology. J. L. Millán
Copyright © 2006 WILEY-VCH Verlag GmbH & Co. KGaA, Weinheim
ISBN: 3-527-31079-7

prominent, the introduction of the $Akp2^{\beta geo}$ allele would enable localization of the PGCs during early stages of development even in the presence of high EAP expression. In turn, Narisawa et al. (1997) generated null alleles of the $Akp2$ and $Akp5$ genes and also examined possible abnormalities in the [$Akp2^{-/-}$; $Akp5^{-/-}$] double-knockout mice. A collaborative side-by-side comparison of the $Akp2^{\beta geo/\beta geo}$ and the $Akp2^{-/-}$ strains (Fedde et al., 1999) revealed that both strains of knockout mice phenocopy the human disease known as infantile hypophosphatasia, although some differences exists between these strains.

The $Akp2^{-/-}$ pups are born with normal appearance and weight, including healthy milk lines, and at this stage are indistinguishable from either wt or heterozygote siblings, but they become recognizable 4–6 days postpartum because of their slightly smaller bodies (Fig. 29). Most of the $Akp2^{-/-}$ mice become progressively exhausted and die at around days 10–12 with a body weight only 30–50% of that of control littermates. This growth retardation occurs despite the ability of neonatal $Akp2^{-/-}$ mice to suckle. Most of the $Akp2^{-/-}$ neonatal mice do not have any detectable AP activity in their serum. Approximately 30% of the $Akp2^{-/-}$ mice, however, have low but detectable levels of AP activity, very likely resulting from IAP expression, since this activity is inhibited by L-phenylalanine. Serum AP levels of $Akp2^{+/-}$ heterozygote mice are consistently 50% of those of wt animals (Narisawa et al., 1997; Fedde et al., 1999).

Severe epileptic seizures are frequently observed in $Akp2^{-/-}$ mice 1–2 days before their death. The seizures are observed in ~50% of the mice and the frequency of the attacks does not appear to be dependent on the genetic background as $Akp2^{-/-}$ with either a higher contribution of C57Bl/6J or of 129/JSv$^{+/+}$ background display a similar pattern. The seizures have various forms of presentation, such as constant running in the cage, high-pitched vocalizations, biting of their tongue and loss of consciousness in a supine position associated with apnea for periods of 30 s or more. These attacks can continue for periods ranging from 30 min to a few hours and occur daily. Body temperature is palpably decreased during these attacks and death often follows a prolonged seizure attack. $Akp2^{-/-}$ mice also show poor coordination and appear disoriented. Nine- to ten-day-old $Akp2^{-/-}$ mice placed

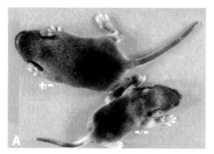

Fig. 29 Macroscopic and behavioral abnormalities in the $Akp2^{-/-}$ mice. A 10-day-old $Akp2^{-/-}$ and an $Akp2^{+/-}$ littermate control are shown. (A) The $Akp2^{-/-}$ mouse is 40–50% smaller that the littermate heterozygote. (B) The same $Akp2^{-/-}$ mouse undergoing a seizure.

on a small (4 × 4 cm) box fall off after an average of 5 s, whereas $Akp2^{+/-}$ or wt animals remain on top for at least 30 s and often 10 min or more. Similarly, when placed on inclined planes, homozygous mice appear not to notice whether they are placed facing upwards or downwards on the slope, whereas control mice would always move up the slope (Narisawa et al., 1997).

Even though TNAP is also expressed during preimplantation development at the blastocyst stage (Section 2.1), the lack of TNAP expression is completely compatible with the survival of embryos to term. Examination of $Akp2^{-/-}$ embryos, ranging in age from E9.5 to E18.5, indicated complete absence of AP activity in the embryonal tissues, but the embryos are morphologically normal. The lack of TNAP activity also appears not to have any adverse consequences on the migratory properties of PGCs and their ability to colonize the gonads and initiate normal gametogenesis (MacGregor et al., 1995). It has been hypothesized that APs may play a role in the processes of cell guidance and migration in a variety of non-mammalian species. Several studies have demonstrated a potential involvement of AP expression both in the migration of the pronephric duct during development in the salamander (Zackson and Steinberg, 1988; Thibaudeau et al., 1993) and in growth cone guidance during axonal outgrowth from the distal limb in the developing grasshopper (Chang et al., 1993). In addition, expression of AP is often restricted both temporally and spatially to populations of cells that are involved with migration or morphogenesis. MacGregor et al. (1995) examined germ cell development between E11.5 and prepubertal stages in the $Akp2^{\beta geo/\beta geo}$ mice. They documented equivalent numbers of germ cells in 14-day-old testes of both the heterozygous and homozygous null mice and found no gross defect in the migration of PGCs during embryogenesis and in the numbers of PGCs in the genital ridges of E10.5 mutant embryos compared with heterozygous or wt littermates (MacGregor et al., 1995). Hence, despite the fact that TNAP clearly is a useful marker to detect PGCs, TNAP function is not required for PGC migration or differentiation. Survival of the $Akp2$ null embryos also occurs even in the complete absence of TNAP in the mouse placenta, which is the tissue showing the highest level of TNAP expression during development. No other compensatory AP activity is detected at any of these stages, suggesting either that an alternative biochemical pathway compensates for the lack of TNAP function or that this isozyme is nonessential for these tissues during embryogenesis and early postnatal stages.

Adult tissues that normally express TNAP (such as the skeleton and kidney) are completely negative in homozygous mice. Gross anatomical examination of $Akp2^{-/-}$ mice revealed significant alterations in several organs. The spleens of $Akp2^{-/-}$ mice are paler and smaller than those of controls even after correcting for body size differences. The $Akp2^{-/-}$ mice have almost no body fat and very reduced muscle structure. Their small intestine consistently contains large amounts of gas, never observed in control heterozygous animals, suggesting impaired intestinal function. The cortex of the thymus of $Akp2^{-/-}$ mice is thinner than that of the littermate heterozygous controls and contains increased number of apoptotic cells. The spleen of $Akp2^{-/-}$ mice shows a reduction in cell mass in the outer layer and an altered ratio of red to white pulp compared with heterozygous controls.

The nerve roots emerging from the spinal cord and descending within the dura in $Akp2^{-/-}$ animals were found to be consistently thinner than those from heterozygous control mice, even after correcting for their size difference. The average ratio relating the area of the nerve roots to that of the spinal cord was 9.1–3.9% for nine different $Akp2^{-/-}$ mice compared with 19.5–4.7% for corresponding sections from nine littermate $Akp2^{+/-}$ or wt controls at the L2, L3, L5 and L6 position. This difference was not observable in the thoracic region. Necropsies performed on $Akp2^{-/-}$ mice that had died after severe seizure attacks revealed bleeding in the thoracic cavity in about 50% of the mice. Bleeding in lung tissue and in the cranial ventricle was also observed (Narisawa et al., 1997).

A comparison of skeletal preparations of embryos and newborns, stained with Alizarin Red and Alcian Blue, revealed no differences between the $Akp2^{-/-}$, $Akp2^{\beta geo/\beta geo}$ and wt mice (Fedde et al., 1999). At postnatal day 6, all genotypes had indistinguishably mineralized skeletons with obvious metaphyses. Secondary growth centers (epiphyses) had not yet appeared at the knees. However, the staining of 8-day-old $Akp2^{-/-}$ bones clearly showed poor mineralization in the parietal bones, scapulae, vertebral bones and ribs. The parietal bones were thinner and the scapulae and vertebrae showed areas of hypomineralization. By day 9, when secondary ossification centers at the knees were just beginning to appear in the control tibia (and forepaws), they were absent in the $Akp2^{-/-}$ and $Akp2^{\beta geo/\beta geo}$ mice (Fedde et al., 1999; Tesch et al., 2003). The epiphyses in the wt mice were well mineralized by day 10, but absent in the $Akp2^{-/-}$ mice. At subsequent ages the $Akp2^{-/-}$ and $Akp2^{\beta geo/\beta geo}$ mice had indistinct metaphyses, hypomineralized condyles and persistent absence of the secondary ossification centers of the tibias (Fedde et al., 1999). Evidence of spontaneous fractures was found in the fibulae. Occasionally, only broken incisors without other obvious abnormalities were observed. Osteoblasts with abnormal morphology were observed on the surfaces of trabeculae and in the periosteal region of the diaphysis of $Akp2^{-/-}$ mice but not of $Akp2^{+/-}$ controls. The shortening of the trabeculae in the homozygous mice was also confirmed with Von Kossa staining for calcium. At the electron microscopic level, up to 50% of the mature osteoblasts in the parietal bones of $Akp2^{-/-}$ mice contained abnormal vacuoles.

Tesch et al. (2003) assessed the mineral characteristics (small-angle X-ray scattering, quantitative backscattered electron imaging) and examined bones of $Akp2^{\beta geo/\beta geo}$ mice by light microscopy and immunolabeling. Their results showed a reduced longitudinal growth and a strongly delayed epiphyseal ossification in the null mutant mice. They found disturbances in mineralization pattern, in that crystallites were not orderly aligned with respect to the longitudinal axis of the cortical bone. However, a great variability in the mineralization parameters was noticed from animal to animal. Also, immunolabeling of osteopontin (OPN) revealed an abnormal distribution pattern of the protein within the bone matrix. In wt mice, OPN was predominantly observed in cement and reversal lines, whereas in the null mice, OPN was also randomly dispersed throughout the nonmineralized matrix, with focal densities. In contrast, they found the distribution pattern of osteocalcin to be comparable in mutant and control animals. The authors concluded that ablation of *Akp2* gene results not only in hypomineralization of the skeleton, but also in a severe disorder of the mineral crystal alignment pattern in the corticalis of growing long bone in association with a disordered matrix architecture, presumably as a result of impaired bone remodeling and maturation.

As in patients with hypophosphatasia, there was a striking elevation of urinary PP_i levels in the $Akp2^{-/-}$ and $Akp2^{\beta geo/\beta geo}$ mice. Also, elevated levels of urinary PEA and a striking accumulation of plasma PLP were detected in the $Akp2^{-/-}$ mice (Fedde et al., 1999). Plasma levels of the dephosphorylated PLP product, pyridoxal (PL), were low in the $Akp2^{\beta geo/\beta geo}$ mice (days 10–14), despite PL therapy. This is in contrast to all but a subset of especially severe hypophosphatasia patients (Whyte et al., 1985, 1988; Chodirker et al., 1990). PL levels in the $Akp2^{-/-}$ mice were normal (without vitamin supplementation) at 6 days of age. Each of the major findings in the study by Fedde et al. (1999) was observed in both the $Akp2^{\beta geo/\beta geo}$ and $Akp2^{-/-}$ mouse colonies. However, the severity of the hypophosphatasia phenotype did differ somewhat between these mutant strains. In the $Akp2^{\beta geo/\beta geo}$ mice, reductions in body weight, growth rate and plasma PL concentrations and increased urinary PEA levels were greater than in the $Akp2^{-/-}$ mice. However, fractures seemed more prevalent in the $Akp2^{-/-}$ than in the $Akp2^{\beta geo/\beta geo}$ mice. Nevertheless, the gene inactivation procedures themselves were not expected to underlie differences between the knockout strains, as they both resulted in undetectable levels of *Akp2* message expression. The mutant alleles of *Akp2* have been confirmed to be null for AP activity in both mouse models (MacGregor et al., 1995; Waymire et al., 1995; Narisawa et al., 1997). Hence it is likely that differences in the background strain composition of the two colonies are responsible for the variability in the mutant phenotypes. Although the ES cell lines used to transmit the targeted mutation through the germline were similar (129/Sv versus 129/J for $Akp2^{\beta geo/\beta geo}$ and $Akp2^{-/-}$, respectively), the chimeric animals were bred with C57BL/6 females ($Akp2^{\beta geo/\beta geo}$) or 129/SvJ females ($Akp2^{-/-}$). Accordingly, the genetic constitutions of the resulting mouse strains were different. The strain composition (genetic background) of the $Akp2^{\beta geo/\beta geo}$ breeding colony was ~87% C57BL/6 and 13% 129/Sv. For the $Akp2^{-/-}$ colony, it was 25% C57BL/6 and 75% 129/J. Several papers have documented that genetic background can have a marked influence on phenotype after gene disruption in mice. For example, Goldstein et al. (1985) found about a 20-fold difference in TNAP activity in lungs from CBA/J and C57L/J inbred strains and this difference was inherited additively with a heritability of 0.84. They also found that the IAP activity varied over 100-fold between A/J and DBA/1J strains. Therefore, it is likely that the greater incidence of fractures and more severe skeletal deformities in the $Akp2^{-/-}$ mice resulted either from their higher proportion of 129/J and/or reduced C57BL/6 strain composition. Indeed, such subtle differences in parental allele combinations could also underlie some of the variability observed in clinical expression of hypophosphatasia in affected siblings (Whyte, 1994, 1995).

8.1.2
Dental Abnormalities in $Akp2^{-/-}$ Mice

Early loss of deciduous teeth is a clinical manifestation of hypophosphatasia (Section 7.1.1). TNAP is expressed at high levels in odontoblasts of the periodontium and a positive correlation exists between the levels of TNAP and the amount of acellular cementum formed. Beertsen et al. (1999) analyzed $Akp2^{\beta geo/\beta geo}$ mice between 6 and 25 days of age and reported a 2–3 day delay in the eruption of incisors in the oral cavity and also the onset of mineralization of the mantle dentin in the roots of the developing molars. However, they found that dentin and enamel formation in the homozygous mutant mice showed a normal pattern except for some localized hypoplasias. Their most conspicuous finding was the identification of a defective formation of acellular cementum along the molar roots, which instead of forming in a continuous layer was deposited as very thin and irregularly shaped patches around the bases of the periodontal ligament fibers (Beertsen et al., 1999). Similarly, in collaboration with Dr Huw F. Thomas, University of Texas Health Science Center, San Antonio, TX, USA, we have examined mandibular first molars of $Akp2^{-/-}$ and $Akp2^{+/-}$ mice at 7 and 24 days of age by light and electron microscopy. We found noticeable differences in the acellular cementum, root dentin and enamel matrix of $Akp2^{-/-}$ samples compared with littermate heterozygote controls (Fig. 30). Whereas $Akp2^{+/-}$ roots showed normal acellular and cellular cementum formation, no acellular cementum was observed in $Akp2^{-/-}$ roots. $Akp2^{-/-}$ teeth also showed a delay in root dentin formation that manifested as an apparent lack of mineralization.

8.1.3
Deficient Mineralization by $Akp2^{-/-}$ Osteoblasts *In Vitro*

Early studies by Fallon et al. (1980) had indicated that the stereospecific inhibition of TNAP activity prevented *in vitro* cartilage calcification. These authors used the anthelmintic drug L-tetramisole (levamisole) (Section 5.1.2.2) and its inactive isomer, D-tetramisole (dexamisole) on *in vitro* calcification studies of rachitic rat cartilage. They found that 50 mM and greater concentrations of L-tetramisole virtually abolished TNAP activity and prevented *in vitro* cartilage calcification, while preserving the structural integrity of the matrix vesicles (MVs). Those conclusions have now been further validated with TNAP null osteoblasts. Wennberg et al. (2000) evaluated the ability of $Akp2^{+/-}$ and $Akp2^{-/-}$ primary osteoblasts to form and mineralize bone nodules *in vitro* in media supplemented with ascorbic acid, using β-glycerophosphate as P_i source. At different time points (days 4, 6 and 8), cultures were fixed and stained with the von Kossa procedure to visualize mineralized nodules. Staining of post-confluent cultures of $Akp2^{-/-}$ osteoblasts showed that these cells were able to form cellular nodules, typical of long-term calvarial osteoblast cultures. However, in contrast to cultures of wt or heterozygous osteoblasts, mineralization by $Akp2^{-/-}$ osteoblasts was never initiated (Fig. 31). Calcium measurements further confirmed the lack of mineral deposition in these $Akp2^{-/-}$ cultures.

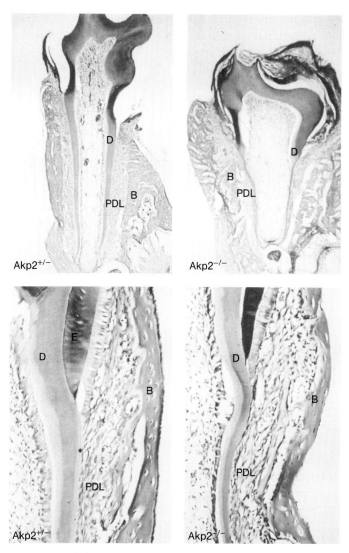

Fig. 30 Dental defects in Akp2$^{-/-}$ mice. Light micrographs of 18-day-old Akp2$^{+/-}$ and Akp2$^{-/-}$ mandibular first molars. Note the presence of a thin layer of acellular cementum (*) in the control but not in the Akp2$^{-/-}$ root surface. PDL, periodontal ligament; B, alveolar bone; D, dentine; E, enamel.

The number and size of the nodular structures did not vary significantly between the different genotypes, only mineralization of the nodules appeared to be affected by the lack of TNAP. Of particular interest is the finding that initiation of mineralization was delayed in the Akp2$^{+/-}$ heterozygous osteoblast cultures compared with wt osteoblasts. This was correlated with a delayed increase in the levels of TNAP activity in Akp2$^{+/-}$ in comparison with wt osteoblasts. The extent of bone mineral deposition in these cultures was confirmed by quantifications of deposited calcium. These results were compatible with the von Kossa staining data, showing clear phenotypic differences between wt, $Akp2^{+/-}$ and $Akp2^{-/-}$ osteoblasts concerning bone nodule mineralization. Calcification of $Akp2^{-/-}$ osteoblast cultures could be restored by exposure to conditioned media from wt osteoblast cultures and also by adding purified soluble recombinant human TNAP to the $Akp2^{-/-}$ osteoblast cultures. As in control cultures, the deposition of mineral in these cultures was restricted to bone nodules. In contrast, neither heat-inactivated recombinant TNAP nor enzymatically inactive mutants of TNAP, such as [R54C]TNAP or [V365I]TNAP (Section 7.1.2), were able to induce mineralization (Wennberg et al., 2000). These data suggest that a certain level of TNAP activity has to be reached for calcium deposition to be initiated and indicate that a moderate reduction in the levels of expression of TNAP protein and enzyme activity may be sufficient to impair the mineralization process and cause dominant, albeit subtle, phenotypic abnormalities in some cases of hypophosphatasia.

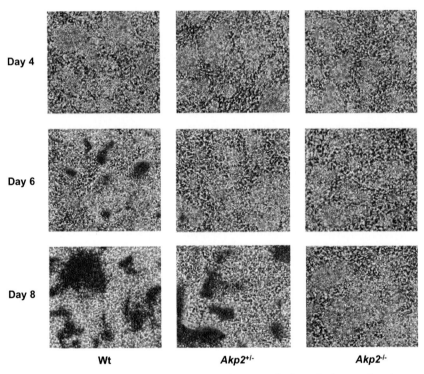

Fig. 31 Bone nodule formation and mineralization of primary calvarial osteoblasts. Von Kossa staining was used to detect the presence of mineral deposits in bone nodules formed during culture of wt, $Akp2^{+/-}$ and $Akp2^{-/-}$ osteoblasts. Stainings were performed after 4, 6 and 8 days of culture. Cell nuclei have been counterstained in red. The results shown are representative for three individual experiments. Taken from Wennberg et al. (2000) and reproduced with permission from the *Journal of Bone and Mineral Research*.

Since extravesicular ATP-dependent extension of mineral deposition is impaired in MVs from patients with perinatal hypophosphatasia (Anderson et al., 1997), Johnson et al. (2000) assessed the ability of $Akp2^{+/-}$ and $Akp2^{-/-}$ osteoblast-derived MV fractions to precipitate calcium. In doing so, aliquots of the MV fractions were treated with the detergent Triton X-100 (0.1%), which is capable of enhancing ATP-initiated MV mineralization via MV membrane perturbation (Hsu and Camacho, 1999). In the absence of detergent treatment, $Akp2^{-/-}$ osteoblasts and control osteoblasts released MV fractions that were comparably able to precipitate calcium in an ATP-dependent manner. However, treatment of the MV fractions with Triton X-100 unmasked a significant defect in the ATP-dependent calcium-precipitating ability of $Akp2^{-/-}$ osteoblast-derived MV fractions relative to wt MV fractions. The defect directly correlated with the genotype, because it was greater in MV fractions derived from $Akp2^{-/-}$ mice than from $Akp2^{+/-}$ mice. Furthermore, the PP_i content of MV fractions was 4-fold greater in cultured $Akp2^{-/-}$ than wt calvarial osteoblasts (Johnson et al., 2000). The authors also found co-localization of nucleosidetriphosphate pyrophosphohydrolase (NPP) and TNAP on the MVs and established that the specific activity of NPP1 was 2-fold greater in MV fractions of wt osteoblasts than in $Akp2^{-/-}$ cells. This downregulation can be explained by a direct effect of extracellular PP_i on the expression of the NPP1 gene (*Enpp1*) (Section 8.1.4.2). Interestingly, however, transfection of the TNAP cDNA significantly increases the NPP1 activity in the osteoblast MV fraction. Hence TNAP attenuates NPP1-induced PP_i generation that would otherwise inhibit MV-mediated mineralization, whereas TNAP also paradoxically regulates NPP1 distribution to osteoblast MVs by a yet unknown mechanism (Johnson et al., 2000).

8.1.4
Metabolic Pathways Affected in $Akp2^{-/-}$ Mice

8.1.4.1 Neuro-physiological Abnormalities

Vitamin B_6, an important nutrient that serves as a cofactor for at least 110 enzymes, can be found in three free forms (or vitamers), i.e. pyridoxal (PL), pyridoxamine (PM) and pyridoxine (PN), all of which can be phosphorylated to the corresponding 5′-phosphated derivatives, PLP, PMP and PNP (Coburn et al., 1996; Jansonius, 1998). PLP serves as a coenzyme in reactions involving catabolism of various amino acids and decarboxylation reactions necessary for generation of the neurotransmitters dopamine, serotonin, histamine, γ-aminobutyric acid (GABA) and taurine. PLP, as the cofactor of ornithine decarboxylase, regulates polyamine synthesis, thereby affecting phospholipid, nucleic acid and protein synthesis. Through its participation in methionine metabolism, vitamin B_6 influences the metabolism of phosphatidylcholine and polyunsaturated fatty acids and homocysteine. PLP also affects the expression and action of steroid hormone receptors and modifies immune function through actions on T and B lymphocyte responses. About 70% of the vitamin B_6 in the body is located in muscle, where it is primarily associated with glycogen phosphorylase. The phosphorylation of PL,

PM and PN is catalyzed by PL kinase. PLP and PMP are interconvertible through aminotransferases or PMP/pyridoxine 5′-phosphate oxidase. Removal of the phosphate group is a function of AP, primarily the TNAP isozyme (Whyte, 1995). Since only dephosphorylated vitamers can be transported into the cells, decreased TNAP activity in hypophosphatasia results in marked increases in plasma PLP (Whyte et al., 1985, 1995).

Administration of pyridoxal, a hydrophobic form of vitamin B_6 that can easily traverse biological membranes, temporarily suppress the epileptic seizures (Waymire et al., 1995) and reverses the apoptosis in the thymus and the abnormal morphology of the lumbar nerve roots in Akp2 null mice (Narisawa et al., 2001). However, the Akp2 null mice still show a 100% mortality rate before weaning and their demise is often preceded by epileptic seizures. Vitamin B_6 is an important coenzyme in the biosynthesis of the neurotransmitters GABA, dopamine and serotonin and is likely to be important for the normal perinatal development of the central nervous system. It is known that vitamin B_6 depletion causes epilepsy in rats accompanied by reduced PLP levels in the brain. In general, the epilepsy is believed to involve the GABAergic synapses. Glutamic acid decarboxylase (GAD) catalyzes the synthesis of GABA, requiring PLP as a cofactor, and knockout mice lacking one of the GAD isozymes, GAD65, developed spontaneous epilepsy in the adult stage (Kash et al., 1997). Waymire et al. (1995) showed that the $Akp2^{\beta geo/\beta geo}$ mice had reduced brain levels of GABA and hypothesized that the epileptic seizures observed in these mice result from GAD dysfunction due to shortage of PLP. Increased plasma PLP has also been observed when TNAP activity was reduced in moderately zinc-deficient rats (Wan et al., 1993). Vitamin B_6 deficiency in suckling rats has been found to result in a decrease in brain sphingolipids since sphingosine synthesis is also a PLP-dependent reaction (Stephens and Dakshinamurti, 1975). Narisawa et al. (1997, 2001) observed an abnormal morphology of the lumbar nerve roots in both $Akp2^{-/-}$ and vitamin B_6-deficient wt mice during the suckling stage when myelination is rapidly occurring. It is possible that impaired myelination of nerve cells in the brain and nerve roots descending within the dura may be a contributing factor to the development of epileptic attacks in these mice. Furthermore, the mechanisms that cause the vitamin B_6-dependent epilepsy seems to correlate with postnatal development, since no seizures were observed in the vitamin B_6-depleted adult mice. Since vitamin B_6 administration only temporarily suppresses the epileptic seizures, and fails to rescue the Akp2 null mice from their sudden death, it is possible that the lack of TNAP function during embryonic development may contribute to this 100% penetrant phenotype. The nerve roots of postnatal mice do not express TNAP; however, in the embryonic stages between E8.5 and E13.5, TNAP is expressed in the pons, medula oblongata, cranial nerves and spinal nerves (Narisawa et al., 1994). The localization of TNAP activity in cross-sections of the spinal cord of E10.5–E13.5 embryos corresponds to the regions called intermediomedial tract and intermediolateral tract of the spinal cord, which are the locations where autonomic fibers project. The most likely cause of death in the $Akp2^{-/-}$ mice appears to be apnea, which often occurs in conjunction with the epileptic seizures. Necropsy shows large amounts of blood in the lungs, a sign of suffocation and intracranial hemorrhage. There was no evidence of bacterial infection in the histological analysis and any disorder in blood clotting is unlikely insofar as these mice had adequate numbers of platelets and their blood clotted normally. Recently, Fonta et al. (2004) found a specific and strong TNAP activity in the neuropile, matching the pattern of thalamo-cortical innervation in layer 4 of the primate sensory cortices (visual, auditory and somatosensory). Such a pattern is also evident in rodents and carnivores, making TNAP a powerful marker of primary sensory areas. Remarkably, TNAP activity is regulated by sensory experience, as demonstrated by monocular deprivation paradigms in monkeys. The distribution of TNAP activity matches that of the glutamate decarboxylate, the GABA-synthesizing enzyme found in presynaptic terminals. These authors' electron microscopic data indicate that TNAP is found at the neuronal membranes and in synaptic contacts and they propose that TNAP may be a key enzyme in regulating neurotransmission and could therefore play an important role in developmental plasticity and activity-dependent cortical functions (Fonta et al., 2004).

8.1.4.2 The Function of TNAP in Bone Mineralization

Although abnormalities in the metabolism of PLP explain many of the abnormalities of infantile hypophosphatasia (Narisawa et al., 2001), they are not the basis of the abnormal mineralization that characterizes this disease. In bone, TNAP is confined to the cell surface of osteoblasts and chondrocytes, including the membranes of their shed MVs (Bernard, 1978; Morris et al., 1992) where, by an unknown mechanism, TNAP is highly enriched compared with the plasma membrane (Morris et al., 1992). Deposition of hydroxyapatite during bone mineralization initiates within the lumen of MVs (for a review, see Anderson et al., 2005a). It has been proposed that the role of TNAP in the bone matrix is to generate the P_i needed for hydroxyapatite crystallization (Majeska and Wuthier, 1975; McComb et al., 1979; Fallon, et al., 1980). GPI-anchored TNAP appears to also control the quality of the calcium mineral, at least *in vitro* (Harrison et al., 1995). However, TNAP has also been hypothesized to hydrolyze the mineralization inhibitor PP_i to facilitate mineral precipitation and growth (Moss et al., 1967; McComb et al., 1979; Rezende et al., 1998). When PP_i is present at near physiological concentrations, in the range 0.01–0.1 mM, PP_i has the ability to stimulate mineralization in organ-cultured chick femurs (Anderson and Reynolds, 1973) and also by isolated rat MVs (Anderson et al., 2005b), whereas at concentrations above 1 mM, PP_i inhibits calcium phosphate mineral formation by coating hydroxyapatite crystals, thus preventing mineral crystal growth and proliferative self-nucleation. Given that at physiological pH the K_m of TNAP for PP_i is about 0.5 mM (Di Mauro et al., 2002), in this range of physiological extracellular PP_i concentrations most PP_i would be hydrolyzed to orthophosphate (P_i) by TNAP, thus providing P_i for incorporation into nascent mineral. Therefore, PP_i has a dual physiological role: it can function as a promoter of mineralization at low concentrations, but also as an inhibitor of mineralization at higher concentrations.

Anderson et al. (2004) carried out a detailed characterization of the ultrastructural localization, the relative amount and ultrastructural morphology of bone mineral in tibial growth plates and in subadjacent metaphyseal bone in the $Akp2^{-/-}$ mice. Alizarin Red staining, micro-computed tomography (micro-CT) and Fourier Transform infrared imaging spectroscopy (FT-IRIS) confirmed a significant overall decrease in mineral density in the cartilage and bone matrix of $Akp2^{-/-}$ mice. High-resolution transmission electron microscopy indicated that mineral crystals were initiated, as is normal for the first stage of mineralization, within MVs of the growth plate and bone of TNAP-deficient mice (Fig. 32). However, mineral crystal proliferation and growth (second stage of mineralization) was inhibited in the matrix surrounding MVs, as is the case in the hereditary human disease hypophosphatasia (Anderson et al., 1997). These data suggested that hypomineralization in TNAP-deficient mice results primarily from an inability of initial mineral crystals within MVs to self-nucleate and to proliferate beyond the protective confines of the MV membrane. This failure of the second stage of mineral formation may be caused by an excess of the mineral inhibitor pyrophosphate (PP_i) in the extracellular fluid around MVs. Indeed, Johnson et al. (2000) studied osteoblast-derived MVs isolated from the $Akp2^{-/-}$ mice and found that MVs lacking TNAP activity show greatly increased levels of PP_i and reduced activity of the PP_i-producing enzyme NPP1. Furthermore, overexpressing TNAP in cells led to an

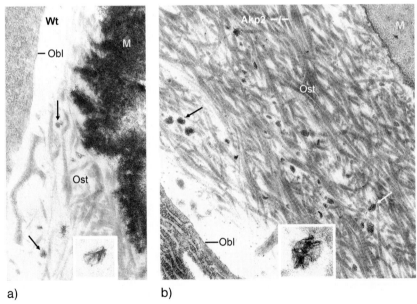

Fig. 32 Electromicrographs of wt and $Akp2^{-/-}$ metaphyseal bones. Normal uncalcified osteoid (Ost) layer (a) versus widened osteoid layer in TNAP-deficient tibial metaphyseal bone (b). A few intact matrix vesicles, containing apatite-like needles (indicated by arrows and shown at higher magnification in insets, are present in the uncalcified osteoid of both TNAP wt and TNAP-deficient tibias. M, mineralized bone matrix; Obl, osteoblast; Ost, osteoid. Taken from Anderson et al. (2004) and reproduced with permission from the *American Journal of Pathology*.

increased expression of NPP1 and reduced levels of PP$_i$ in the MVs. These data suggested that TNAP, by an enzyme activity-dependent mechanism, acts to control mineralization in part by modulating the content of the mineralization inhibitor PP$_i$ in osteoblasts and osteoblast-derived MVs.

Recent studies have provided compelling proof that a major function of TNAP in bone tissue consists in hydrolyzing PP$_i$ to maintain a proper concentration of this mineralization inhibitor to ensure normal bone mineralization. The nucleotidetriphosphate pyrophosphohydrolase activity of NPP1 and the transmembrane

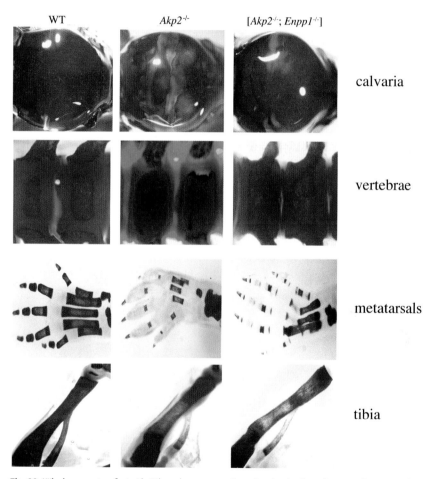

Fig. 33 Whole mounts of wt, *Akp2$^{-/-}$* and [*Akp2$^{-/-}$*; *Enpp1$^{-/-}$*] skeletal tissues. Whole mounted calvaria, spine, metatarsals and tibias, stained for calcium with Alizarin Red and unmineralized osteoid with Alcian Blue from wt, *Akp2$^{-/-}$* and [*Akp2$^{-/-}$*: *Enpp1$^{-/-}$*] double-knockout mice. Although double deficiency of TNAP and NPP1 restored mineralization in the calvaria and spine and to a lesser extent in the metatarsal bones of the feet, the phalangeal bones and the tibias of double-knockout mice still displayed moderate to severe hypomineralization in comparison with wt controls. Taken from Anderson et al. (2005) and reproduced with permission from the *American Journal of Pathology*.

PP$_i$-channeling protein ANK are responsible for supplying the larger amount of PP$_i$ to the extracellular spaces. Mice deficient in NPP1 (*Enpp1$^{-/-}$*) or ANK (*ank/ank*) have decreased levels of extracellular PP$_i$ and display soft tissue ossification. *Enpp1$^{-/-}$* mice develop features essentially identical with to the previously described phenotype of the tiptoe walking (ttw/ttw) mice (Okawa et al., 1998; Sali et al., 1999). These include the development of hyperostosis, starting at ~3 weeks of age, in a progressive process that culminates in ossific intervertebral fusion and peripheral joint ankylosis, in addition to Achilles tendon calcification. The *ank/ank* mice have also been characterized as a model of ankylosis (Ho et al., 2000; Nurnberg et al., 2001). Given the high endogenous levels of extracellular PP$_i$ characteristic in the *Akp2$^{-/-}$* mice, a reasonable hypothesis to test was that affecting the function of either NPP1 or ANK would have beneficial consequences on hypophosphatasia by reducing the amounts of extracellular PP$_i$ in the *Akp2$^{-/-}$* mice and conversely that affecting the function of TNAP should also ameliorate the soft tissue ossification in the *Enpp1$^{-/-}$* and *ank/ank* mutant animals. Indeed, experiments to rescue these abnormalities by combining two mutations, i.e. [*Akp2$^{-/-}$*; *Enpp1$^{-/-}$*] and [*Akp2$^{-/-}$*; *ank/ank*], revealed that both double mutants led to improvements in hypophosphatasia and also in the ossification of the vertebral apophyses (Hessle et al., 2002; Harmey et al., 2004).

First, the calvarial bones and the vertebral apophyses of the [*Akp2$^{-/-}$*; *Enpp1$^{-/-}$*] mice displayed a clearly normalized mineralization (Fig. 33), whereas a partial correction was observed in the metatarsal bones, but not in the phalanges. Ten-day-old *Enpp1$^{-/-}$* mice have precipitated mineral in every lumbar vertebral apophyses (100%), whereas in the corresponding areas of age-matched *Akp2$^{-/-}$* mice mineral deposits are virtually absent (6%). More than half the lumbar vertebrae of the [*Akp2$^{-/-}$*; *Enpp1$^{-/-}$*] double-knockout mice had mineral deposits, a ratio comparable to that found in the wt mice. Each functional allele of *Akp2* and *Enpp1* contributed to the mineralization status of the spine as even carriers (heterozygotes) of *Akp2* and *Enpp1* null mutations were affected at this anatomical location (Hessle et al., 2002). Importantly, double-homozygous mice live for up to 25 days without vitamin B$_6$ administration, whereas *Akp2$^{-/-}$* mouse never survive longer than 14 days in the absence of vitamin B$_6$ supplementation. The administration of vitamin B$_6$ to the [*Akp2$^{-/-}$*; *Enpp1$^{-/-}$*] double-knockout mice does not increase their lifespan any further. All single- and double-knockout mice examined so far that involve the Akp2 null allele appear to succumb at about day 25 of age (100% penetrance). The authors then examined the ability of [*Akp2$^{-/-}$*; *Enpp1$^{-/-}$*] double-knockout primary osteoblasts to mineralize bone nodules *in vitro*. Similarly to the *Akp2$^{-/-}$* primary osteoblasts (Section 8.1.3), the [*Akp2$^{-/-}$*; *Enpp1$^{-/-}$*] osteoblasts cultured with β-glycerophosphate as the principal P$_i$ source showed an absence of mineralization. However, incubating primary osteoblasts with sodium phosphate as the free P$_i$ source, a condition that better mimics the *in vivo* environment, allowed for the detection of differences in the mineralization ability of osteoblasts of differing genotypes. *Akp2$^{-/-}$* osteoblasts cultured for 21 days precipitated significantly less mineral than wt osteoblasts, whereas osteoblasts of *Enpp1$^{-/-}$* mice precipitated significantly more mineral than control cells. A normal mineralization capacity was restored in the [$Akp2^{-/-}$; $Enpp1^{-/-}$] osteoblasts. The authors then quantified the amount of PP$_i$ in the MVs isolated from the cultured osteoblasts. MVs from $Akp2^{-/-}$ osteoblasts had increased amounts of PP$_i$, whereas those from the $Enpp1^{-/-}$ osteoblasts had decreased levels of PP$_i$. The MVs derived from the [$Akp2^{-/-}$; $Enpp1^{-/-}$] osteoblasts exhibited PP$_i$ concentrations that were comparable to those of wt MVs.

In a subsequent paper, Anderson et al. (2005b) reported that while the calvarial bones and the vertebral apophyses of the [$Akp2^{-/-}$; $Enpp1^{-/-}$] mice displayed a clearly normalized morphology, as did the metatarsal bones, neither the femur nor the tibia from these double-knockout mice appeared rescued (Fig. 33). These data suggested that the simultaneous ablation of the $Akp2$ and $Enpp1$ genes, although affecting a dramatic amelioration of mineralization deficits in certain bones, do not show significant rescue of mineralization abnormalities in the long bones. Micro-CT analysis of the upper tibial growth plate, cortex and metaphysis of wt $Akp2^{-/-}$, $Enpp1^{-/-}$ and [$Akp2^{-/-}$; $Enpp1^{-/-}$] tibias revealed that in fact $Enpp1^{-/-}$ tibias display a moderately to severely reduced mineral content in both the upper growth plate and subjacent bone (Anderson et al., 2005b). Given that $Akp2^{-/-}$ mice display severe hypomineralization and that $Enpp1^{-/-}$ tibial growth plates are also under-mineralized, it is not surprising that in [$Akp2^{-/-}$; $Enpp1^{-/-}$] double-knockout tibias, mineral content was even more reduced in comparison with $Enpp1^{-/-}$ single-knockout tibias. Thus, the rescue of hypophosphatasia abnormalities by abolishing NPP1 function is site-specific and extends to the calvaria, spine and incompletely in the metatarsals, in [$Akp2^{-/-}$; $Enpp1^{-/-}$] double-deficient mice. However, hypomineralization of the long bones remains, despite the dramatic amelioration of the mineral deficits at other sites and the systemic correction of circulating PP$_i$ levels. This is probably due to the relatively higher endogenous levels of NPP1 activity in the cranium, vertebrae and ligaments, compared with long bones, so that the genetic ablation of the NPP1 gene would result in a greater correction of extracellular PP$_i$ concentrations at those skeletal sites (Anderson et al., 2005b).

Harmey et al. (2004) cross-bred $Akp2^{+/-}$ heterozygote mice to ank/ank mutant mice to obtain [$Akp2^{-/-}$; ank/ank] double-mutant mice. Similarly to [$Akp2^{-/-}$; $Enpp1^{-/-}$] mice, the degree of mineralization in the spines of [$Akp2^{-/-}$; ank/ank] mice showed an improvement in the levels of abnormal mineralization (80%) in comparison with the $Akp2^{-/-}$ single-knockout (10%) and the ank/ank mutant mice (100%). In $Akp2^{-/-}$ animals, the level of mineralization was ameliorated by mutation of just one Ank allele in these mice as the [$Akp2^{-/-}$; Ank/ank] mice showed normalized levels of mineralization, ~70%, i.e. control levels. In the [$Akp2^{-/-}$; ank/ank] double-mutant animals there remained a higher degree of mineralization than in control mice, but the percentages were lower than in the ank/ank mutant animals. Hence that crossbreeding $Akp2^{-/-}$ mice to ank/ank mice results in a partial rescue of the abnormal mineralization of both single-mutant mice. The profile of both intra- and extracellular PP$_i$ levels from calvarial osteoblasts isolated from the respective knockout or mutant mice over the course of a 21-day bone nodule assay was also examined. Interestingly, whereas normalization of extracellular PP$_i$

levels was observed in [$Akp2^{-/-}$; ank/ank] osteoblasts and MVs, the intracellular PP_i concentrations in these mice remained abnormal.

Interestingly, the expression of yet another mineralization inhibitor, i.e. osteopontin (OPN), was found to be highly elevated in $Akp2^{-/-}$ mice (Harmey et al., 2004) whereas it was decreased in both the $Enpp1^{-/-}$ and the ank/ank osteoblasts (Johnson et al., 2003). In vitro experiments on wt osteoblasts treated with exogenous PP_i revealed an increase in OPN expression and decreased NPP1 and ANK expression. These data support a direct regulation of OPN expression by NPP1 and ANK, mediated by PP_i. Importantly, both PP_i and OPN levels were corrected in [$Akp2^{-/-}$; $Enpp1^{-/-}$] and [$Akp2^{-/-}$; ank/ank] double-knockout mice. Figure 34 shows the strict correlation between levels of extracellular PP_i and of serum OPN in all genotypes. As described earlier (Section 8.1.2), Tesch et al. (2003) reported that OPN was found throughout the nonmineralized bone matrix, occasionally with high focal densities in the vicinity of mononucleated cells. It seems possible that this focal density results from increased transcription and translation of the Opn gene by osteoblasts as a result of the increased local concentration of extracellular PP_i. However, it is still unclear how the osteoblasts respond to changes in the extracellular levels of PP_i.

Figure 35 depicts the most up-to-date model regarding the concerted action of TNAP, NPP1 and ANK in regulating extracellular PP_i and OPN concentrations and thus in controlling hydroxyapatite deposition. Hypophosphatasia in the $Akp2^{-/-}$ mice arises from deficits in TNAP activity, resulting in an increase in PP_i

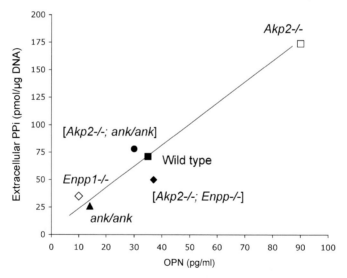

Fig. 34 Correlation between serum PP_i and OPN levels. The elevated levels of PP_i in $Akp2^{-/-}$ mice cause a secondary increase in OPN, whereas the decreased PP_i concentrations in $Enpp1^{-/-}$ and ank/ank mice result in depressed OPN levels. There is a strict correlation between serum PP_i and OPN levels. The double-knockout [$Akp2^{-/-}$; $Enpp1^{-/-}$] and [$Akp2^{-/-}$; ank/ank] mice both show normalized PP_i levels that also result in correction of OPN levels.

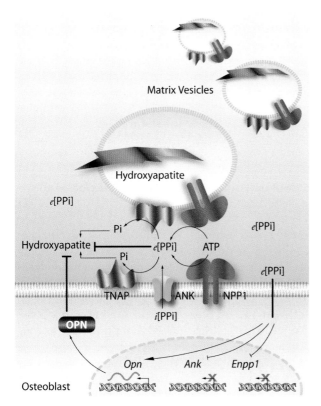

Fig. 35 Concerted action of TNAP, NPP1 and ANK in regulating extracellular PP_i and OPN levels. Both NPP1 and ANK raise extracellular levels of PP_i while TNAP is required for depletion of the PP_i pool. Both TNAP and NPP1 are functional in matrix vesicles whereas ANK is not. Therefore, NPP1 plays a more crucial role in PP_i production than ANK. As a result, the absence of NPP1 in $Enpp1^{-/-}$ mice results in a more severe phenotype than in *ank/ank* mice. A negative feedback loop exists in which PP_i, produced by NPP1 and transported by the channeling action of ANK, inhibits expression of the *Enpp1* and *Ank* genes. In addition, PP_i induces expression of the *Opn* gene and production of OPN, which further inhibits mineralization. In the absence of TNAP, high levels of PP_i inhibit mineral deposition directly and also via its induction of OPN expression. The combined action of increased concentrations of PP_i and OPN causes hypomineralization. In the absence of NPP1 or ANK, low levels of PP_i, in addition to a decrease in OPN levels, lead to hypermineralization. This model clearly points to NPP1 and ANK as therapeutic targets for the treatment of hypophosphatasia. Similarly, targeting TNAP function can be useful in the treatment of hypermineralization abnormalities caused by altered PP_i metabolism. This figure was originally published by Harmey et al. (2004) and is reproduced here with modifications and with permission from the *American Journal of Pathology*.

levels and a concomitant increase in OPN levels; the combined inhibitory effect of these molecules leads to hypomineralization. In contrast, an NPP1 or ANK deficiency leads to a decrease in the extracellular PP_i and OPN pools, thereby allowing ectopic soft tissue ossification. The hypomineralization defects in $Akp2^{-/-}$ mice, along with elevated PP_i and OPN levels, are ameliorated by ablation of either the *Enpp1* or *Ank* gene. Conversely, ablating the function of the *Akp2* gene causes improvement of the abnormalities in the *Enpp1*$^{-/-}$ and *ank/ank* mutant mice via the resulting increase in the concentrations of two inhibitors of mineralization, PP_i and OPN. Thus, NPP1 and ANK represent possible therapeutic targets to correct abnormally elevated levels of PP_i such as those found in hypophosphatasia and TNAP appears as a rational target molecule for the treatment of soft tissue ossification abnormalities due to decreased levels of extracellular PP_i (Section 13.4).

8.1.4.3 Co-expression of TNAP and Fibrillar Collagens Restricts Calcification to Skeletal Tissues

The production by osteoblasts of an inhibitor of extracellular matrix (ECM) mineralization such as PP_i suggests a model whereby the removal of an inhibitor rather than the synthesis of an inducer of mineralization would explain why ECM mineralization occurs in bone. TNAP is necessary for bone mineralization, but this is most likely not sufficient for ECM mineralization, otherwise other organs expressing it such as liver and kidney would mineralize. The apparent contradiction between TNAP function that takes place in bone and TNAP expression that is not bone specific suggests that the spatial restriction of ECM mineralization to bone could be explained by a dual genetic requirement. Murshed et al. (2005) have shown that co-expression in osteoblasts of TNAP and Type I collagen is necessary and sufficient to induce ECM mineralization in bone. Type I collagen is favored as another necessary molecule because mineralization occurs along collagen fibrils (Bachra and Fischer, 1968; Glimcher, 1998) and the expression pattern of TNAP and Type I collagen is fully consistent with this model, since although the *Akp2*, $\alpha1(I)$collagen and $\alpha2(I)$collagen genes are expressed in several tissues, the only tissues in which they are co-expressed are bones and teeth, two mineralizing tissues, in which they are specifically co-expressed in osteoblasts and odontoblasts, respectively (Murshed et al., 2005). Beertsen and Van den Bos (1992) reported that bIAP covalently bound to slices of guanidine-extracted demineralized bovine dentin and implanted subcutaneously on the rat skull rapidly accumulated hydroxyapatite (HA) crystals and proposed that AP plays a crucial role in the induction of hydroxyapatite (HA) deposition in collagenous matrices *in vivo*. Nakamura et al. (2004) showed that TNAP activity was located not only in the plasma membrane of fibroblasts, but also in the collagen fiber bundles and fibrils in the periodontal ligament and further suggested that the bond between TNAP and the collagen fibers is dependent on the GPI anchor.

The dual genetic model proposed by Murshed et al. (2005) has two implications. The first is that the extent of ECM mineralization should vary according to the ability of osteoblasts to synthesize Type I collagen. To test if this was the case, the authors used a rat osteoblastic cell line, the ROS 17/2.8 cell line that does not express fibrillar collagens and mouse wt osteoblasts. The ECM surrounding wt osteoblasts easily mineralized in the presence of P_i; in contrast, the ECM surrounding ROS 17/2.8 did not. However, ROS cells permanently transfected with a *Col2a1* expression vector encoding Type II collagen and cultured in presence of β-glycerophosphate were able to mineralize their ECM. These data indicated that a fibrillar collagen network is necessary for ECM mineralization. However, a collagenous network alone is not sufficient to induce bone mineralization since the ECM surrounding *Akp2* null osteoblasts, which is rich in Type I collagen, does not mineralize. A second implication of this model is that ectopic expression of TNAP in Type I collagen-expressing cells should induce an ECM mineralization very similar to that seen in bone, whereas ectopic expression of TNAP in mesenchymal cells that do not express fibrillar collagen genes should not. This hypothesis was tested in cell culture and *in vivo*. First, Murshed et al. (2005) ectopically expressed TNAP in NIH3T3 fibroblasts that express Type I collagen genes but are not surrounded normally by a mineralized ECM. TNAP-expressing NIH3T3 cells were then cultured in the presence of PP_i, the substrate of TNAP. Under these culture conditions, the ECM surrounding the TNAP-expressing NIH3T3 cells became mineralized while the ECM surrounding NIH3T3 cells transfected with an empty vector did not (Fig. 36).

To test this hypothesis *in vivo*, Murshed et al. (2005) generated transgenic mice expressing TNAP either in the dermis, a skin layer rich in Type I collagen, or in the epidermis, a skin layer that does not contain fibrillar collagen. The α2(I) collagen-TNAP transgenic mice expressed TNAP in skin fibroblasts and produced a functional TNAP as determined by Fast Blue staining. The α2(I)-TNAP mice developed a dramatic mineralization of their skin ECM. This ECM mineralization consisted of hydroxyapatite crystals and occurred along collagen fibers, as is the case in bone (Fig. 36). In contrast, expression of TNAP under control of the keratin 14 promoter in keratinocytes, a cell type that does not secrete fibrillar collagen, did not lead to mineralization of epidermis ECM. Further analysis of the α2(I)-TNAP mice showed ECM mineralization in other locations also, such as the arteries and sclera of the eye, two other tissues rich in Type I collagen and where the transgene is also expressed. Taken together, these data are consistent with the hypothesis that the co-expression of genes encoding TNAP and fibrillar collagens is necessary and sufficient to induce ECM mineralization in bone and other tissues.

The discussion above should have made it clear that the extracellular P_i/PP_i ratio is of fundamental significance for bone ECM mineralization, as is the presence of fibrillar collagen in skeletal tissues. Thus, in the bone ECM, while the extracellular P_i concentration is fairly constant, TNAP's enzymatic degradation of PP_i controls the P_i/PP_i ratio to favor proliferation of hydroxyapatite crystals outside the MVs and also along collagen fibrils. But why are $Akp2^{-/-}$ mice born with a mineralized skeleton and still contain hydroxyapatite crystals inside their MVs? Anderson and colleagues have documented the presence of hydroxyapatite crystals in the MVs of hypophosphatasia specimens both in humans (Anderson et al., 1997) and in mice (Anderson et al., 2004) (Section 8.1.3). As TNAP sits on the outer surface of the MV membrane, there is no TNAP-mediated hydrolysis of PP_i inside the MVs. Therefore, it is likely that another enzyme is responsible for either cleaving PP_i or elevating the intravesicular concentration of P_i so as to achieve a P_i/PP_i

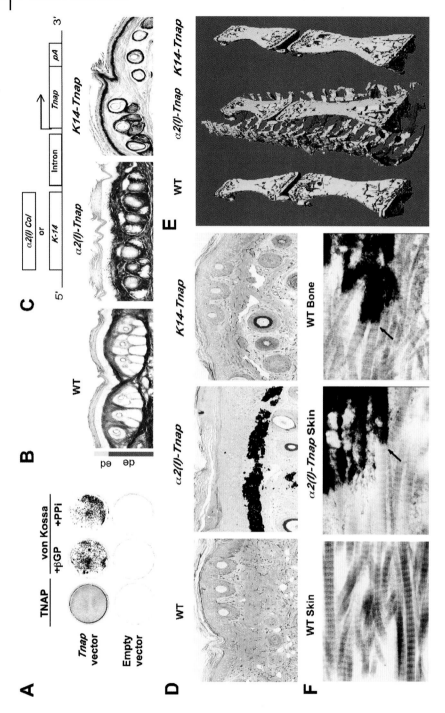

◀ **Fig. 36** Co-expression of TNAP and Type I collagen is required for ECM mineralization. (A) NIH3T3 cells transfected with a TNAP cDNA expression vector produced TNAP as shown by Fast Blue staining. ECM surrounding these cells mineralized when cultured in presence of βGP or PP_i. No ECM mineralization was seen in the case of empty vector transfected NIH3T3 cells. (B) Van Gieson staining (pink) of skin showing collagen in the dermis but not in the epidermis. (C) Transgene constructs for dermis and epidermis specific expression of TNAP (top). Fast Blue staining (blue) showing TNAP activity in the epidermis of K14-*Tnap* mice and dermis of $\alpha 2(I)$-*Tnap* mice (bottom). (D) Von Kossa staining showing massive mineral deposition in dermis of $\alpha 2(I)$-*Tnap* mouse, while no mineral deposition was seen in the epidermis of K14-*Tnap* mouse ($n = 6$) (E) Micro-CT analysis of the mineralized tail of an $\alpha 2(I)$-*Tnap* mouse (middle). Similar analysis with WT (left) and K14-*Tnap* (right) are presented for comparison. (F) Electron micrograph showing mineral deposition along collagen fibrils in the $\alpha 2(I)$-*Tnap* mouse. Taken from Murshed et al. (2005) and reproduced with permission from *Genes and Development*.

ratio conducive for initial seed crystal deposition during the first stage of MV-mediated calcification. Farquharson's laboratory has shown that PHOSPHO1, a soluble phosphatase present in MVs (Stewart et al., 2003; Houston et al., 2004), has specificity for PEA and phosphocholine (Roberts et al., 2004). It seems probable that PHOSPHO1 may play the important role of increasing the P_i/PP_i ratio inside MVs and, thus, would control the first step of initiation of hydroxyapatite crystal deposition inside MVs. This hypothesis, however, remains to be tested experimentally. Figure 37 shows electron micrographs that document the different steps of mineralization, spelling out which molecules are known or hypothesized to be involved in the initiation of mineralization.

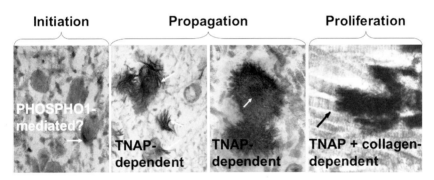

Fig. 37 Mechanisms of initiation of mineralization. The first step (initiation step) mediates the deposition of seed crystals of hydroxyapatite (HA). It is possible that the function of PHOSPHO1 may be involved in this process. This step is independent of the function of TNAP as it is unaffected in *Akp2*$^{-/-}$ mice. Subsequently, the hydroxyapatite (HA) crystals grow beyond the confines of the MV membrane (propagation step). Clearly, this step is dependent on the function of TNAP as it is affected in *Akp2*$^{-/-}$ mice. Finally, the growing hydroxyapatite (HA) crystals continue to expand along collagen fibrils in the extracellular matrix (ECM mineralization). This step is dependent on the activity of TNAP and the presence of fibrillar collagens (Type I, II or X).

8.1.4.4 Other Organs Affected in $Akp2^{-/-}$ Mice

It is clear that PLP and PP_i are physiological substrates of TNAP and that abnormalities in their metabolism cause the epileptic seizures and hypomineralization, respectively, in hypophosphatasia. Interestingly, several hypophosphatasia mutations were found to affect PLP catalysis to a larger extent than PP_i catalysis (Section 7.1.2), providing a clue as to one mechanism to explain phenotypic variability in hypophosphatasia, i.e. some individuals suffer from epileptic seizures while displaying mild bone abnormalities, whereas others have the reverse manifestations (Whyte, 1995). Although pyridoxal administration improves the lifespan of the $Akp2^{-/-}$ mice, they still succumb to the disease at about day 25 of age. Even [$Akp2^{-/-}$; $Enpp1^{-/-}$] mice, which display corrected skeletal abnormalities, die at about this same time point (Section 8.1.4.1). The ultimate cause of death in the mice has not yet been identified. At the time of death the animals have considerable respiratory distress and episodes of apnea. It is possible that anatomical disturbances in the central nervous system resulting from lack of TNAP activity during embryonic development that are not correctable by vitamin B_6 supplementation account for the late epilepsy and death (Narisawa et al., 2001). Alternatively, a report has pointed to the potential role of TNAP in dephosphorylating AMP to produce adenosine, needed for proper mucociliary clearance and inflammatory responses in the airways that help prevent lung infections (Picher et al., 2003). It is possible that abnormalities in adenosine production in the lungs of $Akp2^{-/-}$ mice contribute to their demise and this possibility should be tested experimentally. There is also some anecdotal evidence that PEA may be epileptogenic in hypophosphatasia patients (Takahasi et al., 1984). The significant elevations of endogenous PEA levels in the $Akp2^{-/-}$ mice could cause this complication as the mice age, which would then not be correctable by vitamin B_6 administration.

Meanwhile, the reasons for the increased excretion of PEA in hypophosphatasia remain unclear. One possibility is that PEA is a natural substrate of TNAP (Whyte et al., 1995), but the putative metabolic pathways involved have not been elucidated. An alternative explanation may relate to abnormalities in the function of O-phosphorylethanolamine phospholyase (PEA-P-lyase) (Fleshood and Pitot, 1969, 1970). PEA-P-lyase is an enzyme reported to require PLP as cofactor and since TNAP is crucial for the normal metabolism of PLP (Narisawa et al., 2001), it is possible that it is the insufficient amount of PLP inside the cells that leads to suboptimal activity of PEA-P-lyase, which in turn leads to increased excretion of PEA in hypophosphatasia. Although one study has addressed this issue (Grøn, 1978) and did not find support for this hypothesis, there is merit in exploring further this possible metabolic pathway as being affected in hypophosphatasia. Of interest in this regard is a study involving members of a large kindred with adult hypophosphatasia in which PEA and phosphoserine levels in their urine correlated inversely with both total and liver TNAP activity in their serum, but not with the activity of bone TNAP (Millán et al., 1980). Those findings suggest that altered hepatic metabolism is responsible for the increased urinary excretion of PEA, and perhaps phosphoserine, in hypophosphatasia. The role of TNAP in the liver has not been approached experimentally using the $Akp2^{-/-}$ models, primarily because of the premature death of these animals. However, it is expected that once models of adult hypophosphatasia, displaying less severe disease, become available, other organs possibly affected by a deficiency in TNAP function will be examined in detail.

Although TNAP is present in the fetal small intestine (Komoda et al., 1986), the adult mouse small intestine contains no immunoreactive TNAP (Hoshi et al., 1997). The adult colon, however, does express this enzyme, which is secreted from the colonocyte as part of the SLPs (Sections 7.3.1 and 8.2) containing surfactant protein A and phosphatidylcholine (Eliakim et al., 1997). These SLPs are increased in the small intestine by fat feeding (Mahmood et al., 1994) and by over-expression of the IAP cDNA, but their function in the colon is completely obscure. The $Akp2^{-/-}$ mouse model provided an opportunity to examine the effects of TNAP absence on the content of SLP in small bowel and colon and on other aspects of intestinal structure (Shao et al., 2000). Microscopically, the mucosa of the small intestine and colon in $Akp2^{-/-}$ mice appeared somewhat thin but otherwise normal. Examination for SLP by immunocytochemistry in wt and heterozygous mice revealed strong cellular staining with less staining in the lamina propria and a small amount of reactive luminal debris in the small intestine. In the stomach, most staining was found in the luminal debris over the apical surface of the cells. In the colon, staining was found in the luminal debris and in the lamina propria, but was weaker over the mucosal cells themselves, with occasionally strong staining individual cells. In $Akp2^{-/-}$ animals, the patterns were the same in the stomach and colon. In the small intestine, SLP immunoreactivity was markedly diminished in the cells and in the lamina propria. The expected presence of IAP was confirmed in these tissues by immunocytochemistry. Transmission electron microscopy of the small intestine showed small phagolysosomes (residual bodies) in the apical portion of the cell. In the homozygous $Akp2^{-/-}$ animals, these structures were markedly dilated and filled with debris. Examination of the colon revealed the striking appearance of osmium staining droplets within the mitochondria of colonocytes. These structures seemed to be identical with droplets of neutral fat. The rest of the appearance of the colonocyte was normal. Nerve roots in $Akp2^{-/-}$ animals appear to be significantly diminished in size, consistent with previous findings (Narisawa et al., 1997). Such loss of neuronal tissue might be reflected in a loss of neurons in the enteric nervous system, which might explain the gaseous distention identified in the small intestine of homozygous animals that died spontaneously. When the ganglia of the small intestine were examined, the neuronal content of the ileum in the submucosa and muscular layers was similar to the wt. However, in the submucosa and muscular layers of jejunum, the number of neurons/circumference was clearly less than in the wt controls (Shao et al., 2000).

TNAP has also been found expressed on murine B-lymphocytes activated by either polyclonal mitogens or T helper cells, where TNAP expression was linked to B cell differentiation and antibody secretion (Marquez et al., 1989). The mouse lymphocyte surface alloantigen, Ly-31, defined by monoclonal antibody N1.10 (IgG2b, k) was found to detect an allotypic determinant of mouse TNAP (Dairiki et al., 1989). Similarly, a MAb G-5-2, that recognizes a 76 kDa molecule preferentially expressed on the surface of pre-B and plasma cells, was found to recognize specifically TNAP (Marquez et al., 1990). Nevertheless, recent data appear to indicate that B-lymphocyte differentiation is not affected in $Akp2^{-/-}$ mice (Rickert et al., personal communication). However, it remains to be investigated whether or not $Akp2^{-/-}$ mice display any phenotypic abnormality when their immune system is challenged.

8.2
Phenotypic Abnormalities in $Akp3^{-/-}$ Mice

The striking increase in lymph and serum levels of IAP after a fatty meal in rodents (McComb et al., 1979; Young et al., 1981; Alpers et al., 1990) and, in turn, the dramatically decreased of IAP upon starvation (Hodin et al., 1994) prompted studies to unravel the mechanism whereby IAP anchored in the apical membrane of the enterocyte could enter the lymph and serum. IAP is anchored to the external surface of the apical brush border via a glycosyl-phosphatidylinositol (GPI) anchor (Section 4.3.2) (Engle et al., 1995b) and the absence of a transmembrane sequence made especially puzzling the appearance of IAP in the serum. Sussman et al. (1989) reported that IAP is secreted bidirectionally from villous enterocytes and that the secretion of a particulate AP with extracellular release from the membrane can account for the appearance of the IAP isozyme in both the serum and the lumen. Using tannic acid as mordants to fix membranous structures during processing for electron microscopy, membranes were found on the surface of the enterocyte that were enriched in IAP (DeSchryver-Kecskemeti et al., 1989). These membranes appeared to have morphological characteristics similar to those of pulmonary surfactant, namely occurring in swirls but without cross-linking features. Membranes isolated from the rat small intestine were found to contain surfactant proteins B and D, whereas those membranes from rat and human small intestine also contained surfactant protein A (Eliakim et al., 1989, 1997). These membranes were isolated from the surface of stomach, small bowel, colon and bladder and were enriched in AP activity. They were enriched for phospholipids, with a cholesterol/phospholipids ratio only half of that found in apical brush-border membranes (Eliakim et al., 1989). Moreover, the major phospholipid fatty acid was palmitate, as found in pulmonary surfactant. These membranes also lowered surface tension about half as well as did pulmonary surfactant. For all these reasons, the membranes were called surfactant-like particles (SLPs). Similar particles were isolated from the apical surface of human intestine (Mahmood et al., 1993) and are probably present throughout the gastrointestinal tract.

If SLPs represent the mechanism mediating the secretion of IAP from the enterocytes, one would expect the kinetics of IAP and SLP appearance outside the enterocyte to be coordinated. Antibodies were raised against the SLP membrane, which were then adsorbed against brush-border membranes, so that they no longer recognized IAP. Following IAP activity and SLP protein content after feeding of corn oil to rats provided the necessary correlation. Feeding corn oil from 0.5

to 2.0 mL provided a dose-dependent increase in SLPs isolated from the surface of the cells, and also in the serum (Eliakim et al., 1991). Furthermore, the appearance in serum and lumen of IAP was coordinately regulated. Zhang et al. (1996) found that some IAP is secreted from the cell associated with the SLPs consistent with a role SLPs in transepithelial transport of triacylglycerols in the enterocyte. SLPs were isolated from the apical surface of the cells, yet no cytoplasmic membranes were seen approaching the apical membrane when tissue was examined by electron microscopy after fat feeding. Therefore, the possible movement of SLP membranes through the tight junction was explored using the colonic cell line Caco-2. Cytochalasin D increased by 2–3-fold both Caco-2 monolayer permeability and the appearance of SLP (IAP activity and SLP protein) in the apical medium (Engle et al., 1995a). Quantitative image analysis of electron micrographs showed that the SLP content increased 4–10-fold on the apical surface of the cells, whereas the total cell particle content was unchanged. Hence SLP appears to be secreted first across the basolateral membrane and some of that secreted SLP is transported to the apical surface/intestinal lumen through the tight junctions (Fig. 38). Subsequently, the validity of IAP as a marker for SLP secretion was confirmed using prelabeling of the Caco-2 membranes with [^{3}H]palmitate and following their secretion (Engle et al., 2001). Once SLP is secreted it remains for a finite time on both the apical surface of the cell and in the basolateral compartment. To understand the reason for this residence time, interactions of SLP with apical and basolateral membranes and with extracellular matrix proteins were measured using rat SLP with solid-phase binding assays and gel overlays (Mahmood et al., 2002). SLP bound more tightly to basolateral membranes, but also bound to apical membranes. Binding of SLP on the surfaces of the cell may enable it to mediate its extracellular functions more efficiently.

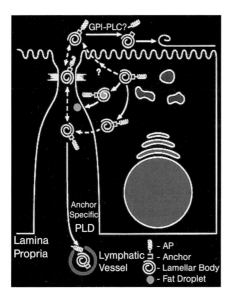

Fig. 38 IAP secretion from the enterocyte. Following absorption of long-chain fatty acids from dietary triacylglycerols, intracellular triacylglyerols are formed and surrounded by a membrane (lamellar body or SLP) enriched in IAP, which is attached to the membrane via a glycosylphosphatidylinositol (GPI) anchor. These structures are seen in the apical portion of the cell, near the basolateral membrane and in the intercellular space, where the lipid droplet and membrane become dissociated. About 25% of the SLP migrates to the apical cell surface via opened tight junctions, where it is bound and remains probably for the lifespan of the mature enterocyte. GPI-PLC secreted by the gall bladder may release some of the IAP in the lumen. The rest of the SLP migrates to the lymphatic vessels and in the lymphatic/blood IAP is removed by the action of GPI-PLD, which is abundant in serum.

The coordinated expression of IAP and SLP was explored further by isolating the lamina propria, representing the compartment that received secreted IAP from the basolateral surface of the cells. Fat feeding increased lamina propria content by 2–3-fold and increased 2-fold the specific activity of IAP relative to protein in the membrane (Yamagishi et al., 1994b). AP activity fell in the Golgi apparatus with the same kinetics as it peaked in the enterocyte and lamina propria (3 h), followed by the serum (3–5 h) and the luminal washings (5 h). Radiolabeled SLP had a short half-life in serum (4.5 min), with the liver accounting for 50% of the uptake (Yamagishi et al., 1994a). IAP became separated from the SLPs in the lamina propria by the action of GPI-PLD (Section 4.3.2), but this effect was minimized by perfusing the intestine with saline to prevent serum contamination (Eliakim et al., 1990a). Moreover, IAP was still attached to SLP on the surface of the enterocyte. Hence the likely progression *in vivo* is that the IAP–SLP complex is secreted from the enterocyte and the components are not separated until they appear in the serum, where GPI-PLD activity is very abundant. After secretion, the lipid particle is seen in the intercellular spaces, surrounded by IAP detected by gold-labeled antibody (DeSchryver-Kecskemeti et al., 1991).

A test of the relationship between SLP and fat absorption involved the use of the non-ionic detergent Pluronic L-81 that produces a decrease in the secretion of chylomicrons but not in fatty acid absorption (Tso et al., 1984). After addition of the detergent to the corn oil feed, AP activity fell by ~60% in both the intestinal mucosa and in the serum, and also in the lamina propria (Mahmood et al., 1994). The total particle protein content also fell in these compartments, but only by about half as much as did AP activity, consistent with the observation that fat feeding increased the specific activity of SLP-associated IAP. By electron microscopy, enterocytes from Pluronic-treated animals showed a 2–3-fold increase in large cytoplasmic lipid droplets and the appearance of lamellae at the poles of the droplets, along with a marked decrease in the number of SLPs overlying the apical brush–border. Immunoelectron microscopy confirmed the presence of IAP in these capped structures adjacent to the lipid droplets, consistent with the identification of the lipid droplet-associated lamellae with SLPs. These changes, produced by an agent that inhibited chylomicron transport, were consistent with a role for SLP in transepithelial triacylglycerol transport in the enterocyte. However, Nauli et al. (2003) found that the nonabsorbable fat olestra is unable to stimulate lymphatic IAP secretion and also that the hydrophobic surfactant Pluronic L-81, which blocks chylomicron formation, fails to inhibit this increase in lymphatic IAP secretion. These results suggest that it is the lipid uptake into the mucosa and/or re-esterification to form triacylglycerols, but not the formation of chylomicrons, that is necessary for the stimulation of the secretion of IAP into the lymph. Hence IAP could be involved in production or function of SLP either directly or indirectly.

To test this hypothesis and to examine the consequences of ablating IAP function *in vivo*, Narisawa et al. (2003) inactivated the murine IAP gene (*Akp3*). Disruption of *Akp3* expression was confirmed by RT-PCR, Northern blotting, Western blotting and immunohistochemistry whereas expression of the syntenic EAP

(*Akp5*) gene was not affected. Despite the absence of IAP expression, mice heterozygous or homozygous for the *Akp3* null mutation did not show any abnormalities in appearance, behavior and fertility after having been bred for several generations. Histomorphological examination of tissues revealed a similar normal appearance of all internal organs known to express IAP. Furthermore, no apparent fat absorption changes were observed in the $Akp3^{-/-}$ mice under regular laboratory diet.

However, abnormalities were noted in homozygous $Akp3^{-/-}$ mice fed with high-fat diets either by forced oil feeding or long-term high-fat feeding (Fig. 39). For forced oil feeding experiments, pre-fasted mice were gavaged with corn oil and the intestines collected and fixed at various time points (ranging from 0 to 48 h) following oil feeding. Tissue sections were examined after hematoxylin–eosin staining and Oil Red O staining, which specifically stains fat. Narisawa et al. (2003) observed time-dependent changes in the rate of disappearance of fat drop-

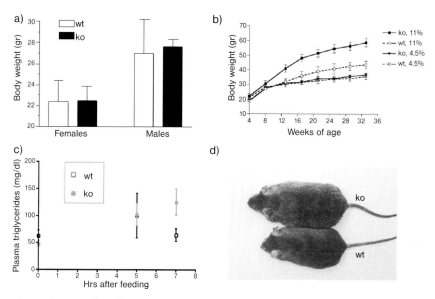

Fig. 39 Plasma triglycerides and body weight of control and $Akp3^{-/-}$ (knockout; KO) mice under different feeding conditions. (a) Triglyceride levels in plasma from the mice subjected to forced oil feeding were measured at 0, 5 and 7 h following gavage. After 7 h, the triglycerides levels were significantly higher in $Akp3^{-/-}$ mice than in WT controls. (b) Body weight of $Akp3^{-/-}$ and control WT mice at 8 weeks of age fed a regular diet containing 4.5% fat. Average values were obtained from 7–11 mice in each group. The body weight of male mice is higher than those of female mice in both KO and control mice; however, there is no statistical difference between the control and the $Akp3^{-/-}$ mice subjected to a regular diet. (c) Body weight of control and $Akp3^{-/-}$ male mice under a regular fat diet (4.5%) or high-fat diet (11%) feeding regime initiated at 4 weeks of age. Note the difference in weight gain in $Akp3^{-/-}$ mice subjected to high-fat diet compared with control mice. (d) Difference in body size between $Akp3^{-/-}$ and WT male mice at 30 weeks of age fed long-term with a high fat diet. Taken from Narisawa et al. (2003) and reproduced with permission from *Molecular and Cellular Biology*.

lets in the duodenal mucosa of control and $Akp3^{-/-}$ mice. At 0 h following forced high-fat feeding, there was no difference between $Akp3^{-/-}$ mice and wt littermate controls. Similar numbers of fat droplets were also localized at the apical region of enterocytes in $Akp3^{-/-}$ and wt mice 3 h after fat feeding. However 5 h after feeding, fat droplets were no longer observable in $Akp3^{-/-}$ mice whereas they were still present in significant numbers in the enterocytes of wt mice. This apparent accelerated transport of lipids through the intestine of $Akp3^{-/-}$ mice was accompanied by an elevation in plasma triglycerides 7 h after gavage (Fig. 39). For long-term high-fat feeding, $Akp3^{-/-}$ and wt control mice where fed either a normal 4.5% fat or an 11% fat mouse chow. Control mice fed a regular fat diet revealed a consistent difference in body weight gain between males and females. Long-term observation showed a different rate of weight increase for $Akp3^{-/-}$ versus controls when fed a high-fat diet from 4 weeks of age. This response was observed in both male and female $Akp3^{-/-}$ mice but was much more dominant in males, as expected from the relative difference in weight as shown in Fig. 39. Furthermore, this increase in body weight was not dependent on the time at which the high-fat diet was initiated, as similar curves were obtained if high-fat feeding began at 4, 6 or 9 weeks of age. Figure 39 also shows the relative difference in size achieved by a wt and an $Akp3^{-/-}$ mouse at 30 weeks of age under a high-fat diet.

Hence the study by Narisawa et al. (2003) suggests that IAP participates in a rate-limiting step regulating fat absorption. The data suggest that SLPs and IAP are involved in the transcytotic movement of lipid droplets following a meal, but that they slow it down (and perhaps direct it), possibly by protein–protein interactions. What are the data that suggest that IAP and/or SLPs might be involved in transcytosis? First, IAP binds to other proteins, including serum immunoglobulin G (Shinozaki et al., 1998), enterocyte calbindin (Leathers and Norman, 1993) and the hepatic asialoglycoprotein receptor (McComb et al., 1979). No direct assessment of IAP binding to intracellular proteins has been made to date. Second, SLP contains other proteins, including one known to be involved in transcytosis, cubilin (Mahmood et al., 2003). IAP and cubilin are not only present in SLPs isolated from the luminal surface of the cell, as in all the studies reported previously, but were also present in intracellular SLPs. After fat feeding, both cubilin and IAP co-localized to the intracellular regions surrounding lipid droplets (Mahmood et al., 2003). Third, these recent studies have also confirmed the existence of a membrane surrounding the intracellular lipid droplet in the rat. Thus, SLP appears to surround the lipid droplet and contains proteins that could modify the movement of the lipid droplet through the cell. The accumulated data provide support for the hypothesized role for SLPs and IAP in fat absorption. Indeed, the IAP knockout model was created to test the hypothesis that these factors were related to fat absorption. The original hypothesis (that IAP/SLPs were essential for transcytosis) appears to be proven incorrect by the results of the knockout experiments, but the fact that they play a role has been confirmed. As noted above, none of the data achieved to date are inconsistent with the revised hypothesis, i.e. that IAP and/or other SLP components bind to intracellular proteins and retard the movement of the fat droplet across the enterocyte. The role of IAP to retard the rate of fat absorption could be viewed as logical for animals that ingest high-fat diets. The activity of AP increases steadily from the concentrations found in the blind eel to the highest levels in mammals and the properties that are characteristic of IAP isozyme were established late in evolution and only in mammals (Komoda et al., 1986).

Despite the absence of IAP in the intestine, $Akp3^{-/-}$ mice displayed rather subtle phenotypic abnormalities. Recently, re-examination of the $Akp3^{-/-}$ mouse intestines showed that despite the lack of IAP, they still had considerable IAP-like activity in the duodenum and ileum, particularly after forced oil feeding (Narisawa et al., personal communication). This expression appears to result from the new Akp6 locus recently identified in the mouse genome, but further studies will be needed to understand better the possible compensatory role that Akp6 expression may have in the intestine and what phenotypic abnormalities would result from inactivation of the Akp6 gene. There is precedent for the existence of multiple IAP genes. Alpers' laboratory cloned two IAP mRNAs, i.e. IAPI and IAPII, from a rat intestinal cDNA library and found that only IAPII appeared to be involved in mediating SLP transport in the rat intestine (Lowe et al., 1990; Engle and Alpers, 1992; Tietze et al., 1992). In humans, PLAP and IAP have been found to be co-expressed in the intestine (Behrens et al., 1983). Furthermore, a remarkable level of complexity with at least seven IAP-like genes being expressed was also uncovered in the bovine intestine (Weissig et al., 1993; Manes et al., 1998). These isozymes varied widely in their kinetic parameters and turnover number (Section 5.3). Furthermore, if more than one of the IAP genes are expressed in the same cell, heterodimers are likely to form, which, as we have discussed, will display noncooperative allosteric behavior (Section 5.2), where the stability and the catalytic properties of each monomer are controlled by the conformation of the other subunit. Given that the mouse intestine also expresses EAP (Hahnel et al., 1990), it would seem that the mouse system displays a considerable degree of complexity with respect to AP expression, but the functional interplay of the Akp3, Akp5 and Akp6 genes and their enzyme products remains to be established.

It is interesting that of the four human AP genes, only abnormalities in the TNAP gene have been associated with a genetic disease, i.e. hypophosphatasia. There have been no reported cases of a lack of IAP, PLAP or GCAP associated with any syndrome or genetic defect. This might mean that deficiencies in these genes are embryonic lethal, but this seems unlikely given the mild phenotypes observed after inactivating the orthologous genes in mice, i.e. Akp3 and Akp5. However, the experimental data in this section provide a clue as to where to look for possible abnormalities, i.e. the phenotypic abnormalities seen in the $Akp3^{-/-}$ mice suggests that polymorphisms or missense mutations in the human IAP-like genes (*ALPI, ALPP* and *ALPP2*) may predispose to obesity, a condition of great public health concern and cost to the medical establishment in the USA. Furthermore, it has been clearly established that there is a marked post-prandial increase in serum IAP levels in humans following ingestion of a fat-containing meal (Day et al., 1992a; Domar et al., 1991, 1993; Matsushita et al., 1998). However, this increase is strictly dependent on the ABO genotype and secretor status of the individual, i.e. nonsecretors (of blood group antigens) had serum IAP levels 20% of those of secretors after fat feeding (Section 9.3). Among secretors, the numbers were lowest for type A and highest for O and B. These data show a relationship between IAP secretion and fat feeding in humans. The data on the $Akp3^{-/-}$ mice and these published observations already suggest an experimental model to test fat absorption in the presence of low and higher IAP secretion into blood in humans, i.e. if IAP reflects SLP secretion, we would predict that there would be more fat absorption in those individuals with lowest IAP secretion (i.e. type A, nonsecretors). There are currently no published studies on the relationship between obesity, blood group and secretor status and intestinal alkaline phosphatases (IAPs), but the available experimental data clearly speak for an association of these genes as participating in the development of the multifactorial condition referred to as obesity.

8.3
Phenotypic Abnormalities in $Akp5^{-/-}$ Mice

The embryonic isozyme of AP (EAP) is one of the genes transcribed early in the mouse embryo (Hahnel et al., 1990). Since the initial demonstration of AP activity in mammalian embryos (McComb et al., 1979), most experiments have focused on the time of onset and the localization of AP activity in preimplantation embryos by using cytochemistry, electron microscopy (Izquierdo et al., 1980) and immunocytochemistry (Ziomek et al., 1990). In 1989, it was revealed that the preimplantation isozyme of AP in mouse is biochemically different from those of all the other known mouse isozymes of AP (Kim et al., 1989; Lepire and Ziomek, 1989). Identification of EAP at the RNA level in preimplantation embryos (Hahnel et al., 1990) and characterization of the genomic structure of its gene (Manes et al., 1990) established a framework for future experiments to determine the function of this enzyme.

Work on the cloning and expression of the mouse AP genes had suggested that the mouse $Akp5$ gene was the ortholog of the human $ALPP$ and $ALPP2$ genes, encoding the PLAP and GCAP isozyme, respectively (Hahnel et al., 1990; Manes et al., 1990) and Narisawa et al. (1997) undertook the generation and characterization of the $Akp5^{-/-}$ mice as an opportunity to clarify the role of these human isozymes. Inactivation of the $Akp5$ gene in embryo-derived stem (ES) cells and the generation of double-knockout cells by gene conversion revealed a subtle phenotype in vitro. As shown in Fig. 40, whereas $Akp5^{+/-}$ D3 cells display a normal morphology comparable to untargeted D3 cells, $Akp5^{-/-}$ cells showed a ball-like morphology. Colonies of undifferentiated D3 cells or $Akp5^{+/-}$ cells form bi- or triple layers at the center of the cell colony and monolayers at the periphery where the cells are often flat, showing a clear cytoplasm. The $Akp5^{-/-}$ cells, however, formed colonies with fewer flat cells. These $Akp5^{-/-}$ cell colonies detached easily from the tissue culture plastic. In order to confirm that the round morphology in $Akp5^{-/-}$ cells was caused by the lack of expression of EAP and not to an artifact of manipulating the cells, the authors reintroduced the *Akp5* gene, subcloned in the mammalian expression vector pSVT7, into *Akp5*$^{-/-}$ cells by co-transfection with the hygromycin-resistant gene under control of the phosphoglycerolkinase (*Pgk*) promoter. Hygromycin-resistant *Akp5*$^{-/-}$ cells overexpressing EAP recovered the ability to spread (Fig. 40) seen in wt D3 and *Akp5*$^{+/-}$ cells and also in D3 cells overexpressing EAP. Clones that were only hygromycin resistant but did not express EAP retained the round morphology. It was of interest that *Akp5*$^{-/-}$ ES cells showed morphological changes *in vitro*. Given that APs contain a loop in the crown domain that may mediate interactions between APs and ECM proteins (Tsonis et al., 1988; Bossi et al., 1993), including collagen (Section 4.1.6), the ball-like morphology of *Akp5*$^{-/-}$ cells suggested that EAP may also have a role in mediating the interaction between the cell surface of early embryonic cells and the ECM. In this context, the overexpression of EAP forced the cells to differentiate into a type of adherent monolayer supporting a possible role of EAP in cell attachment and/or migration. Despite this clear *in vitro* phenotype, however, the resulting homozygous *Akp5*$^{-/-}$ mice failed to show any obvious phenotypic abnormalities.

Since EAP is the most strongly expressed AP isozyme in preimplantation embryos (Hahnel et al., 1990; Lepire and Ziomek, 1989), either an embryonic lethality or gross abnormalities in early development would have been expected following inactivation of the EAP gene. However, *Akp5*$^{-/-}$ pups were born in the expected Mendelian ratios and adult *Akp5*$^{-/-}$ mice appeared healthy and reproduced normally (Narisawa et al., 1997). That study, however, did not focus on the preimplantation period, when the *Akp5* gene is largely transcribed and was performed on mice with a mixed genetic background. Therefore, a subsequent study by Dehghani et al. (2000) examined whether the absence of an active *Akp5* gene in animals with a relatively homogeneous background (96% 129P3) would affect the development of the preimplantation embryo. The average gestation length and litter size of first pregnancies were compared between control and *Akp5*$^{-/-}$ mice. The mean gestation length was 12.72 h longer in the homozygous *Akp5*$^{-/-}$ mice than in the control mice, i.e. 496.8 ± 15.6 versus 484.08 ± 12.48 h. In addition, the absence of a functional *Akp5* gene caused a significant decrease in the litter size from 6.29 ± 1.1 in control mice to 4.64 ± 1.6 in *Akp5*$^{-/-}$ mice. There was no correlation between the two variables of gestation length and litter size in either the control or the *Akp5*$^{-/-}$ mice, suggesting that the prolonged gestation did not result from a smaller litter size. The authors also found a significant difference between the number of cells in embryos of control and *Akp5*$^{-/-}$ mice recovered 93 h after injection of human chorionic gonadotrophin (Fig. 41a). While the numbers of embryos recovered were comparable between the two groups, the *Akp5*$^{-/-}$ embryos were mostly at the morula stage, whereas control embryos were mostly developing a blastocoele at this time point.

To investigate further the reasons for the differences in gestation length and litter size between control and *Akp5*$^{-/-}$ mice, Dehghani et al. (2000) studied the progression of embryos through the preimplantation stages *in vitro*. This was done by following individual embryos and by comparing wt and *Akp5*$^{-/-}$ groups for percen-

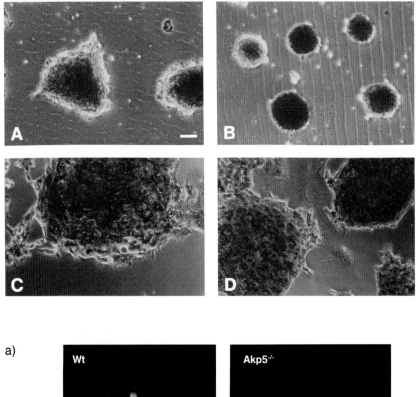

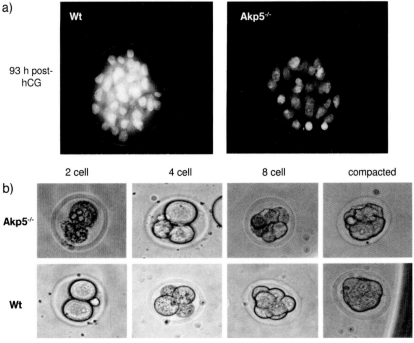

Fig. 40 D3 ES cell clones fixed and stained for AP activity. The red color indicates AP activity. (A) $Akp5^{+/-}$ clone. Most of the detectable AP activity is TNAP, which is co-expressed with EAP. Cells in the peripheral region of the colonies are flatter and AP-negative. (B) $Akp5^{-/-}$ clone displaying round morphology with few of the peripheral flat cells. (C) $Akp5^{-/-}$ clone over-expressing EAP under control of the SV40 promoter, stained after heat treatment at 65 °C for 30 min to destroy the TNAP activity. The observed AP activity is EAP. The cells have recovered the wt behavior as adherent monolayers. (D) Parental D3 cells over-expressing EAP under control of the SV40 promoter were stained after the same heat treatment.

Fig. 41 Phenotypic abnormalities in $Akp5^{-/-}$ embryos. (a) Representative pictures of embryos collected 93 h after human chorionic gonadotropin (hCG) injection and stained with DAPI to facilitate cell counts. Left panel, a control embryo (scored with 35 nuclei indicating the presence of 35 cells); right panel, an $Akp5^{-/-}$ embryo (scored with 27 nuclei indicating the presence of 27 cells). (b) Comparison of degenerating $Akp5^{-/-}$ embryos to normal control embryos. The embryos were individually cultured from the two-cell stage in KSOM/AA medium. Scale bar = 20 µm. Taken from Dheghani et al. (2000) and reproduced with permission from *Developmental Dynamics*.

tage of embryos in a defined morphological stage at defined intervals. Fewer $Akp5^{-/-}$ embryos survived to the blastocyst stage and those that did survive developed significantly more slowly than the 129P3 control embryos through compaction. However, after 68 h, the surviving $Akp5^{-/-}$ embryos developed more slowly than the control embryos from the four-cell to morula stage. The highest percentages of $Akp5^{-/-}$ 6–8 cell, compacted and blastocyst stage embryos were found at 4, 24 and 24 h later than the same stages for the control group. In addition, an increased percentage of degenerated embryos were observed in the $Akp5^{-/-}$ group with a plateau at 96 h with 62% embryo mortality (Fig. 41b). At that time point most control *in vitro*-derived embryos were compacted and only 3.7% were degenerated (Dehghani et al., 2000). The authors concluded that the presence of an active EAP isozyme is beneficial for preimplantation development of the mouse embryo and its absence leads to fewer blastocysts *in vitro*, delayed parturition and reduced litter size *in vivo*. However, the mechanism by which the enzyme may be exerting this embryonic effect remains to be elucidated.

Part III
AP Expression in Health and Disease

9
APs as Physiological and Disease Markers

9.1
Clinical Usefulness of TNAP

9.1.1
TNAP as a Marker of Bone Formation

TNAP has been used for many years as a biochemical marker of bone turnover, specifically bone formation, and for monitoring the treatment of patients with metabolic bone disease. Bone is formed by osteoblasts, which initially are derived from local mesenchymal stem cells. With the right stimulation, precursor stem cells undergo proliferation and differentiate into preosteoblasts and then into mature osteoblasts. Osteoblasts express all of the differentiated functions required for bone formation and synthesize the matrix constituents necessary for osteogenesis, a process that involves cell proliferation, extracellular matrix deposition, maturation and calcification. TNAP functions as an ectoenzyme attached to the osteoblast cell membrane by a GPI anchor (Section 4.3.2). Bone AP is most likely released into the circulation as a result of cellular turnover and programmed cell death (apoptosis) of bone-forming cells (Farley and Stilt-Coffing, 2001) and circulates in an anchorless (soluble) form after conversion of the membrane-bound to the soluble form (Raymond et al., 1994; Anh et al., 2001) as a result through the action of endogenous GPI-PLC or GPI-PLD, or both. Although both phospholipases are present in humans, only GPI-PLD is abundant in the circulation (Low et al., 1986; Davitz et al., 1987; Low and Prasad, 1988; Anh et al., 1998). Clearance of bone AP and other glycoproteins from the circulation occurs via uptake by the liver, more specifically by the galactose receptor, also known as the asialoglycoprotein receptor (Ashwell and Harford, 1982). Differences in carbohydrate structure and the content of sialic acid in bone or liver TNAP isoforms, IAP and PLAP results in various clearance rates and half-lives in the circulation from 7.5 h for IAP to 7 days for PLAP (McComb et al., 1979). These differences will also affect their serum levels and the composition and heterogeneity of the individual forms in serum. It is difficult to extrapolate reported half-life values of bone AP, ranging from 1.1 to 4.9 days (McComb et al., 1979; Posen and Grunstein, 1982; Whyte et al., 1982b, 1984; Farley, 1995), which are determined from a small group of individuals, to a general value applicable for every individual. An additional factor

Mammalian Alkaline Phosphatases: From Biology to Applications in Medicine and Biotechnology. J. L. Millán
Copyright © 2006 WILEY-VCH Verlag GmbH & Co. KGaA, Weinheim
ISBN: 3-527-31079-7

that could contribute to the large variation between reported half-lives of bone AP could be the existence of different bone AP isoforms, with a possibility of different clearance rates. However, it is generally reported that the half-life for bone AP is ~2 days (McComb et al., 1979; Moss, 1987; Calvo et al., 1996). From the reported clearance rates and half-lives of serum TNAP and of bone AP specifically, a constant activity over the day would be expected, i.e. no circadian variation, as indeed has been reported (McComb et al., 1979; Van Straalen et al., 1991; Tobiume et al., 1997). However, some studies have shown a circadian variation of TNAP (Schlemmer et al., 1992; Pedersen et al., 1995) and also of bone AP (Nielsen et al., 1990a). Greenspan et al. (1997) reported a diurnal variation of 10–20% of the mean value of bone AP for about half (43%) of the studied individuals. However, studies on circadian variation are at significance variance with each other, so until conclusive evidence to this effect is reported, one must conclude that there is no significant circadian variation for bone AP. Seasonal variation in serum TNAP, however, has been documented with lower values occurring in summer (Devgun et al., 1981) and specifically serum bone AP has a significant seasonal variation, which is the major contributor to the seasonal variation of serum TNAP (Woitge et al., 1998). These changes are directly related to variations in the hormonal regulation of skeletal homeostasis by parathyroid hormone and vitamin D (Woitge et al., 1998).

Measurement of bone AP activity in serum can provide an index of the rate of osteoblastic bone formation (Farley and Baylink, 1986; Leung et al., 1993). It has-been difficult, however, to analyze the bone-specific fraction of AP in serum owing to the co-existing liver-specific fraction of AP, which has similar physical and biochemical characteristics, e.g. the same primary structure. A large number of techniques and methods (Section 16) have been developed for the separation and quantification of the bone and liver isoforms, such as heat inactivation (Moss and Whitby, 1975), various chemical inhibitors (e.g. L-hArg and levamisole) (Lin and Fishman, 1972; Van Belle, 1976), wheat-germ lectin precipitation (Rosalki and Foo, 1984), electrophoresis (Van Hoof et al., 1988), isoelectric focusing (Griffiths and Black, 1987), HPLC (Magnusson et al., 1992) and different immunoassays (Bailyes et al., 1987). The development of commercial immunoassays for routine assessment of the bone-specific TNAP isoform (Broyles et al., 1998; Gomez et al., 1995; Hill and Wolfert, 1990; Panigrahi et al., 1994) has indeed increased the availability and use of bone AP, particularly as a biochemical marker of bone turnover for monitoring therapy in osteoporotic patients (Kleerekoper, 1996; Miller et al., 1999; Riggs, 2000).

The most commonly used approach for the visual identification of the different AP isozymes and isoforms is electrophoresis. Pretreatment of the sample with neuraminidase (Moss and Edwards, 1984) or addition of wheat-germ lectin to the electrophoresis support (Rosalki and Foo, 1984; Crofton, 1992) yields increased resolution between the bone and liver TNAP isoforms. Moreover, unusual AP forms can be identified with electrophoresis, which favors electrophoresis over heat inactivation and inhibition techniques. Another electrophoretic technique that is being used because of its high resolving power is isoelectric focusing (Sinha et al., 1986; Griffiths and Black, 1987; Root et al., 1987). Wallace et al.

(1996) presented a gel with 12 major bands, in addition to some sub-bands of AP activity. Two bone bands (bands 5a and 5b) were identified and both bands were resistant to GPI-PLC treatment, indicating that these bands could correspond to the two major bone AP isoforms B1 and B2 identified by HPLC analysis (Magnusson et al., 1992). Using electrophoretic mobility, sialic acid content, inhibition properties, heat lability, molecular mass and binding to lectins as criteria, Crofton (1987) examined the isozyme profile of plasma AP in preterm and term neonates. She found that term infants had two AP isozymes in plasma, namely, bone- and fetal-type IAP. Liver-type TNAP, PLAP- and adult-type IAP were not detected in any of the plasma samples. Bone-type TNAP levels increase during the first 6 weeks of life in both preterm and term infants and this increase is very marked in babies fed with nonsupplemented (no calcium or phosphate added) baby formulas (Crofton and Hume, 1987). In healthy adults, the bone and liver TNAP isoforms constitute ~95% of the total serum AP activity with similar quantitative levels (Magnusson et al., 1997). However, serum TNAP activities can be up to 5-fold higher during skeletal growth since the balance of cellular activity is in favor of net bone formation, hence bone AP is more prominent than liver AP during childhood and adolescence (Van Hoof et al., 1990; Magnusson et al., 1995a). At least six different AP isoform peaks can be separated and quantified by weak anion-exchange HPLC in serum from healthy individuals: three bone AP isoform peaks (B/I, B1 and B2) and three liver AP isoform peaks (L1, L2 and L3) isoforms (Fig. 42a) (Magnusson et al., 1992, 1993). In healthy adults, the three bone isoforms, B/I, B1 and B2, account on average for 4, 16 and 37%, respectively, of the total serum AP activity (Magnusson et al., 1999). In serum, the minor fraction B/I is not a pure bone isoform as it co-elutes with the IAP isozyme and is composed, on average, of 70% bone and 30% IAP.

Some biochemical differences between the bone AP isoforms have been reported (Magnusson et al., 2002b); however, these isoforms are similar with respect to freeze–thaw stability, instability during concentration by vacuum centrifugation, solubility, heat inactivation, concanavalin A precipitation, reaction kinetics and inhibition by L-Phe, L-hArg and levamisole. The bone AP isoforms differ with respect to sensitivity to precipitation with wheat-germ lectin, i.e. B1 and B2 have more (or more reactive) sialic acid residues compared with B/I. Native gradient gel electrophoresis gave estimated molecular weights of 126, 136 and 141 kDa for the B/I, B1 and B2 isoforms, respectively. These differences could be explained by the differences in the content of sialic acid residues since desialylation of B1 and B2 with neuraminidase decreased the apparent molecular weight to 127 kDa for both B1 and B2, that is, to approximately that observed for B/I (126 kDa). No change in molecular weight was observed for the B/I isoform after treatment with neuraminidase (Magnusson and Farley, 2002). Farley and Magnusson (2005) recently found that the inhibition of either N-linked glycosylation or oligosaccharide synthesis for GPI-anchor addition could affect the synthesis and the distribution of skeletal AP, but not the kinetics of the AP reaction.

It has been reported that the sialic acid residues present in bone AP affect the immunoreactivity of MAbs against bone AP (Magnusson et al., 2002a). It is of

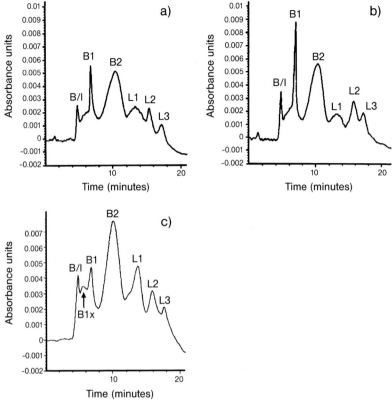

Fig. 42 TNAP isoforms in serum. HPLC separation of serum AP. (a) A serum AP isoform profile from a healthy female, 46 years of age, with total AP 2.6 μkat L^{-1}. One μkat corresponds to 60 u. Peaks, retention times and activities are, in order of elution: B/I, 4.77 min, 0.09 μkat L^{-1}; B1, 6.85 min, 0.46 μkat L^{-1}; B2, 10.45 min, 0.93 μkat L^{-1}; L1, 13.33 min, 0.57 μkat L^{-1}; L2, 15.49 min, 0.37 μkat L^{-1}; and L3, 17.12 min, 0.18 μkat L^{-1}. (b) A serum AP isoform profile from a growth hormone deficient female, 47 years of age, with total AP 2.9 μkat L^{-1}. Peaks, retention times and activities are, in order of elution: B/I, 4.79 min, 0.13 μkat L^{-1}; B1, 6.94 min, 0.66 μkat L^{-1}; B2, 10.36 min, 1.03 μkat L^{-1}; L1, 13.24 min, 0.40 μkat L^{-1}; L2, 15.67 min, 0.43 μkat L^{-1}; and L3, 17.21 min, 0.25 μkat L^{-1}. (c) A serum AP isoform profile from a female patient with severe renal insufficiency and on chronic dialysis therapy, 70 years of age, with total AP 3.6 μkat L^{-1}. Peaks, retention times and activities, in order of elution, are: B/I, 4.88 min, 0.16 μkat L^{-1}; B1x, 5.40 min, 0.19 μkat L^{-1}; B1, 6.85 min, 0.45 μkat L^{-1}; B2, 10.02 min, 1.25 μkat L^{-1}; L1, 13.92 min, 0.79 μkat L^{-1}; L2, 16.06 min, 0.49 μkat L^{-1}; and L3, 17.72 min, 0.27 μkat L^{-1}. The figures were kindly provided by Dr Per Magnusson. Data taken from Magnusson et al. (1997, 1999) and reproduced with permission from the *Journal of Bone and Mineral Research*.

interest that preparations of bone AP from human osteosarcoma (SaOS-2) cells are currently used as calibrators in these immunoassay kits. A higher proportion of the bone AP isoform B/I (39–80%), compared with human serum, has been reported in these calibrators (Magnusson et al., 2001). In serum from healthy subjects, B/I has a lower activity than either the B1 and B2 isoforms and represents, on average, only 4% of the total AP activity measured (Magnusson et al., 1999). Some differences have been reported between these bone AP immunoassays in groups of patients that express different ratios of the bone AP isoforms, which could be explained by differences between the antibodies used in these immunoassays (Woitge et al., 1996; Price et al., 1997; Magnusson et al., 1998a). The ISOBM TD-9 workshop recently evaluated the properties, specificity and target epitopes of 19 coded MAbs against TNAP (Section 6.3). These studies were intended to determine the nature and location of the epitopes and provided immunological data about the different TNAP isoforms. Data from that workshop showed that some of these MAbs can to some extent distinguish between the circulating bone AP isoforms B/I, B1 and B2 and that desialylation with neuraminidase increased the affinity for the MAbs and decreased the differences in cross-reactivity between the isoforms. These observations suggest that both the immunoaffinity and the bone AP isoform specificity may also be dependent on the amount/number of terminal sialic acid residues (Magnusson et al., 2002a).

Some anatomical differences in the skeletal content of the bone AP isoforms have also been investigated in human femora from three sites: (i) cortical and (ii) trabecular bone from the diaphysis and (iii) trabecular bone from the greater trochanter. Trabecular bone, from both sites, had higher bone AP activities compared with cortical bone. Cortical bone had about 2-fold higher activities of B1 compared with B2 and trabecular bone had about 2-fold higher activities of B2 compared with B1 (Magnusson et al., 1999, 2002b). Ten-fold higher activities have been reported for the bone AP isoforms B1 and B2 during childhood and adolescence in comparison with adults (Magnusson et al., 1995a). The circulating peak levels for total TNAP and bone AP isoforms follow reported peak levels for adolescent growth hormone secretion. The adolescent growth spurt is very divergent in boys and girls, with girls beginning 2 years earlier and boys having a greater spurt. The circulating levels of the bone AP isoforms also vary independently during the pubertal growth spurt. Girls aged 15–16 years have higher levels than boys for the calculated ratio B2/B1 owing to a more rapid decline of B2 compared with the B1 bone AP isoform after puberty (Magnusson et al., 1995a).

Ross et al. (2000) demonstrated, in a prospective study, that increased levels of serum bone AP is significantly associated with an increased risk of osteoporotic fracture in postmenopausal women. In a clinical study of patients at risk of osteoporosis, due to treatment with oral corticosteroids, and in patients at risk of increased bone synthesis because of treatment with cyclosporin, a decrease of bone AP during corticosteroid treatment and an increase of bone AP during cyclosporin treatment could be demonstrated (Van Straalen et al., 1991). Clinical differences have been reported between the bone AP isoforms in disease states such as growth hormone deficiency (Fig. 42b) (Magnusson et al., 1997), hypophosphatasia, X-linked hypophosphatemia (Magnusson et al., 1992), stress fracture, Paget's bone disease (Magnusson et al., 1995b), celiac disease (Jansson et al., 2001), chronic renal failure (Magnusson et al., 2001) and metastatic bone disease (Magnusson et al., 1998b).

Decreased B1 activity has been reported after 2 weeks of insulin-like growth factor I (IGF-I) administration and after 1 month of growth hormone therapy, followed by an increase after 3 months (Magnusson et al., 1997). The B2 isoform was not influenced by IGF-I administration, but was similarly increased after 3 months of growth hormone therapy. It was proposed that the initial decrease of B1 could be an effect of endocrine IGF-I action mediated by growth hormone. Different responses of B1 and B2 during IGF-I and during growth hormone therapy suggest different regulations of the bone AP isoforms *in vivo* (Magnusson et al., 1997). However, the significance of these finding in terms of function and clinical utility of the bone AP isoforms remain to be determined. A fourth bone AP isoform, identified as B1x, has also been identified in serum from patients with chronic renal failure and in extracts of human bone tissue (Fig. 42c) (Magnusson et al., 1999, 2001). The possible existence of a circulating kidney AP isoform was ruled out since bilaterally nephrectomized patients also had measurable activities of B1x. Renal bone disease, common in chronic renal failure, is a multifactoral disorder of bone remodeling encompassing a spectrum of histologically classified disorders ranging from high-turnover osteitis fibrosa through adynamic osteopathies to low-turnover bone disease. Accurate classification of the type of renal bone disease is difficult and usually achieved through bone histomorphometry after iliac crest biopsy. However, less invasive approaches such as biochemical markers of bone turnover, e.g. bone AP, could be of potential value to clinicians as early indicators of diagnostic and/or pharmacological efficacy (Ureña and de Vernejoul, 1999).

Clearly, a greatly reduced level of serum TNAP (and of total AP activity) is a primary indication to rule out hypophosphatasia. One report has documented the use of a MAb to detect TNAP on a chorionic villus sample taken in the first trimester of pregnancy. Very low activities of the TNAP isozyme indicated an affected fetus and diagnosis of hypophosphatasia was confirmed by ultrasound scan at 15 weeks gestation and by AP measurement in amniotic fluid supernatant and fetal serum (Warren et al., 1985). Furthermore, Henthorn and Whyte (1995) successfully assessed a fetus at risk for lethal infantile hypophosphatasia using amniocyte DNA and allele-specific oligonucleotide (ASO) probes for two missense mutations in the *ALPL* gene that had been discovered in a sister who died at 8 months of age from the disease. The mother was known to carry the 747 (cDNA) G>A transition, whereas her husband and 5-year-old daughter, who were also healthy, carried the 1309 A>T transversion (Section 7.1.2). Amniocytes, obtained at 16 weeks' gestation, provided genomic DNA for polymerase chain reaction (PCR) amplification of the appropriate TNAP gene exons. Orimo et al. (1996) also obtained a prenatal molecular diagnosis during the first trimester of pregnancy in a Japanese woman whose first child (the proband) had been a compound heterozygote for infantile hypophosphatasia. The authors examined chorionic villus DNA samples obtained at 10 weeks of gestation for the base substitutions detected in the proband DNA using PCR, RFLP and ASO analysis. The genotype of the fetus turned out to be the same as that of the proband.

9.1.2
TNAP and Bone Cancer or Bone Metastasis

Osteosarcomas display high serum TNAP levels and these levels are significantly higher in metastatic disease than in patients with localized disease (Bacci et al., 1993). However, bone-type TNAP activity did not allow Van Hoof et al. (1992) to distinguish between nonmalignant bone disease and bone metastasis. The negative predictive value of bone-type TNAP in the diagnosis of bone metastasis was low, but its positive predictive value was high and superior to that of total AP measurements (Liu et al., 1996) where it also has prognostic value (Bacci et al., 2002). Elevation of serum TNAP has also been reported in clear cell chondrosarcoma of bone (Ogose et al., 2001).

More than 70% of patients with breast cancer and a~84% of patients with prostate cancer develop bone metastasis at the time of autopsy (Jacobs, 1983). Bone metastasis is generally classified as either osteolytic or osteoblastic. Breast cancers often cause osteolytic bone metastasis whereas prostate cancers are usually osteoblastic, although bone metastasis commonly contains both osteolytic and osteoblastic components. Serum levels of TNAP increase in cases of osteoblastic activity including osteoblastic bone metastasis (Cooper et al., 1994; Imai et al., 1992; Wolff et al., 1996, 1997, 1999). Patients with prostate cancer and skeletal metastases have increased bone AP activities, particularly increased B2 activities corresponding to ~75% of the total serum AP activity, significantly higher when compared with the value of 35% found in the control group of healthy men (Magnusson et al., 1998b). The increased B2 levels could be explained by the fact that secondary bone tumors often develop in the vertebral column, each vertebra containing a larger fraction of trabecular bone with a higher turnover rate than cortical bone. Hence the increased bone AP isoform B2 activities in patients with skeletal metastases could be related to tumor invasion of the trabecular bone compartment (Magnusson et al., 1998b). Early studies indicated that the total serum AP levels were of value as an earlier predictor of response to therapy than X-rays or bone scans and more reliable than the prostatic acid phosphatase profile in monitoring patients with prostate cancer and bone metastases (Mackintosh et al., 1990). Morote et al. (1996) developed a two-monoclonal antibody radioimmunoassay (Section 16.6) allowing an accurate measurement of the serum bone TNAP isoform with the goal of comparing the clinical performance of bone TNAP and prostate-specific antigen (PSA) in patients with untreated prostate carcinoma and to determine whether or not TNAP can provide valuable additional information to PSA regarding the degree of skeletal extension in patients with prostate carcinoma. Bone AP was more efficient than PSA in the prediction of positive bone scans and its level was significantly related to the magnitude of skeletal involvement (Morote et al., 1996). Both total serum AP and the bone AP were increased significantly in patients with prostatic carcinoma with bony metastasis. The level of bone AP had a higher positive predictive value and specificity for bony metastasis than the level of total serum AP, but a lower negative predictive value and sensitivity than total AP or PSA (at PSA levels >20 ng mL^{-1}) (Chen et al., 1997). Using the Tandem-R Ostase (Section 16.6), bone AP was found to be a valuable marker for clinical response evaluations to use in the serial follow-up of patients with metastatic prostate cancer. However, this marker correlates well with PSA level as a quantitative assessment of the bone scan only after the PSA level is >50 ng mL^{-1} (Murphy et al., 1997).

Androgen deprivation produces an increase in the bone AP serum concentration in prostate cancer patients. A major increase seems to be produced during the first year of follow-up and thereafter this increase is reduced by around 50% annually (Morote et al., 2002). In turn, lower serum levels of the bone AP seems to correlate with a better response to anti-androgen withdrawal. Moreover, a level of bone AP higher than 50 ng mL^{-1} would predict the absence of response (Morote and Bellmunt, 2002). A flare in serum total AP activity post-orchidectomy has also been shown to be of negative prognostic value for progression-free survival in patients with prostate cancer. Measurement of AP activity within 4 weeks of castration represents a useful adjunct in assessing which prostate cancer patients, undergoing androgen ablation, may benefit from additional early chemotherapy (Pelger et al., 2002). However, serum bone AP should not be used as a marker for the diagnosis of osteoporosis in men with prostate cancer and this serological test cannot replace bone densitometry as a diagnostic tool in these cases (Morote et al., 2003).

Plasma levels of bone and liver TNAP isoforms have also been useful markers to diagnose and monitor bone and liver metastases in patients with breast cancer. In a study by Mayne et al. (1987), the bone-specific TNAP isoform was increased in 21 of 50 patients (42%) with radiologically confirmed bone metastases, whereas total AP activity was increased in only 10 of 50 patients (20%); liver AP activity was raised in 12 of 25 patients (48%) with liver metastases. All patients with liver metastases had bone metastases. Bone AP activity was significantly higher in patients with symptomatic bone disease. Isoform determination provided additional information that would have changed patient management in five of 20 patients who were monitored serially. The authors concluded that measurement of AP isozymes, although less sensitive than imaging procedures, can assist in screening for, and in early detection of, a high proportion of bone and liver metastases and can provide useful objective evidence of their response to treatment (Mayne et al., 1987). Using an immunoradiometric assay, Reale et al. (1995) studied 145 female patients, 97 with radically operated breast cancer and 48 with benign mammary cysts, in order to evaluate the correlation of serum levels with the metabolic process of bone rearrangement in patients with bone metastases. The study showed that skeletal TNAP, having high specificity (86.48%) and sensitivity (78.6%) for early progression (the average anticipation time compared with scintigraphic detection was 101 days) could represent a valid marker for bone metastases in association with mucinous markers in the follow-up of patients operated for breast cancer. In addition, dynamic serum determination of skeletal AP could be a valid help in monitoring the efficacy of therapy in patients with bone progression. In breast cancer, however, metastases to bone mostly cause increased bone resorption, i.e. they lead to osteolytic lesions, both from direct effects of the tumor itself and through osteoclastic activation.

The diagnosis and follow-up of bone metastatic cancer patients usually relies on skeletal X-ray and bone scintigraphy. Thus, markers of bone remodeling, i.e. bone AP, are often measured in conjunction with markers of bone resorption, such as N-telopeptide of type I collagen (NTx). Kanakis et al. (2004) applied two solid-phase immunoassays used for the determination of bone-specific TNAP isoform and NTx in serum of breast cancer post-menopausal women with bone metastasis and healthy individuals. The data showed elevated levels for both markers, indicating a high rate of bone degradation in breast metastatic cancer (Kanakis et al., 2004). Thus, the bone fraction of plasma TNAP activity has some value in the identification of bone metastasis during the evaluation and follow-up of breast cancer (Stieber et al., 1992; Moro et al., 1993) and prostate cancer (Curatolo et al., 1992) patients. Furthermore, leukocyte AP (TNAP isozyme) has been shown to serve as a useful indicator in cases of advanced lung cancer (Walach and Gur, 1993). However, in most instances TNAP levels signal metabolic changes secondary to the malignant process and as such the TNAP isozyme cannot be considered a typical cancer marker.

9.1.3
TNAP Expression in Cholestasis

The hepatocyte is another cell type that expresses TNAP, although at very low levels under normal physiological conditions. The hepatocytes have an apical membrane facing the bile canaliculus, a basal membrane facing the sinusoid and a lateral membrane connecting those to other surface domains. Patients with cholestasis have greatly elevated tissue and serum levels of liver-derived TNAP, which can be found in three forms: soluble (released from its GPI anchor), bound to membrane fragments or in lipoprotein-X complexes. In humans with cholestasis there is evidence for two sources of liver AP in serum, one derived from sinusoidal (basolateral) and the other from canalicular (apical) hepatic membranes (De Broe et al., 1985) The canalicular AP may appear in serum because of regurgitation from the bile, a fluid in which bile salts cause release of enzymes without cell lysis. In addition to AP released by detergent action, AP is also found in large particles in serum. These forms separate on density gradients at densities equal to that of the cell basolateral membrane. The absence of AP-containing particles in bile makes the sinusoidal (basal) membrane domain the likely site for membrane shedding into serum. Early studies had indicated that both the circulating particulate AP and the so-called lipoprotein-X were detected in 25% of patients with cholestasis (McComb et al., 1979) and that soluble AP was associated with lipoprotein-X in many of these patients. Further studies suggested that this lipoprotein-X was composed of phospholipid and serum proteins, particularly albumin (McComb et al., 1979). The AP–lipoprotein-X complex was demonstrated as an abnormally moving band in the agar-agarose gel pattern in patients whose sera contained lipoprotein-X (Weijers, 1980). The position of the abnormally moving band in the agar-agarose gel AP stained pattern and the percentage of AP activity in the complex in relation to the total serum AP activity depended on the serum lipoprotein-X concentration. Weijers et al. (1980) purified the serum AP–lipoprotein-X complex from several patients suffering from intrahepatic and extrahepatic cholestasis and demonstrated that the complex had a relative molecular mass of at least 669 000. Crofton and Smith (1981a,b) confirmed that high molecular weight AP had similar properties in serum and bile and the association with lipoprotein-X was demonstrated by precipitation of both molecular forms by an antiserum specific for lipoprotein-X. Evidence for both tight junction passage and transcellular regurgitation was reported to explain the presence of lipoprotein-X in serum, but it was not clear whether this AP arose from the sinusoidal or canalicular membrane. In addition, it was not certain whether it was released by the action of a GPI-PLC, serum GPI-PLD, bile salt detergent action or a combination of these factors. Alpers et al. (1990) summarized the various hypotheses concerning the secretion of hepatic AP during cholestasis: AP appears in serum in various combinations of forms: free (or soluble), particulate on a plasma membrane-like fragment or particulate on lipoprotein-X. AP may reach the serum either by direct release from the sinusoidal membrane, by transcytosis or by regurgitation from the bile through tight junctions. The soluble form may result from detergent action, enzymatic release or a combination of the two. At the present time it is not possible to postulate a single mechanism for the secretion of hepatic AP. Wolf (1990) reported on a fast-moving liver isoform, also known as high molecular weight isoform, that represented fragments of hepatic cell plasma membranes with various membrane-bound enzymes localized on the surface. High molecular weight TNAP isoform was present in patients' serum samples in conditions associated with intrahepatic and extrahepatic cholestasis and primary or metastatic hepatic malignancy (Wolf, 1994). Kihn et al. (1991) also concluded that the high molecular weight AP in serum most often originated from fragments of hepatic plasma membranes. Bile acids probably facilitate the enzymatic release of TNAP from the sinusoidal surface of hepatocytes (Solter et al., 1997). In fact, the bile acids taurine-conjugated cholic acid and chenodeoxycholic acid can cause up-regulation of TNAP mRNA in HuH7 and HepG2 hepatoma cell lines (Khan et al., 1998). Elsafi et al. (1989) concluded that the increase in AP activity in serum from rats with bile duct obstruction and cirrhosis mainly had a hepatocytic origin. In a clinical study, Hadjis et al. (1990) measured the levels of serum AP in eight patients with bile duct obstruction limited to one lobe of the liver. Although an initial rise of enzyme concentration was documented in every patient, unrelieved biliary obstruction was associated with a gradual return of AP to normal values. The return to normal levels coincided with the development of atrophy of that part of the liver deprived of its bile drainage. It appears that normal serum AP levels can be expected with advanced obstructive biliary disease. Biliary AP, another isoform of liver-type TNAP, has been found in the serum of patients with biliary obstruction and metastatic liver cancer. The demonstration of serum biliary AP, in particular in nonjaundiced patients with high serum AP, may indicate the presence of tumor in the bile duct (Bhudhisawasdi et al., 2004). Van Hoof et al. (1992) reported that the positive predictive value of liver and high molecular weight AP was higher than that of total AP in detecting liver metastasis, but liver and high molecular weight AP did not permit the differentiation between malignant and nonmalignant liver disease. Total AP activity was of slightly more value in ruling out liver metastasis. Monitoring the liver-type TNAP level was also indicated to be useful to estimate postoperative liver failure as measured by changes in bilirubin following hepatectomy (Osada and Saji, 2004). The biliary isoform of liver-type TNAP was found to be a highly sensitive indicator of liver dysfunction also in cases of cystic fibrosis (Schoenau et al., 1989).

Little is known about the role of TNAP in liver physiology or in liver diseases (Section 7.2). In control liver, TNAP activity is low and localized in the bile canalicular plasma membranes. The total parenchymal activity increases about 3-fold after the induction of experimental cholestasis and is considered to be a compensatory mechanism in order to enhance the excretion of bile salts from hepatocytes (Frederiks et al., 1990). Since endotoxin is usually elevated in patients with liver damage, Xu et al. (2002) suggested that TNAP might contribute to protection from injury by a mechanism involving neutralization of endotoxin. In a very interesting and complete study, Deng et al. (1996) purified circulating liver plasma membrane fragments (LPMF) from human serum by means of a MAb against leucine aminopeptidase. TNAP is bound to these LPMF through a GPI anchor and Deng et al. refer to it as membrane-bound liver AP (Mem-LiAP). Low concentrations of Triton X-100 or high bile salt concentrations released GPI anchor-bearing LiAP (Anch-LiAP) from purified LPMF; once released, Anch-LiAP was slowly and progressively converted to hydrophilic dimeric LiAP or soluble LiAP (Sol-LiAP), free from its GPI anchor. Low levels of GPI-PLD activity were measured in the pure LPMF that apparently were released by the action of detergents and contributed to the spontaneous conversion of Anch-LiAP to Sol-LiAP. In the absence of detergents, GPI-PLD had little effect on Mem-LiAP, both in purified form and in serum. In serum, and also in purified conditions, only a small range of detergent of bile salt concentrations permitted the conversion of Mem-LiAP to Sol-LiAP. The authors proposed a model for the release in the circulation of Mem-LiAP, Anch-LiAP and Sol-LiAP, involving both LPMF-associated GPI-PLD and liver sinusoid bile salts (Fig. 43). Quoting from their paper: "LiAP normally circulates in human serum as Sol-LiAP, its dimeric hydrophilic form. In pathological conditions, particularly in cholestasis, hydrophobic forms can be released into the circulation: Mem-LiAP combined with GPI-anchored and/or lipoprotein bound LiAP. The formation of Sol-LiAP by GPI-PLD-mediated anchor degradation requires free Anch-AP. Bile salts activate membrane-associated GPI-PLD by releasing both GPI-PLD and Anch-LiAP from LPMF. In general, Mem-LiAP largely predominates over Anch-LiAP in cholestatic sera. Sol-LiAP in serum appears to be the product of GPI-PLD activity. In serum, GPI-PLD activity is inhibited by bicarbonate and by complexes of Anch-LiAP with serum components. Only a small range of detergent bile salt concentrations permits the conversion of Mem-LiAP to Sol-LiAP, explaining the ability of Mem-LiAP and Anch LiAP to retain their structure in the presence of circulating bile salts and GPI-PLD". To-date, this paper by Deng et al. (1996) remains the most comprehensive study concerning the release of liver AP in health and disease.

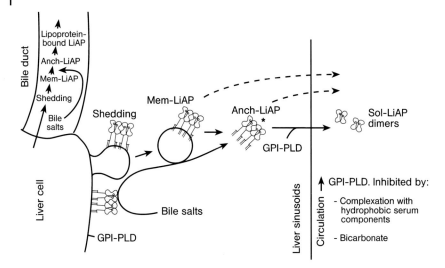

Fig. 43 Model hypothesizing possible interactions of bile salts and GPI-PLD with Mem-LiAP and Anch-LiAP under physiological conditions. Taken from Deng et al. (1996) and reproduced with permission from the *American Journal of Physiology*.

9.1.4
TNAP in Other Conditions

Alterations in TNAP levels can also be informative in other pathologies. Total, liver-type and bone-type TNAP activities are significantly higher in untreated hyperthyroid patients than in the treated group or a healthy control population. Bone-type TNAP was more frequently and more markedly abnormal than liver-type TNAP (Tibi et al., 1989). This increase in serum TNAP activity frequently seen in patients with hyperthyroid Graves' disease may partly be due to a direct effect of thyroid hormone on osteoblasts or osteoblast-like cells (Kasono et al., 1988). TNAP has also been found elevated in patients with rheumatoid arthritis, where the elevation was found to be due to the liver isoform in 39% of the cases and to the bone isoform in 22% of the cases. In women, both total AP and the levels of the bone AP correlated with the number of joints involved in rheumatoid arthritis (Cimmino et al., 1990). Changes in the AP profile were also observed in these patients (Cimmino and Accardo, 1990). The usefulness of measuring bone AP in patients with Paget's disease has also been well documented (Deftos et al., 1991). Higher TNAP activities and P_i concentrations in saliva have been associated with rampant caries (Gandhy and Damle, 2003). Vogelsang et al. (1996) measured the serum levels of AP in patients with Crohn's disease using agarose electrophoresis to identify changes in the liver AP, fast liver AP (another name for high molecular weight liver AP), bone AP and IAP. They concluded that total AP activity seems to be influenced by inflammatory activity mainly via fast liver AP that could be a marker of disease activity (Vogelsang et al., 1996). Reduced AP activity presumably derived from TNAP in cerebrovascular endothelial cells in neonates born prior to 28 weeks gestation have been taken as signs of immature cerebral white matter afferent cells which compromises vascular function (Anstrom et al., 2002). Challa et al. (2004) suggested that the combination of 100 µm thick celloidin sections and AP enzyme histochemistry of the vascular endothelium offers a greatly enhanced, 3D morphological perspective and reveals intricate details of the vasculature of the brain, and that this technique might be useful for the study of tumor angiogenesis and to evaluate vascularity in experimental and human brain tumors after various therapies (Challa et al., 2004). Histochemical detection of TNAP has been found to be informative in cases of primary human brain tumors, including meningiomas and neurinomas (Murakami et al., 1993). In fact, deletion of chromosome 1p and loss of expression of TNAP indicates progression of meningiomas. It was concluded that TNAP is a useful marker enzyme for the loss of a putative regulatory (tumor suppressor) gene on chromosome 1p or that *ALPL* itself might represent a new tumor suppressor gene homozygously inactivated in meningiomas (Muller et al., 1999; Niedermayer et al., 1997).

TNAP is also expressed in cells of the hematopoietic lineage. In the normal hematopoietic system, the post-mitotic neutrophilic granulocyte is the only cell type that expresses leukocyte AP under basal conditions. However, TNAP is a membrane marker for activated B cells (Garcia-Rozas et al., 1982). Culvenor et al. (1981) examined 19 cell lines of the B lymphoid lineage, including Abelson pre-B, B lymphoma and plasma cell tumor lines, and all but one had substantial TNAP activity. In contrast, nine T lymphoid and nine nonlymphoid hematopoietic lines examined had very low activity. Hence TNAP seems to provide a useful marker for tumor lines of the B lymphoid lineage and for their plasma membranes. Marquez et al. (1989) reported that a B-cell proliferative response elicited either by high concentrations of rabbit anti-IgM antibodies or by LPS in the presence of phorbol esters is characterized by the lack of both antibody secretion and expression of TNAP activity. In contrast, B cells stimulated to differentiate into Ig-secreting cells by B cell differentiation factors, almost in the absence of a proliferative response, express high levels of TNAP activity, as do those that are LPS-stimulated (Marquez et al., 1989). Later, using a MAb G-5-2 that recognizes murine TNAP, the authors concluded that TNAP was also expressed in some T-cells (Marquez et al., 1990). Analysis of the TNAP present in the cytoskeletal fraction in fully differentiated B lymphocytes, X63 myeloma cells and Sp2/O hybridoma cells by Pezzi et al. (1991) revealed that during the course of B-lymphocyte activation, TNAP shifted from a soluble to a Triton-insoluble form and changes in the phosphorylation of Triton-insoluble proteins with molecular weights of 120, 100, 90, 75, 34 and 31 kDa were detected, coinciding with the appearance of the TNAP in this fraction. Despite these observations, the function of TNAP in leukocytes and lymphoid cells remains unknown, although a role in programmed cell death (apoptosis) has been suggested. Souvannavong et al. (1995, 1997) reported that expression of TNAP by the 7TD1 B-cell hybridoma was amplified by ultraviolet irradiation and this effect occurred both in cycling and in apoptotic cells. Furthermore, levamisole treatment almost totally abrogated apoptosis induced by ultraviolet irradiation at doses that failed to affect 7TD1 cell survival and the authors suggested that TNAP could play a role in the signaling cascade that mediates apoptosis in irradiated cells. Along these lines, Hui et al. (1997b) showed that high-level TNAP expression in rat fibroblasts induced collagen phagocytosis and apoptosis.

Precursors of the myelomonocytic pathway, represented by leukemic cells isolated from several cases of chronic myelogenous leukemia in its stable and blastic phase and acute myelogenous leukemia, are devoid of a TNAP transcript (Dotti et al., 1999). However, TNAP represents a specific and restrictive marker for the terminal maturation of the neutrophilic granulocytes (Rambaldi et al., 1990; Garattini and Gianni, 1996). In this regard, elevations in the levels of neutrophil AP have been documented in mothers of trisomy 21 children (Grozdea et al., 1984, 1988, 1991; Vergnes et al., 1988). An increase in neutrophil TNAP in the parents of trisomy 21 children has also been reported (Grozdea et al., 1980). Neutrophil AP score is currently readily available in most laboratories and may also be helpful in the diagnosis of hypophosphatasia (Iqbal et al., 2000).

9.2
Clinical Usefulness of PLAP in Normal and Complicated Pregnancies

PLAP is synthesized in the syncytiotrophoblast after the 12th week of pregnancy and is shed into the maternal circulation. The serum PLAP concentrations increase from low basal levels as pregnancy proceeds and the increased production correlates with the growth of the placenta and the increased synthesis of estrogen (McComb et al., 1979). Okamoto et al. (1990) described the expression of PLAP during the course of pregnancy and the amount of PLAP mRNA and its activity in normal placental villi. Both PLAP and its mRNA were found in placentas as early as 7 weeks of gestation and both continued to increase throughout pregnancy, but they showed different patterns of increase. The amount of PLAP mRNA began to increase dramatically around 13 weeks and continued gradually until term. PLAP activity per gram of villi increased gradually to the 13th week and then markedly after the 20th week. PLAP levels in sera from pregnant women were also measured and showed a pattern similar to that of PLAP activity per gram of villi. The rate of enzyme enrichment in the maternal blood follows a logarithmic course. The continuous increase in the expression of PLAP throughout pregnancy suggests that PLAP may play a role in feto-maternal metabolism and placental differentiation.

It was recognized early that deviations from the "prediction curves" of serum PLAP in pregnancy could indicate fetal disorders and death *in utero* (McComb et al., 1979). However, despite the convenience in sampling and measurement, PLAP levels during normal pregnancy are generally considered too widely scattered to define useful normal ranges. The introduction of "prediction curves" by Yamaguchi and Shimozato (1978), however, suggested that placental function tests could be used as screening tests during the third trimester of pregnancy. Several complications such as "small-for-date" babies, severe and mild toxemia and diabetes were associated with changes in the levels of PLAP (Yamaguchi and Shimozato, 1978; Holmgren et al., 1979). Using a specific monoclonal antibody-based immunoassay, PLAP was measured in plasma samples of 117 women who delivered an infant of birth weight <2.5 kg. PLAP values greater than twice the normal median were found in 32% of maternal plasma samples from low birthweight cases in one series and in 35% in another series, whereas in normal outcome controls the corresponding value was 8% (Brock and Barron, 1988; Best et al., 1991). Furthermore, women with elevated PLAP levels are at increased risk for preterm delivery (Meyer et al., 1995).

A relationship between the occurrence of two PLAP variants and fetal disorders has also been proposed. Beckman et al. (1972a) found that in extracts of chorionic tissue from spontaneous abortions the frequency of the Pl^2 allele (Section 4.2) of the *ALPP* gene was significantly higher than in tissue from a series of induced abortions and from live births. Although the number of investigated cases was small ($n = 51$), this may suggest the presence of a prenatal mechanism working against the Pl^2 allele. Beckman and Beckman (1975) also found a significant increase of previous abortions in couples whose pregnancies gave rise to placentas with the D variant (variant 18, a rare variant with low electrophoretic mobility, Section 4.2). These data suggest that an optimal PLAP structure–function is correlated with proper development and that, possibly, prenatal selection mechanisms are acting against certain "defective" alleles of PLAP. The proportion of newborns with a low birth weight (below the 10th percentile) was also found to be lower in infants with the Pl^1 allele of PLAP than in those with other PLAP alleles, especially among nonwhites (Amante et al., 1996). High or increasing plasma total AP activity levels at 26 weeks, but not 19 weeks, was significantly associated with subsequent preterm delivery and a lower birth weight (Goldenberg et al., 1997). Specific antibodies to PLAP have been detected in placental extracts bound to placental membrane preparations. Therefore, PLAP appears to be immunogenic in at least some pregnancies (Freeman et al., 2001). In other reports, IgG–PLAP complexes (Grozdea et al., 1988a) have been detected in maternal serum of 11 cases of anti-Rh-complicated pregnancies.

Congenital Chagas disease, due to the intracellular parasite *Trypanosoma cruzi*, is associated with premature labor, miscarriage and placentitis. PLAP activity decreases in serum and placental villi from term Chagasic women. *In vitro*, *Trypanosoma cruzi* invades a wide variety of mammalian cells by a process that is still poorly understood. Trypomastigotes adhere to specific receptors on the outer membrane of host cells before intracellular invasion, causing calcium ion mobilization and rearrangement of host cell microfilaments. Sartori and colleagues propose a pathogenetic role for PLAP in congenital Chagas disease. They suggest that PLAP could be involved in the internalization of *T. cruzi* in HEp2 cells or trophoblast cells, via activation of a tyrosine kinase receptor and rearrangement of actin microfilaments (Sartori et al., 2002, 2003). More recently, Mezzano et al. (2005) have shown that a significant reduction of PLAP expression was detected immunologically in infected diabetic and normal placental villi cultured under hyperglycemic conditions of 71 and 81%, respectively, compared with controls. A significant decrease in PLAP specific activity was also registered in homogenates and in the culture media from both infected diabetic and normal placentas under hyperglycemic conditions (of about 50–70%) and in Chagasic ones (of about 87%), when compared with controls. The authors conclude that PLAP might be involved in parasite invasion and diabetic and hyperglycemic placentas could be more susceptible to *T. cruzi* infection.

The presence of PLAP has also been demonstrated immunohistochemically in frozen sections of human endometrium (Davies et al., 1985a). The enzyme is present in glandular epithelium, but is found most commonly in the surface epithelial layer throughout the menstrual cycle. It has also been demonstrated in malignant endometrial epithelium in eight out of 12 patients (Davies et al., 1985a). Increased activity of thermostable AP with the biochemical properties of PLAP have been reported in peritoneal fluid obtained from patients with untreated endometriosis (Kang et al., 1990).

9.3
IAP Expression in Relation to ABO Status, Fat Feeding and Other Pathologies

It has been clearly established that there is a marked post-prandial increase in serum IAP levels in humans following ingestion of a fat-containing meal (Domar et al., 1991, 1993; Day et al., 1992a; Matsushita et al., 1998). The post-prandial rise in serum IAP activity is significantly greater following a long-chain fatty acid meal than following a medium-chain fatty acid meal (Day et al., 1992a). In turn, IAP levels are dramatically decreased upon starvation (Hodin et al., 1994). The increase in serum IAP following a fatty meal is strictly dependent on the ABO genotype and secretor status of the individual, i.e. nonsecretors (of blood group antigens) had serum IAP levels 20% of that of secretors after fat feeding (McComb et al., 1979; Domar et al., 1991). Among secretors, the numbers were lowest for type A and highest for type O and B.

Deng et al. (1992b) investigated the biophysical properties of human IAP in duodenal fluid and serum. They found that IAP could be released by the enterocyte into duodenal fluid as a mixture of three isoforms. A proportion of the enzyme was found associated with triple-layered membrane vesicles (which they called vesicular IAP) and although occasionally free hydrophilic IAP dimers were present, the remaining enzyme usually consisted of a mixture of hydrophobic IAP dimers and more complex hydrophobic IAP structures of larger size, both entities identified as "intestinal variant" AP (VAR IAP). The hydrophobicity of VAR IAP appeared to stem exclusively from its attached GPI anchor (Section 4.3.2). Both vesicular IAP and VAR IAP were converted to hydrophilic enzyme upon removal of the GPI tail by GPI-PLD present in duodenal fluid. The IAP released into the vascular bed consisted mainly of VAR IAP; vesicular IAP was absent. The enzyme characteristics of VAR IAP partially purified from duodenal fluid and from serum were identical. In plasma, VAR IAP appeared to associate with lipoprotein complexes and was thus protected from further degradation by plasma GPI-PLD. Such complex formation may explain why, in the serum of a healthy reference population, VAR IAP was more abundant than hydrophilic dimeric IAP (Deng et al., 1992).

Nevertheless, raised serum levels of IAP have so far been of little diagnostic value. Clinical conditions including inflammatory diseases of the bowel, such as Crohn's disease or ulcerative colitis, would be natural candidates for a clinical application of such measurements, but no elevations of IAP have been found in those patients (Domar et al., 1988). In fact a small decrease in serum IAP levels was observed in patients with these diseases, probably owing to increased shedding into the intestinal lumen rather than into the bloodstream. Intestinal ischemia necessitates rapid re-establishment of blood flow to prevent irreversible anoxic tissue damage. However, reperfusion leads to additional injury as a consequence of the generation of oxygen free radicals. Studies by Sisley et al. (1999) utilizing a rat *in vitro* lipid peroxidation model demonstrated that the generation of free radicals resulted in the inactivation of only the intestinal brush-border IAP isozyme, with no effect on other membrane-bound digestive enzymes. Brush-border membrane enzymes were assayed in mucosal extracts from intestines with ischemia versus reperfusion. IAP activity was significantly decreased in both the canine and human reperfusion models, with no change in specific activities of sucrase, maltase and γ-glutamyl transpeptidase, indicating that IAP measurements may permit quantitative assessments of therapeutic interventions in human intestinal reperfusion injury.

Concerning other diseases, the MAb AAP-1 (Section 6.2), specific for the IAP isozyme, was used to develop an immunoassay for amniotic fluid samples. Brock et al. (1984) used this assay to investigate a panel of 124 control second-trimester amniotic fluids and 21 fluids with a one in four risk of a cystic fibrosis fetus. Eight of 10 affected cases had values below an arbitrary cut-off of one-third median, whereas all the nonaffected cases were above this level (Brock et al., 1984). Tibi et al. (1988) determined the serum AP isozyme profile in patients with type 1 and type 2 diabetes and in a nondiabetic control group. As expected, IAP activity in both diabetics and the control group was significantly higher in BO secretors than A secretors or ABO nonsecretors. The authors found no difference in the levels of IAP between type 1 and type 2 diabetics, but the diabetics had a significantly higher activity of IAP than the corresponding blood group/secretor status category of the control group. Liver-type TNAP was also significantly higher in the diabetics than the control group, but no elevations were found for the bone isoform. Because of the suggestion that IAP was elevated in the serum of patients with chronic renal failure, Alpers et al. (1988) studied the serum of 42 patients undergoing hemodialysis with elevated enzyme activity. Using a sensitive and specific electroimmunoassay for IAP, 26 of 42 serum samples were positive, compared with three of 25 samples obtained from hospitalized patients with elevated phosphatase activity. The fractional amount of the IAP isozyme was also higher, ranging from 1.5 to 41% of the total serum AP, compared with 0.1–1.2% in control sera. Kidneys removed during transplantation or post mortem contained a membranous AP with immunological activity identical with IAP in five of six patients. This enzyme accounted for 8–21% of the total kidney AP activity. By morphology, the immunoreaction was localized to the apical membranes of the collecting tubules. The authors concluded that an elevation in IAP isozyme levels rather than the presence of early metabolic bone disease or hepatic disease should be considered in renal failure patients with mildly elevated (up to 50% over normal) total serum AP (Alpers et al., 1988). IAP is a specific and sensitive marker for alterations of the S3 segment of the human proximal tubule, the preferred part for several nephrotoxins. Nuyts et al. (1992) found that IAP excretion was clearly increased in mercury-exposed workers, compared with other parameters, indicating that the determination of IAP could be a useful screening test for renal effects in occupational mercury exposure. Increased levels of serum IAP activities have also been observed in jaundiced patients with intrahepatic disease and subnormal values in those with post-hepatic disease. However, the diagnostic value of IAP was not as good as that derived with other techniques (Kuwana and Rosalki, 1991).

9.4
Complexes of APs and Immunoglobulins

Several reports have documented the existence of AP–immunoglobulin complexes in patients' sera (Buttery et al., 1980). Crofton et al. (1981) studied 31 patients in whose sera an immune complex between AP and immunoglobulin G had been detected. Twenty-three of those patients (74%) had a disease with either an autoimmune etiology or associated with circulating immune complexes or autoantibodies. In all those cases, AP was attached to the F(ab')$_2$ region of the immunoglobulin molecule and not to the Fc region. The authors concluded that the complex was an immune complex formed by antibody-antigen reaction in the circulation (Crofton, 1981). Maekawa et al. (1985) detected an abnormal band of AP activity by electrophoresis in the serum of a patient with liver cirrhosis and the authors showed it to be a complex between liver-type TNAP and immunoglobulin A (IgA) of the lambda type, probably as a result of an antibody–antigen reaction. Another patient with cholestatic liver disease was reported to have two macromolecular complexes in serum with molecular weights of 640 000 and 420 000, which consisted of a complex of liver-type TNAP with kappa-type immunoglobulin A (Wenham et al., 1983).

Another electrophoretically slow-moving AP-antibody complex was found in the serum of a patient with osteomalacia. The molecular weight of this complex was estimated to be 500 kDa and was composed of IAP in complex with IgG kappa light chain (Kohno et al., 1983). Nakagawa et al. (1983) reported a macromolecular AP in the serum of a patient suffering from myasthenia gravis complicated with thymoma and was shown by immunoelectrophoresis to be bound to immunoglobulins A and G. The IgG appeared to be immunoreactive to both IAP and PLAP and in fact was able to interfere with PLAP catalytic activity. Mader et al. (1994)

studied the patterns of AP-binding proteins in patients with diverse inflammatory diseases. The AP-binding protein was identified as IgG on Western blots and in ELISA using human IgG-specific antibodies. It was shown that this IgG binds to AP from both calf (bovine) and human intestine but not to human TNAP or bacterial AP. Moderate reaction was also observed with human PLAP. In another report, patients suffering from acute bacterial infections had specific IgG autoantibodies to IAP expressed in high titer (Kolbus et al., 1996).

9.5
Hyperphosphatasia

Several cases of hyperphosphatasia have also been documented, due to elevated levels of TNAP, IAP or PLAP. Most of these elevations, however, appear to have no pathological consequences and often no apparent underlying cause (Wolf, 1995). For example, a study of 11 members of the same family after an incidental detection of raised bone-type TNAP in one of them without any apparent underlying cause suggested a probable autosomal dominant pattern of inheritance (Cirera Nogueras et al., 1982). In another report, five children, aged 16–38 months, were found to have serum TNAP levels of skeletal origin 7–30 times the upper limit of the reference range (Steinherz et al., 1984). No other abnormalities or explanations for the unusual enzyme levels were found and the enzyme levels later returned to the reference range. These and other authors (Griffiths et al., 1995) have concluded that benign, transient hyperphosphatasemia should be recognized as a clinical entity since awareness of this condition would curtail extensive unnecessary evaluation.

A persistent elevation of IAP was found in a boy whose only symptom was transient periodic fatigue observed at home, but not apparent during hospitalization. His blood type was O, RH$^+$, Le (a$^-$, b$^+$) and he was a secretor of H-substance, but his serum IAP levels remained unchanged after fat loading (Nathan et al., 1984). In two other cases, increased activities of IAP have been reported in the absence of disease (Bentouati et al., 1990; Lieverse et al., 1990). Onica et al. (1989) described a family with an inherited persistent elevation of serum AP activity in the absence of malignant disease, observed for at least 15 years. Isozyme studies revealed that this increased activity was due to an enzyme which showed similarities to serum PLAP having the properties high heat stability, moderate inhibition by EDTA and lack of interaction with wheat-germ lectin. The enzyme was less sensitive than PLAP to inhibition by L-Phe, L-Trp, L-Leu, L-leucylglycylglycine and L-phenylalanylglycylglycine. The enzyme also differed from PLAP in its electrophoretic mobility, isoelectric heterogeneity and apparent molecular weight. In a later report, they further described that the enzyme reacted with the H7, but not with the C2 MAb (Section 6.1). The authors suggested that this upregulated enzyme might correspond to a rare phenotype of PLAP (Onica et al., 1990).

10
Neoplastic Expression of PLAP, GCAP, IAP (Regan, Nagao, Kasahara) and TNAP Isozymes

10.1
Some History

PLAP was one of the first enzymes recognized as an oncofetal protein by Fishman and colleagues (Fishman et al., 1968a,b). It was first identified in the serum of a patient (Regan) with a squamous cell carcinoma of the lung and its functional characteristics were indistinguishable from those of the placental enzyme by several catalytic, immunological and structural criteria (Fishman et al., 1968a; Greene and Sussman, 1973). A complicating factor was the presence of the so-called "PLAP-like" enzyme in some samples. The origin of the term "PLAP-like" dates back to when an AP (termed "Nagao isozyme" after the name of the patient) with all the characteristics of a rare variant of PLAP, the D-variant, was detected in the serum of a patient with pleuritus carcinomatosa (Nakayama et al., 1970). The D-variant of PLAP (Section 4.2) has several characteristics that distinguish it from other PLAP phenotypes: inhibition by L-Leu and EDTA and a slow mobility in starch gel electrophoresis (Beckman and Beckman, 1968; Inglis et al., 1973; Doellgast and Fishman, 1976). Re-evaluation of many serum samples that had previously been characterized as containing the "Regan isozyme" positively revealed that up to 50% of these exhibited L-Leu inhibition and slow migration in starch gels. Many of these Nagao isozyme-containing samples were derived from sera of patients with seminoma of the testis. Interestingly, a heat-stable, L-Leu-sensitive AP was also found in trace amounts in normal testis (Chang et al., 1980), comprising 0.3–4.6% of total AP. Wei and Doellgast (1981) used anti-PLAP polyclonal antibodies, that had been rendered allozyme-specific by extensive cross-adsorption, to detect antigenic differences between the Nagao isozyme and the D PLAP allozyme. They argued that the Nagao isozyme and the D variant were closely related but distinct enzymes. Also, Millán et al. (1982b), using the allozyme specific MAb F11, concluded that the Nagao isozyme was likely the product of a distinct locus. Now it is clear that those samples containing the L-Leu-inhibitable PLAP-like Nagao AP were expressing the GCAP isozyme, not the PLAP isozyme (Millán and Manes, 1988).

Numerous associations have been reported between the expression of GCAP and PLAP and malignancy. Thus, tumors that express these markers can be broadly divided into two groups: (a) those with an enhanced production of the iso-

Mammalian Alkaline Phosphatases: From Biology to Applications in Medicine and Biotechnology. J. L. Millán
Copyright © 2006 WILEY-VCH Verlag GmbH & Co. KGaA, Weinheim
ISBN: 3-527-31079-7

zyme normally expressed in that tissue (eutopic expression) and (b) those showing expression of one or more isozymes not characteristic of the normal tissue (ectopic expression). Benham et al. (1981b) conducted a search for expression of heat-stable PLAP isozyme in 19 unselected human tumor cell lines, known not to be HeLa. That study suggested that the PLAP locus may be expressed in at least low levels in a much higher proportion of tumors and tumor cell lines than previously recognized and the authors suggested that the so-called "ectopic" synthesis of PLAP in tumor cells may not necessarily be due to derepression of a structural locus which is completely unexpressed in normal adult tissues but represents an enhancement of expression in malignancy, or there may be clonal expansion of a particular cell type that normally expresses the PLAP gene at a high level (Benham et al., 1981b). These authors also examined the AP activity in eight independent cell lines derived from human testicular germ cell tumors (Benham et al., 1981a). Seven out of eight of the lines had high TNAP activity, but also low levels of a heat-stable, PLAP-like AP. We now know that the presence of the PLAP-like, Nagao isozyme in cancer samples results from the tumoral expression of the GCAP gene. In the case of testicular cancers, this tumoral expression represents an overexpression of a gene normally expressed in that tissue, i.e. eutopic expression, rather than a re-expression of a developmental gene not normally expressed in that tissue, i.e. ectopic expression (Jeppsson et al., 1984a). Similarly, as described in Section 2.2, Povinelli and Knoll (1991) have documented that even though expression of the *ALPP* gene is predominant in the syncytiotrophoblast, transcripts from the *ALPP2* are also detectable at about 2% of the level of *ALPP* transcripts. Ovitt et al. (1986) examined the expression of *ALPP* and *ALPP2* in the placenta and in three choriocarcinoma cell lines and concluded that transformation of normal to malignant trophoblast is associated with a switch in expression from PLAP to GCAP. Consistent with this view, two out of four GCAP cDNA sequences available to date have been derived from choriocarcinoma cell lines, JEG3 (Watanabe et al., 1989) and BeWo (Lowe and Strauss, 1990). This expression of GCAP in choriocarcinomas can therefore also be considered as upregulated eutopic expression. The ectopic expression of PLAP has been observed in cancer of the lung, ovary, uterus and gastrointestinal tract (McComb et al., 1979; Loose et al., 1984) and is also found ectopically in ovarian cancer (Vergote et al., 1987; Nozawa et al., 1990) and in some tumors of the pineal gland and thymus. Using a PLAP-specific polyclonal antibody and the immunoperoxidase technique in formalin-fixed, paraffin-embedded sections, Wick et al. (1987) examined 37 germ cell neoplasms and 483 somatic tumors. All germ cell lesions were reactive for PLAP/GCAP, but so were 62 somatic carcinomas, usually in female mullerian, intestinal and lung cancers and less often in carcinomas of the breast and kidney. Malignant mesotheliomas were nonreactive for PLAP, as were carcinomas of the nasopharynx, adrenals, liver, pancreas, stomach, prostate and urinary bladder.

During the last 25 years, many clinical reports have been published concerning the tumoral expression of PLAP and GCAP (the Regan isozyme versus the Nagao isozyme), often with greatly discordant results. Some of the discordances and the ensuing controversies with regard to the usefulness of these isozymes as tumor markers are due to differences in the specificity of the assays that could not distinguish between these closely related gene products. In fact, the most useful application of PLAP determinations, i.e. follow-up of patients with seminoma of the testis, entails measurement of GCAP and not PLAP, but these determinations are still reported as PLAP, since most assays available, with few exceptions, will not differentiate between these tumor markers.

10.2
GCAP as Marker for Testicular Cancer

Testicular germ cell tumors (TGCTs) can be divided into two main histological subtypes: seminoma and nonseminatomous TGCTs (Mostofi and Sobin, 1977). Nonseminatomous TGCTs include embryonal carcinoma, yolk sac tumors, choriocarcinoma and teratoma. All of these are derived from one precursor lesion, carcinoma *in situ* (CIS) (Skakkebaek et al., 1987). Many studies concerning AP expression in TGCTs have been carried out, on serum samples of patients, on tumor tissue samples and on tumor cell lines. The following is a brief discussion of some of those studies, with special attention to the techniques used and the findings.

Indirect immunofluorescence using rabbit antisera against PLAP stained primary testicular tumor cells in cytological smears. Seminomas were usually positive, whereas embryonal carcinomas were occasionally positive. Using MAbs against PLAP/GCAP and IAP, indirect immunoperoxidase staining on seminomas detected PLAP/GCAP in about 90% of seminomas, but none in embryonal carcinoma or interstitial cell tumors. IAP was not detected in those tumors, but strong staining for IAP was found occasionally in teratomas and weak staining for PLAP/GCAP was also found there (Uchida et al., 1981). Using six monoclonal antibodies (MAbs) that had been characterized with regard to their binding to PLAP, GCAP and IAP, PLAP was detected in four of seven seminomas, three of seven embryonal carcinomas and one yolk sac carcinoma. GCAP was identified in two additional seminomas and four embryonal carcinomas. Only one seminoma had no detectable PLAP or GCAP. IAP was found in three teratocarcinomas and trophoblastic giant cells of two seminomas (Paiva et al., 1983). Epenetos et al. (1984) used the H17E2 PLAP MAb and found that all seminomas and malignant teratomas tested gave strong positive labeling while not staining normal testis. Using immunohistochemistry with anti-PLAP polyclonal antibodies, nearly 100% of CIS or seminoma cells stained positive, whereas <50% of cells stained positive, in those two out of three samples in embryonal carcinomas that had staining at all. Two choriocarcinomas stained moderately positive, more so in the syncytiotrophoblast than the cytotrophoblast, whereas immature teratomas with endometrial differentiation, and two testicular lymphomas, were completely negative (Hustin et al., 1987). Koide et al. (1987) found that atypical germ cells could be easily identified in testicular samples using a combination of immunohistochemical detection of PLAP/GCAP and periodic acid Schiff staining for glycogen. Hirano et al.

(1987a) assayed seminoma extracts and normal testicular extracts with MAbs and found that in the normal testis ~90% of the catalytic activity originates from TNAP and the remaining activity is due to trace expression of both IAP (~5%) and PLAP/GCAP (~5%). In homogenates of seminoma tissues, highly increased levels of all three isozymes were identified. TNAP, PLAP/GCAP displayed relative increases of 10–100-fold and IAP 2–10-fold compared with normal testis (Hirano et al., 1987a). Hofmann et al. (1989) also extracted AP from normal testis and TGCTs and were able to distinguish all four AP isozymes by isoelectric focusing. In turn, Hamilton-Dutoit et al. (1990) examined the expression of PLAP and GCAP in frozen sections from a variety of normal and neoplastic human tissues using monoclonal antibodies (MAbs) reactive with PLAP (H317) and GCAP (H17E2). PLAP/GCAP reactivity was seen in normal thymus and fetal and neonatal testis and in 21 out of 22 malignant germ cell tumors, but was also found in normal endocervix, normal Fallopian tube and in 28 out of 167 non-germ cell tumors (particularly in ovarian and proximal gastrointestinal tract tumors).

Those papers documented expression of TNAP and GCAP in normal testis, whereas seminomas and germinal epithelia containing atypical germ cells seemed to contain four isozymes, but nonseminatomous TGCTs such as yolk sac tumors and choriocarcinoma did not seem to have GCAP. As methods improved in their sensitivity and specificity, it became apparent that the expression of APs in CIS and TGCTs seemed to be both eutopic and ectopic. Schär et al. (1997) developed a reverse transcriptase PCR method that clearly distinguished between the RNA transcripts of all four isozyme mRNAs. They compared the pattern of AP expression in 15 germ cell tumors, two germinal epithelia adjacent to seminoma, two cell lines of germ cell tumor origin (Tera-1 and BeWo) and five normal testes. In comparison with normal testes, in all seminomatous germ cell tumors eutopic expression of GCAP and ectopic expression of TNAP were demonstrated. In both samples of pure embryonal carcinoma and in the embryonal carcinoma cell line, the transcription of all four mRNAs was shown. These results indicate that the expression of the isozymes depends on the degree of differentiation of a tumor. The increased expression of TNAP in seminoma was later also reported by Shigenari et al. (1998). In another comprehensive study, testicular cancers of germ cell and non-germ cell origin along with testicular parenchyma with and without CIS were analyzed for the expression of the different AP isozymes (Roelofs et al., 1999). Immunohistochemistry on frozen tissue sections showed that expression of PLAP/GCAP and TNAP in CIS, microinvasive seminoma, seminoma and embryonal carcinoma was rather heterogeneous between samples of the same histology of different patients, and also within single samples (Fig. 44). At the molecular level, seminomas were found to express almost exclusively GCAP in accordance with multiple earlier studies. In embryonal carcinoma, however, the predominantly expressed isozyme was found to be either PLAP or GCAP. It is worth noting that spermatocytic seminoma, which is derived from more differentiated germ cells, possibly spermatogonia B, expresses predominantly GCAP, whereas Leydig cell tumors, which are non-germ cell derived, express predominantly PLAP. As in the study of Schär et al. (1997), Roelofs et al. (1999)

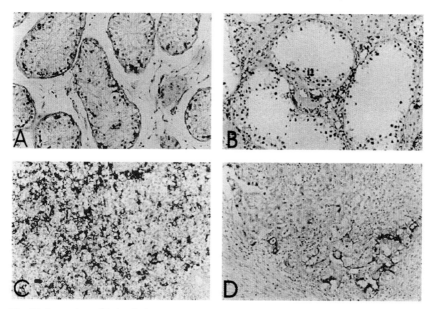

Fig. 44 Expression of PLAP/GCAP in testicular germ cell tumors. Immunohistochemistry for PLAP/GCAP on formalin-fixed, paraffin-embedded tissue sections containing carcinoma-*in-situ* (A), microinvasive seminoma (B), seminoma (C) and embryonal carcinoma (D). Taken from Roelofs et al. (1999) and reproduced with permission from the *Journal of Pathology*.

found that the relative expression of the different TSAP isozymes varies between TGCT samples, possibly depending on the differentiation status. These findings clearly indicate that all AP loci appear be activated in testicular seminomas.

Nevertheless, it is the enhanced eutopic expression of GCAP that appears to have most value in the histopathological diagnosis and classification of seminomas. It is also clear that GCAP is a particularly good marker to diagnose CIS of the testis (Wahren et al., 1979; Lange et al., 1982; Paiva et al., 1983; Jeppsson et al., 1984a), a condition that precedes all cases of TGCTs, except spermatocytic seminoma (Skakkebaek et al., 1987). Progression from CIS to invasive cancer was observed in 70% of the cases after 7 years of observation (Burke and Mostofi, 1988a; Von der Maase et al., 1987). Spontaneous regression of CIS has not yet been reported and it is believed that all cases of CIS will eventually develop into invasive cancer. The diagnosis of CIS is based on routine histological examination of surgical testicular biopsy specimens. Using a polyclonal antibody to PLAP/GCAP, the immunoreactivity of 89 germ-cell tumors for PLAP was as follows: 98% of cases with seminomatous elements were PLAP/GCAP positive and 97% of embryonal carcinomas and 85% of endodermal sinus tumors also showed reactivity. Cytotrophoblastic cells were focally immunoreactive in one of two cases with choriocarcinomatous elements. Staining for PLAP/GCAP was strongest and most diffuse in seminomas (Burke and Mostofi, 1988b). CIS was present in 53 (84%) of 63 specimens that had adjacent seminiferous tubules available for evaluation; PLAP/GCAP was demonstrated in 98% of these. In addition, the germ-cell elements in 11 gonadoblastomas were immunoreactive for PLAP/GCAP. One of 17 cases of undescended testes had CIS cells that were strongly immunoreactive, but the remaining 16 cases were negative. Five dysgenetic gonads without CIS were studied and one was immunoreactive for PLAP/GCAP. Three testicular biopsy specimens from infertile men without CIS were PLAP/GCAP-negative (Manivel et al., 1987). In an immunohistochemical study of 59 routinely processed tissue specimens from 48 adult testes with isolated CIS and of 66 specimens from adult testes without neoplasia, PLAP/GCAP was shown to be a reliable marker of CIS cells preceding the development of a testicular tumor (Bailey et al., 1991; Giwercman et al., 1991). Staining for the presence of PLAP/GCAP is also useful for making the differential diagnosis between classical seminoma and spermatocytic seminoma, which are consistently negative for this marker (Dekker et al., 1992).

Several groups also recommend that PLAP/GCAP serum levels be used as a "serum" marker for seminomas, where it can provide unique clinical information not available through other testicular tumor markers such as AFP and βHCG (Wahren et al., 1979; Goldstein et al., 1980; Lange et al., 1982; De Broe and Pollet, 1988). Detection of PLAP/GCAP in serum from seminoma patients by radioimmunoassay was initially limited by its sensitivity, i.e. $12\,\text{ng}\,\text{mL}^{-1}$ (Wahren et al., 1979). However, subsequent studies utilizing a more sensitive sandwich ELISA (capable of detecting $0.4\,\text{ng}\,\text{mL}^{-1}$) could detect PLAP/GCAP in most sera from seminoma patients, but not those with nonseminatomous germ cell tumors (Lange et al., 1982). Using a solid-phase monoclonal antibody enzyme immunoassay, the utility of evaluating the catalytic activity of PLAP/GCAP in serum as a potential marker for seminoma was evaluated in an 18-laboratory multicenter study. PLAP/GCAP activity was found to be frequently increased ($>100\,\text{mU}\,\text{L}^{-1}$) in preoperative serum samples in patients with seminoma and PLAP/GCAP was clearly more frequently increased than βHCG (De Broe and Pollet, 1988).

However, as described above, the production of PLAP/GCAP, and also of most other tumor markers, is not absolutely specific for the tumor tissue. Nonmalignant tissues such as the human testis (Chang et al., 1980) and the human cervix (Benham et al., 1978; Hording et al., 1990) express GCAP. GCAP is also present in epithelial cells of respiratory bronchioli and alveolar type I pneumocytes (Hording et al., 1990). Although the concentration of enzyme in these tissues is very small, they provide a likely origin for the basal circulatory levels of PLAP-like activity found in both normal males and females (Millán et al., 1985a). Heavy smoking increases serum GCAP levels (Tonik et al., 1983; Koshida et al., 1990, 1991), most probably through enhanced production of GCAP by type I pneumocytes (Section 2.2). Using a sensitive solid-phase immunoassays against PLAP/GCAP using MAb 17E2, two separate studies indicated that elevated serum PLAP/GCAP levels were found in all patients with active seminomas and whereas serum levels of PLAP/GCAP correlated with the course of disease, false-positive results were obtained when evaluating smokers (Epenetos et al., 1985a; Horwich et al., 1985). To study the origin of increased serum PLAP-like activity in smokers, heat-stable alkaline phosphatase (AP) activity was assayed from serum and bronchoalveolar lavage fluid in 83 smoking and nonsmoking patients (Kallioniemi et al., 1987). PLAP-like activity was increased in about 80% of the smokers, independently of the underlying lung disease. Isozyme activities in lavage fluid correlated with serum values. When adjusted for the albumin concentration, mean PLAP-like activity in lavage fluid was almost 1000-fold higher than that in serum, suggesting local synthesis of PLAP-like isozymes in the lungs. Although a direct dose–response effect was not observed, the values in serum and in bronchoalveolar lavage fluid tended to be higher in patients smoking >10 cigarettes daily as compared with patients smoking less. In ex-smokers the results indicated that PLAP-like activity decreased to the level observed in nonsmokers within 5 years after cessation of smoking. PLAP activity was L-Leu sensitive, compatible with the Nagao isozyme in almost all cases. In three patients, the activity was due to the L-Leu-resistant (true PLAP) isozyme. Muensch et al. (1986) reported that smoking was a major factor determining the nonspecific elevation PLAP/GCAP. In 98 healthy nonsmokers, the mean of the enzyme activity was determined as $0.068\,U\,L^{-1}$ compared with a mean of $0.378\,U\,L^{-1}$ in 65 smokers. In view of that finding, the authors re-evaluated the usefulness of PLAP/GCAP as a tumor marker in 286 patients with various neoplasms and a negative smoking history. Of these patients, 23% had elevated values for PLAP/GCAP. When compared with the range of GCAP in normal smokers, only 4.1% of the patients showed elevated values (Muensch et al., 1986). Nielsen et al. (1990b) also evaluated the usefulness of PLAP/GCAP as a tumor marker in 1578 serum samples from 236 patients with seminoma. Smoking habits were known for all but seven patients (22 samples). Smoking was associated with significantly higher mean levels of PLAP/GCAP in disease-free patients ($28.8 \pm 2.1\,U\,L^{-1}$ versus $15.9 \pm 1.3\,U\,L^{-1}$ in nonsmokers). Mean PLAP/GCAP levels were higher in patients with active disease ($78.6 \pm 23.5\,U\,L^{-1}$ in nonsmokers and $47.2 \pm 18.5\,U\,L^{-1}$ in smokers). The median values showed a similar trend. However, there was considerable overlap between the various groups and differences between mean and median values indicated that PLAP/GCAP values were distributed asymmetrically. The specificity and sensitivity were, respectively, 88 and 45% (all patients) and 96 and 47% (nonsmokers). Even after the sensitivity and specificity of PLAP/GCAP had been adjusted for a series of threshold values (normal versus abnormal) with a graphical method, only in nonsmokers did PLAP/GCAP seem useful. Otto et al. (1998) used two-dimensional electrophoresis, ion-exchange chromatography and immunoassay to improve the diagnostic specificity of GCAP for the detection of seminoma. They concluded that assessment of GCAP is hampered by its structural heterogeneity, clearly visualized by two-dimensional electrophoresis that depends on allelic amino acid substitutions, varying sialylation and differential cleavage of the membrane anchor. Despite the fact that allelic variability affects the accuracy of immunological measurements (Section 6.1), immunoassay was found to be the only technique sensitive enough to assess GCAP in serum, although 15% of healthy blood donors were shown to have high GCAP values correlated with smoking. The authors also suggested that GCAP might be utilized for the detection of CIS cells exfoliated into ejaculate as a means for a noninvasive, early diagnosis that presumably will not be hampered by the patient's smoking habits (Otto et al., 1998). GCAP measurements in seminal plasma can also be used as a marker for successfully performed vasectomies. Lewis-Jones et al. (1992) used the H17E2 MAb-based enzyme immunoassay in samples obtained both before and after vasectomy and found that GCAP fell to undetectable levels within 14 days of operation. The important conclusion from these studies is that serum levels of PLAP/GCAP can be very useful indicators of malignant disease and serve as tumor markers provided that their use is confined to nonsmokers. However, in tissues, the immunohistochemical detection of PLAP/GCAP is almost diagnostic in cases of TGCTs.

10.3
Usefulness of PLAP/GCAP in Ovarian Cancer

PLAP activity is also frequently increased in samples from ovarian cancer patients. The concentrations of PLAP in ascites and cyst fluids are markedly higher than in serum and cyst fluid values are also generally higher than in ascites fluid from the same individual. Doellgast and Homesley (1984) found that the median levels of heat-stable enzyme for malignant cyst fluids were 50 times greater than for benign cyst fluids. However, determination of this isozyme in serum did not give a useful index of tumor burden, as metastatic disease did not consistently result in elevated serum enzyme levels. Sunderland et al. (1984) found significant expression of PLAP in ovarian cystadenocarcinomas by immunohistochemistry using MAb NDOG2. The use of a highly specific enzyme–antigen immunoassay based on MAb H317 specific for PLAP and with a sensitivity of $0.07\,\mu g\,L^{-1}$, indicated that 23 of 65 (35%) ovarian carcinoma patients were positive (McLaughlin et al., 1983). Similarly, using another assay based on MAb E6 to PLAP, serum PLAP levels $\geq 0.1\,U\,L^{-1}$ were found in 58% of ovarian cancer patients (Nouwen et al., 1985). PLAP was also detected in extracts from 13 of the 14 tumors investigated (range 2.4–$557\,mU\,g^{-1}$) while the highest PLAP content of normal ovarian tissue was estimated at $1.1\,mU\,g^{-1}$. The neoplastic origin of PLAP was confirmed by immunohistochemistry. In all the tumors, staining for PLAP was observed mainly on the plasma membranes of carcinoma cells but the histological distribution was heterogeneous. PLAP staining, present in one of five normal ovaries, was restricted to germinal inclusion cysts (Nouwen et al., 1985). Similarly, using the H317 MAb, McDicken et al. (1985) detected PLAP in >30% of ovarian cancer patient sera and in most solubilized tumor tissue extracts. They found no association between circulating PLAP levels and either tissue extract levels or immunohistological staining of ovarian tumor tissue sections with H317. These authors also demonstrated the heterogeneity of cellular localization of PLAP within different tumors (McDicken et al., 1985). Mano et al. (1986a) used a radioimmunoassay based on polyclonal antibodies and described that serous adenocarcinoma, endometrioid adenocarcinoma and dysgerminoma contained particularly large amounts of PLAP and that this marker was more frequently detected in tissue than in the serum of ovarian cancer patients and suggested that PLAP may be a useful target in immunodetection and immunotherapy. In another report, Mano et al. (1986b) evaluated 1236 samples from 414 patients with ovarian cancer. The frequencies of elevated enzyme levels for patients with or without evidence of disease were 17.7 and 10.9%, respectively. The true positive rate was highest in serous cystadenocarcinoma, undifferentiated carcinoma and dysgerminoma. However, measurement of the enzyme did not give a useful index of stage of disease, tumor burden or prognosis. The value of the enzyme as an index of successful therapy was limited because half of the patients lost this marker during progression (Vergote et al., 1987). De Broe and Pollet (1988) determined that PLAP activity was increased in preoperative serum samples in 49% of ovarian cancer patients. Using a combined enzyme-linked immunoassay able to measure both serum PLAP activity and concentration in the same microtiter plate using the MAb H17E2, Fisken et al. (1989) evaluated 397 serial samples from 87 patients with epithelial ovarian cancer and concluded that PLAP was only of value in the management of patients with known active disease who are already known to be "marker positive" for this antigen. Employing PLAP-specific immunoassays (no cross-reactivity with GCAP) using the C2 MAb, a similar incidence of PLAP positivity was found in the four major groups of adenocarcinomas of the ovaries. An examination of 116 formalin-fixed ovarian epithelial tumors using the C2 MAb revealed that 51% of the tumors displayed positive immunoreactivity with similar incidence (46–67%) in the four major groups of the adenocarcinomas, i.e. serous, mucinous, endometrioid and mesonephric tumors, but the staining index revealed a more intense staining for the mucinous and mesonephric tumors (2.1 and 2.6) compared with the serous and endometrioid tumors (0.9 and 1.5) (Stendahl et al., 1989; Stigbrand et al., 1990). PLAP has also been reported in dysgerminomas (Nozawa et al., 1990). Smans et al. (1999) used a single-nucleotide primer extension (SNuPE) assay (Section 16.7) to discriminate between expression of PLAP and GCAP in samples from 13 ovarian carcinoma patients. The expression of PLAP isozyme was predominant, but in some cases GCAP was also found (Fig. 45). In fact, the expressed isozyme may differ between the primary tumor and the metastases in the same patient. As shown in Fig. 45, one patient (P7), in whom the primary tumor and two metastases were available, whereas the primary tumor expressed only PLAP, GCAP expression was predominant in the metastases (Smans et al., 1999). These results indicated that antibodies used for immunotargeting or immunotherapy of ovarian tumors should preferably display comparable affinity towards PLAP and GCAP.

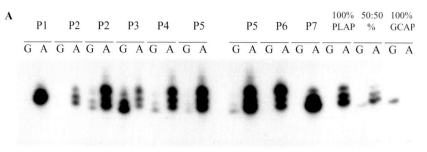

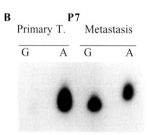

Fig. 45 Expression of PLAP/GCAP in ovarian cancer. Analysis of seven ovarian cancer samples (A) and a primary tumor and its metastasis (B) with the single nucleotide primer extension (SNuPE) assay. Incorporation of [^{32}P]dATP or [^{32}P]dGTP reflects the presence of PLAP and GCAP, respectively. Samples consisting of a defined mixture (100:0, 50:50, 0:100) of PLAP and GCAP RT-PCR product served as controls. Taken from Smans et al. (1999) and reproduced with permission from *International Journal of Cancer*.

10.4
Other Tumors

In addition to its presence in extracts of ovarian carcinoma, carcinomatous tissues from the vulva, endometrium and ovary, PLAP can also be found in tissue extracts of benign conditions of the endometrium and myometrium. Van de Voorde et al. (1985b) detected slightly elevated levels of PLAP in patients with myoma, raised levels in 2/2 patients with endometrial polyps and high values in 2/2 patients with glandulocystic hyperplasia. Moreover, normal endometrium and fragments of normal Fallopian tube also contained fairly high amounts of endogenous PLAP (Van de Voorde, et al., 1985b). Solubilized cervical smears or biopsy material and cervical mucus swabs often contained substantial amounts of PLAP; however, there was no significant difference between the expression levels in patients with pre-invasive and invasive cervical neoplasia, with or without evidence of papilloma virus infection and control groups (McLaughlin et al., 1987). MAb H317 with specificity for PLAP was used to investigate the occurrence of this marker in patients with primary breast carcinoma. All preoperive plasma samples were negative for PLAP in a sensitive solid-phase enzyme immunoassay (McLaughlin et al., 1983), with a lower limit of detection of 0.1 U L^{-1} or ~0.07 µg L^{-1}. In contrast, using a peroxidase-anti-peroxidase staining technique on fixed tissue sections, and enzyme immunoassay on fresh tissue extracts, PLAP could be demonstrated in the carcinomatous tissue of all seven patients investigated (McDicken et al., 1983). PLAP could also be demonstrated in 16 out of 22 gastric carcinomas and in three out of 11 moderate and six out of six severe dysplasias, but not in intestinal metaplasia. There was no correlation between the occurrence of PLAP and histological type of gastric carcinomas (Kralovanszky et al., 1984). Another study using MAbs indicated that the incidence of PLAP positivity was 23% (25 of 107) of all gastric carcinomas (Watanabe et al., 1990). Among gastric carcinomas, a 42% (13 of 31) positivity rate of highly differentiated carcinoma (papillary adenocarcinoma and well-differentiated tubular adenocarcinoma) was significantly higher than that found in poorly differentiated carcinoma (poorly differentiated adenocarcinoma and signet-ring cell carcinoma, five of 41, 12%). The incidence of PLAP positivity was 11% (four of 35) in colorectal carcinoma. In contrast, gastric adenoma, intestinal metaplasia and noncancerous tissue adjacent to cancer did not show staining (Watanabe et al., 1990). Harmenberg et al. (1989, 1991) reported that PLAP also showed rising levels during progression of colorectal carcinoma while liver-type TNAP remained elevated. Goldsmith et al. (2002) found PLAP immunoreactivity in normal human adult and fetal muscle tissue and explored the possible usefulness of PLAP in the diagnosis of soft tissue tumors. A total of 271 tumors that contained myogenic, neural, fibrous, myofibroblastic, lipomatous, neuroepithelial, perivascular and epithelial differentiation were studied. Cytoplasmic PLAP reactivity was detected in all leiomyomas and rhabdomyosarcomas (100%), seven of 15 (46%) leiomyosarcomas, 15 of 19 (79%) desmoplastic small round cell tumors, two of 15 (13%) gastrointestinal stromal tumors, one of eight (13%) Wilms' tumors, one of nine synovial sarcomas (9%) and two of seven (29%) myofibroblastic tumors. However, the usefulness of PLAP as a tumor marker in these kinds of tumors remains limited.

10.5
Immunolocalization and Immunotherapy of Tumors Using PLAP/GCAP as Targets

Polyclonal and monoclonal antibodies (MAbs) against PLAP have been used for tumor immunolocalization of PLAP-producing tumors in both mice and humans. However, it is important to realize that cell lines are able to modulate the expression levels of the markers that one aims at targeting, since this inherent property can confound the interpretation of the data. For example, Hep 2/5 HeLa cells express both PLAP and fetal IAP isozyme, but when grown as solid tumors in nude mice, the levels of PLAP are greatly downregulated, whereas levels of the IAP remain fairly constant (Benham et al., 1981b). In one of the first targeting experiments, Jeppsson et al. (1984b) used the F11 MAb and rabbit polyclonal anti-PLAP antibodies, labeled with ^{125}I and injected i.p. in mice, to detect that harbored developing HEp2 tumors. The distribution of [^{125}I]anti-PLAP in various tissues showed that the labeled antibody was enriched in the tumor with a mean concentration ratio of 7.1 and 6.8 for polyclonal and MAb, respectively. A PLAP-negative rhabdomyosarcoma showed a mean ratio of 1.2. There was a positive correlation between PLAP content and uptake of labeled antibody in the tumors. However, while HEp2 tumor cells in tissue culture showed 100% positivity for PLAP, imprints of the tumor after passage in nude mice only showed 40–50% positivity (Jeppsson et al., 1984b). Jemmerson et al. (1984a) also used the F11 MAb in nude mice where both a PLAP-positive (A431 cells) and a PLAP-negative (SNG cells) tumor were grown. The mice were then injected with radioactively labeled $F(ab')_2$ fragments of the F11 anti-PLAP MAb. The results showed that the MAb localized in the PLAP-positive tumor 10 times or more often better than it localized in the PLAP-negative tumor or in normal mouse tissues. The authors concluded that PLAP appears to be a useful marker for the immunodetection of certain tumors in humans by external scintigraphy. PLAP may similarly be effective as a target for the delivery of toxic reagents to tumor cells *in vivo* using drugs conjugated to PLAP-specific MAbs. Indeed, again using the F11 MAb to PLAP conjugated to the A chain of ricin, Tsukazaki et al. (1985) demonstrated high toxicity and specific uptake of the immunotoxin at low concentration of the antibody on cells in culture and suggested that PLAP could be an effective target for immunotoxin therapy of cancers.

MAb 3F6, $F(ab')_2$ and Fab fragments were evaluated in similar xenograft experiments. MAb 3F6 showed good tumor retention and satisfactory specific/nonspecific ratios. As expected, fragments showed much faster blood clearance rates than the whole antibody. For Fab the *in vivo* instability by 6 h was also demonstrated (Durbin et al., 1988). Similarly, three MAb to PLAP and their Fab and $F(ab')_2$ fragments were evaluated for tumor immunolocalization of human PLAP-producing Hela HEp2 tumors in nude mice. The antibodies and their fragments were labeled with ^{125}I and injected intraperitoneally in mice with developing HEp2 tumors. The animals were followed individually for 14 days with repetitive computerized gamma-camera recordings of the time-dependent antibody uptake in the tumors, decrease in background activity and tumor/background ratio. Excellent radioimmunolocalization was obtained with both the intact PLAP-specific immunoglobulins and their fragments but not with the nonspecific antibodies. No background subtraction had to be used. As much as 15% of the initially injected dose could be visualized in the tumors and for the native MAb up to 80% of the radioactivity in the animals was retained in the tumors after 14 days, a considerably longer observation time than usually reported in such tumor xenograft models. The Fab and $F(ab')_2$ fragments were found to be excreted fast with <5% of the injected dose remaining in the animals after 48 h, but still with positive specific localization to the tumors after an initial high uptake in the kidneys (Stigbrand et al., 1989). Subsequently, the radioimmunotherapeutic potential of the ^{131}I-labeled H7 MAb was investigated in nude mice (BALB/c *nu/nu*) inoculated s.c. with the HEp2 cells. Significant growth inhibition was observed after injection of the radiolabeled H7 antibody, which reduced the tumor growth to only 12% during a 3-week period compared with a growth of more than 100% for the controls (Riklund et al., 1990). In another study, two MAbs against PLAP and liver-type TNAP were administered intravenously to nude mice bearing HEp2 tumors. A biodistribution study showed that the percentage of the injected dose of ^{125}I-labeled anti-PLAP in the xenografts was fairly constant at around 7% until 7 days after injection, whereas the percentage of the injected dose of ^{125}I-labeled anti-TNAP MAb decreased with time as in other tissues which do not contain significant amounts of PLAP or TNAP. On scintiscan, the xenografts in nude mice were distinctly visualized at 7 days after injection of anti-PLAP MAb (Koshida et al., 1996). In another study, HEp2 tumors were clearly visualized as early as 1 day after injection of ^{125}I-labeled H7 (Fig. 46). The remaining radioactivity was located exclusively in the tumors at days 30–81. As much as 12–16% of the injected dose per gram accumulated in the tumors during the first 2 days after injection and remained stable at this high level for ~10 days in all investigated groups (Rossi Norrlund et al., 1997). To evaluate further the ability of an anti-PLAP MAbs to localize to PLAP-expressing tumors, Koshida et al. (1998) established a model of testicular tumor with metastasis to lymph nodes and liver in severe combined immunodeficient (SCID) mice. ^{131}I- or ^{125}I-labeled MAb were simultaneously administered via the intravenous and lymphatic route, respectively. Preferential accumulation of MAb in PLAP-expressing tumors at both the primary and metastatic sites was demonstrated. The percentage of the injected dose of MAb found in the tumor was generally higher when MAb was administered intravenously. Identical tumor/blood ratios

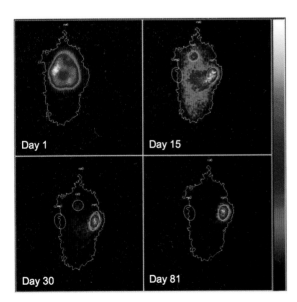

Fig. 46 Radioimmunolocalization of PLAP-expressing tumors. Radioimmunoscintigraphy of human cervix adenocarcinoma HeLa HEp2 tumors in nude mice, using ^{125}I-labeled H7 MAb. Four images of a mouse at days 1, 15, 30 and 81 after a single injection of radiolabeled H7 MAb. The tumor is clearly visualized already on day 1 but the radioactivity is mainly localized in the circulation. After day 30, the radioactivity is found in the tumor and is still detectable in the tumor on day 81. Figure kindly provided by Dr Rossi Norrlund, University Hospital, Umeå, Sweden.

were found with the two routes of administration. These data suggest that intravenous administration of a radiolabeled MAb is superior to lymphatic administration for tumor imaging and radioimmunotherapy (Koshida et al., 1998). In another report, Barka et al. (2000) established an Ehrlich ascites tumor cell line stably expressing PLAP. The mice bearing the PLAP-positive ascites tumor were treated with a mouse MAb to human PLAP or with a control antibody. The average survival of mice treated with the control antibody was 16.4 ± 1.1 days, whereas those treated with anti-PLAP MAb had an average survival of 23.3 ± 5.7 days (Barka et al., 2000).

Considerable effort has been focused on developing adequate tumor models in immunocompetent animals, as opposed to xenografted nude mice, to test immunolocalization and immunotherapy strategies. In one such animal model, the target tumor cells were contained in intraperitoneal diffusion chambers with micropore membrane walls that were permeable to molecules, including the cell specific MAbs but impermeable to cells (Fjeld et al., 1990). Thus, the tumor cells were protected from the host immunocompetent cells. These diffusion chambers were filled with the HEp2 or OHS sarcoma cell line and the MAb preparations injected i.v., i.e. a ^{125}I-labeled Fab fragment of the PLAP-specific antibody H7 or a ^{125}I-labeled F(ab')$_2$ fragment of the sarcoma specific antibody TP-1. Specific targeting of the human tumor cells was demonstrated in both mice and pigs. The target/blood ratios were comparable in the two species, reaching a maximum of about 15 after 24 h with the Fab preparation and a ratio of 25 after 72 h with the F(ab')$_2$ (Fjeld et al., 1990). This same artificial tumor model based on diffusion chambers filled with antigen-coated polymer particles, implanted i.p. in normal immunocompetent mice, was used to estimate the *in vivo* value of the association constant (K_a) in an experimental targeting reaction. The MAb H7 with specificity for PLAP was again chosen for this experiment. Each mouse carried two diffusion chambers, one filled with PLAP-coated particles and a second control chamber with the same amount of uncoated particles. The chambers contained escalating doses of particles, ranging from 0.1 to 16 mg per diffusion chamber, with groups of 6–12 animals per dose level. The day after the implantation, a constant dose of ^{125}I-labeled Fab fragments of H7 was injected i.v. in each mouse. The association constant K_a as measured from the binding data obtained *in vivo* was not significantly different from the value measured *in vitro* when the same target chamber were incubated with the [^{125}I]Fab in test tubes. This indicates that *in vivo* impairment of the antibody avidity is not the reason why a relatively low tumor uptake is generally experienced in immunotargeting studies (Fjeld et al., 1992).

In another approach, Narisawa et al. (1993) generated a series of transgenic mouse lines harboring the entire human GCAP gene linked to progressively longer sequences of flanking DNA. A 5' sequence of 1.7 kb directed GCAP expression to the spermatogenic lineage and to the 8-cell through blastocyst stage of preimplantation mouse embryos. The expression of GCAP in these FVB/N transgenic mice induced a cellular immune tolerance to GCAP. When mouse fibrosarcoma MO$_4$ cells (C3H-derived), stably transfected with the cloned GCAP gene, were injected s.c. in nontransgenic control [C3H × FVB/N] hybrid mice, GCAP positive tumor cells were rejected. However, when GCAP-expressing transgenic [C3H × FVB/N] hybrid mice were challenged with these cells, GCAP-positive tumors developed. Tumors also developed in the transgenic hybrid mice on injection of MO₄ cells transfected with the homologous PLAP cDNA in spite of the presence in PLAP of 10 amino acids that are different from the corresponding residues in GCAP. It was predicted that these transgenic mice should enable the evaluation of the therapeutic potential of bispecific antibodies for T cell recruitment and destruction of GCAP/PLAP-producing tumor cells. In a subsequent study, Smans et al. (1995) used these GCAP-transgenic mice to test the effectiveness of bispecific MAb (BsMAb) therapy of experimental tumors. The anti-PLAP/GCAP/anti-mouse CD3 BsMAb 7E8 × 7D6, previously shown to induce efficient dose-dependent T-cell proliferation and PLAP-positive tumor cell lysis in the presence of recombinant IL-2 (Smans et al., 1991), and the anti-mouse CD3 MAb 7D6 were used in *in vivo* lysis experiments targeting GCAP-positive tumors grown in GCAP-expressing transgenic mice. Mice received intravenous injections twice per week with PBS (group 1) or with 10 μg of the BsMAb 7E8 × 7D6, either alone (group 2) or combined with 1 μg of the anti-CD3 Ab 7D6 (group 3), starting 7 days after the tumor inoculation. A fourth group received a local treatment with mouse splenocytes precoated with 10 μg of 7E8 × 7D6 and 1 μg of 7D6. Between BsMAb injections, groups 2, 3 and 4 received 10 units of recombinant IL-2 (i.v.) each day (Fig. 47). Two weeks of treatment with the BsMAb, either alone or combined with 7D6, resulted in a significant decrease in GCAP-positive tumor cells in groups 2

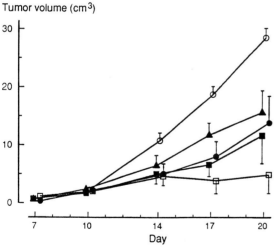

Fig. 47 Bispecific monoclonal antibody therapy of solid tumors in an immunocompetent murine model. Volume of tumors (mean; bars, SEM) developing after the *in vivo* lysis experiment. MO₄ tumor cells (10⁶ GCAP⁻) (○) were injected in the left thigh of all mice and 2.5 × 10⁶ GCAP⁺ MO₄-positive cells were injected in the right thigh of all animals. Group 1, non treated (▲); group 2, treated with 10 μg 7E8 × 7D6 BsMAb i.v. (■); Group 3, treated with 10 μg 7E8 × 7D6 BsMAb and 1 μg 7D6 (●); group 4, treated with 10 μg 7E8 × 7D6 BsMAb, 1 μg 7D6 and preactivated effector cells, s.c. in the right thigh (□). Taken from Smans et al. (1995) and reproduced with permission from *Cancer Research*.

and 3, although tumor volumes were not significantly different. Apparently, the elimination of GCAP-positive cells from the tumor seemed to favor conditions enabling the outgrowth of the few GCAP-negative cells originally present in the tumor inoculate. In contrast, tumor volumes in group 4 (local treatment) were significantly smaller compared with the nontreated group, probably owing to the presence of larger amounts of BsMAb and infiltrated activated T cells capable of secreting cytostatic cytokines such as tumor necrosis factor-α and interferon-γ as compared with groups 2 and 3. This study demonstrated that a BsMAb could specifically concentrate cytotoxic T cells into a solid tumor *in vivo*, with subsequent elimination of the targeted tumor cells (Smans et al., 1995).

However, some GCAP-negative tumor cells were still able to grow, suggesting that BsMAb therapy, when used in a clinical setting, could benefit from targeting several tumor markers to prevent outgrowth of tumor cells lacking a targeted marker. To test this hypothesis, Smans et al. (1999) developed an *in vitro* model based on primary human ovarian carcinoma (OC) cultures and BsMAbs directed against human T cells and several tumor markers, i.e. PLAP, GCAP, folate-binding protein (FBP) and CA19.9. OC cells, isolated from primary tumors, were co-cultured with human peripheral blood mononuclear cells in the presence or absence of various concentrations of BsMAbs against PLAP/GCAP, FBP and CA19.9 administered separately or in combination. Results derived from 18 primary OC samples showed that the combination treatment was better than or equally effective as the best single BsMAb treatment in 60% of cases. Sometimes targeting FBP, PLAP/GCAP or CA19.9 alone was superior to targeting all simultaneously. Combining each BsMAb with a low dose of IL-2 was always beneficial. These results indicated that before using a particular BsMAb in the clinic, it is important to determine the optimal BsMAb for each patient using such an *in vitro* assay on cells from the removed tumor mass (Smans et al., 1999).

Radiolabeled MAbs, with specificity against PLAP/GCAP, have also been used fairly extensively for tumor localization in human patients with different malignancies, with encouraging results. Radiolabeled MAb NDOG2 directed against PLAP was used in the radioimmunodetection of ovarian carcinoma (Davies et al., 1985b). Tumors were successfully visualized in 11 of 15 patients and the abnormalities demonstrated were classified as focal or diffuse. Of 11 patients, eight showed focal abnormalities alone and three had a diffuse abnormality, of which two also showed a focal abnormality. False-positive results may occur not only due to uptake of ^{123}I by gut mucosa and an inadequately blocked thyroid gland but also from activity in an incompletely emptied bladder. A false-negative result occurred due to high background activity in the liver masking a known, discrete tumor deposit (Davies et al., 1985b). The MAb H17E2 against PLAP and GCAP, labeled with ^{111}In, was used in radioimmunoscintigraphy of 15 patients known or suspected to have a TGCTs or carcinoma of the ovary or cervix (Epenetos et al., 1985b). Good images of neoplastic lesions were obtained in most patients with active disease. In one patient with testicular teratoma and elevated HCG, who had a normal CT scan, the labeled antibody located microscopic disease in a lymph node, which was then removed. No false-positive localization was seen in patients with PLAP-negative tumors or sites of inflammation. Subsequently, both MAbs H317 and H17E2 were radiolabeled with ^{123}I or ^{131}I and used in a prospective study of patients with TGCTs (13 cases) or epithelial-origin neoplasms of the ovary (13 cases). The data indicated that a positive antibody scan indicates the definite presence of a tumor, although a negative antibody scan does not always exclude the presence of disease (Epenetos et al., 1986). Positive results were also obtained using the anti-PLAP antibodies HMFG-2 and H17E2 in patients with ovarian carcinomas (Kalofonos et al., 1990). Also, 27 patients with brain gliomas were scanned using ^{123}I-labeled MAbs against epidermal growth factor receptor and PLAP (H17E2). Successful localization was achieved in 18 out of 27 patients. Thus, 10 patients with recurrent grade III or IV glioma who showed good localization of radiolabeled MAb were treated with 40–140 mCi of ^{131}I-labeled antibody delivered to the tumor area intravenously ($n = 5$) or by infusion into the internal carotid artery ($n = 5$). Six patients showed clinical improvement lasting from 6 months to 3 years. No major toxicity was attributable to antibody-guided irradiation (Kalofonos et al., 1989). The same H17E2 MAb to PLAP was subsequently radiolabeled with ^{111}In and ^{123}I and administered intravenously in 33 patients with primary and/or metastatic TGCTs, and also in eight patients who were in complete remission after surgical excision of the tumor (Kalofonos et al., 1990). The presence of a tumor was confirmed and correlated well with conventional diagnostic techniques and, in addition, the antibody scan revealed the presence of active disease in two patients with negative conventional imaging and with elevated serum markers. All patients studied with ^{111}In-labeled MAb had observable concentrations of the radiolabel in the liver (estimated to be ~30% of the administered dose), and also in the kidneys and spleen. The patients studied with the ^{123}I-labeled MAb had observable concentrations in the thyroid gland and the stomach. The best images were seen at 48 and 24 h after the indium- and iodine-radiolabeled antibody respectively. No toxicity was encountered in any of the patients in 4 months of follow-up (Kalofonos et al., 1990). Another study detected a subcutaneous metastasis by external radioimmunoscintigraphy in a patient with gestational choriocarcinoma using the ^{131}I-labeled H17E2 antibody (Athanassiou et al., 1990). The successful detection of occult metastatic deposits in choriocarcinoma, which have become resistant to chemotherapy, may be of value for the potential use of curative surgery in these patients. Two patients with TGCTs were subjected to radioimmunotherapy by using the ^{131}I-radiolabeled anti-PLAP MAb H17E2. Both patients had been previously treated with repeated chemotherapy regimens assisted by autologous bone marrow transplant that were unsuccessful. Radioimmunotherapy was well tolerated and the targeting of multiple neoplastic lesions was satisfactory (Riva et al., 1990). Hence the usefulness of PLAP/GCAP as target antigens for immunodetection and immunotherapy of tumors is rather well established.

10.6
Tumoral Expression of IAP

In the 1970s, Higashino and co-workers reported the presence of an intestinal-like AP variant in patients with hepatocellular carcinoma and named this variant Kasahara isozyme after the first patient's name (Higashino et al., 1972, 1974, 1975a, 1977). The presence of the Kasahara isozyme has been detected in tissue samples from renal cell carcinoma (Higashino et al., 1975a; Hada et al., 1979) and in amnion-derived cell lines (Higashino et al., 1975b). Structurally, the Kasahara variant has been shown to be a heterodimer of PLAP and IAP monomers by the use of isozyme-specific MAbs (Imanishi et al., 1990b). Through cDNA cloning of the FL-amnion enzyme, Higashino et al. (1990) revealed the complete identity of the FL-amnion cDNA with that of adult IAP and have presented mRNA data demonstrating the presence of both PLAP and IAP transcripts in the FL amnion cell line. In another report, the tumor-derived fetal IAP cDNA was shown to be identical in sequence with the adult IAP isozyme gene (Fukui et al., 1997), thus ruling out peptide differences and leaving only variation in carbohydrate content as the basis for phenotypic differences between the fetal and adult IAP isoforms.

Although APs are normally homodimeric molecules, the re-expression in cancer cells of more than one AP isozyme often results in the formation and release into body fluids of heterodimeric enzymes, although the human postnatal intestine also contains heterodimers of IAP and PLAP (Behrens et al., 1983). Ovarian cancer cells often express both PLAP and GCAP (Nozawa et al., 1990; Tholander et al., 1990; Smans et al., 1999) and cell lines derived from these tumors have been shown to express PLAP/GCAP heterodimers (Watanabe et al., 1989; Hendrix et al., 1990). These marker genes respond differently to therapy (van de Voorde et al., 1985a; Mano et al., 1986b; Hirano et al., 1987a; Vergote et al., 1987; Stendahl et al., 1989; Hamilton-Dutoit et al., 1990; Latham and Stanbridge, 1990; Nozawa et al., 1990; Stigbrand et al., 1990). If GCAP and PLAP are expressed in the same cancer cell, heterodimeric PLAP/GCAP molecules are possible, as shown above for the Kasahara isozyme heterodimer, and may be present on the cell surface in ratios that vary according to therapy. Understanding how these heterodimeric molecules differ in their functional and structural properties from the homodimers would help in the accurate quantitation of these markers during the clinical follow-up of cancer patients. It has also been reported that in addition to GCAP, IAP and TNAP are elevated in seminoma tissue. It is suggested that the entire genomic region coding for AP may undergo activation in this circumstance (Hirano et al., 1987a). Despite the common occurrence of heterodimers between TSAP isozymes, only one report has documented the presence of abnormal IAP/TNAP heterodimers in amniotic fluid samples of patients with Down syndrome (Vergnes et al., 2000). The fact that APs can form heterodimers is of structural and possibly functional significance since APs are noncooperative allosteric enzymes where the stability and the catalytic properties of each monomer are controlled by the conformation of the second subunit (Hoylaerts et al., 1997). This means that the properties of the heterodimeric enzymes do not correspond to the weighted average of each homodimeric counterpart.

Several different mechanisms underlying AP expression in tumor cells can be envisioned: (1) A functional involvement of AP isozymes in tumorigenesis; (2) AP expression may represent one factor in a multifactorial etiology; (3) AP expression may represent a close linkage of the AP gene with a disease susceptibility gene; (4) AP may be simultaneously deregulated with the disease susceptibility gene. Alternatively, (5) the expression of AP isozyme genes could be the result of random chromosomal aberrations. Latham and Stanbridge (1990) provided experimental evidence to support the hypothesis that in some cases an AP isozyme gene can be deregulated simultaneously with a disease susceptibility gene. An exclusive correlation was found between the ectopic expression of the IAP isozyme (Kasahara isozyme) and the tumorigenic phenotype in segregants derived from suppressed, nontumorigenic HeLa × fibroblast cell hybrids. This specific association suggested that the loss of a tumor suppressor on chromosome 11 was closely linked to the re-expression of IAP (Latham and Stanbridge, 1990). However, high IAP expression alone was not sufficient to confer rapid tumor growth (Mendonca et al., 1991). The authors concluded that although ectopic IAP expression is unlikely to be functionally relevant to tumorigenicity in these hybrids, the significance of IAP as a tumor marker is still evident from its apparent strong association with a tumor suppressor locus (Latham and Stanbridge, 1992).

Part IV
Uses of APs in Industry and Biotechnology

11
Applications of Recombinant APs

11.1
Expression of Recombinant APs

The expression system that has been by far the most often used to express mammalian APs *in vitro* involves transfecting expression constructs into mammalian cells. The first such report was by Berger et al. (1987b), who transiently expressed the human PLAP cDNA in simian COS cells. The level of PLAP expression was high and it was produced in an enzymatically active form. The bulk of PLAP was associated with the cell membrane, as shown by immunocytochemistry and subcellular fractionation studies. The PLAP produced by these transfected COS cells, like the PLAP of human tissues, was specifically released from the intact cells in a hydrophilic form by GPI-PLC. Soon afterwards, Weiss et al. (1988) used this system to test the first missense mutation, an A162T substitution in exon 6, of the *ALPL* gene that caused hypophosphatasia (Weiss et al., 1988). Narisawa et al. (1990) also used COS-1 cells to examine the consequence of introducing a D91N substitution in PLAP. The mutation abolished catalytic activity but preserved immunoreactivity of PLAP, indicating profound alterations in the active-site environment but no apparent change on the surface of PLAP. That was the first in a large series of structural/functional studies mutagenizing and comparing the behavior of PLAP and TNAP mutations with reference to the crystal structure of ECAP and PLAP. The mammalian cells used for these expression studies were Chinese hamster ovary cells (CHO) for stable expression (Hoylaerts and Millán, 1991; Hummer and Millán, 1991; Bossi et al., 1993; Weissig et al., 1993; Hoylaerts et al., 1997; Manes et al., 1998) and other investigators have used COS-1 cells for transient transfection studies (Orimo et al., 1994; Goseki-Sone et al., 1998; Mornet et al., 2001). Recently, the author's laboratory has resorted to transiently expressing PLAP and TNAP mutant enzymes in COS-1 or COS-7 cells but after replacing the GPI-anchoring sequence of the enzymes with a synthetic epitope tag. The introduction of the eight amino acid FLAG tag sequence (DYKDDDDK) and a premature stop codon at position 489 into the wt TNAP or PLAP cDNA facilitates expression, recovery and purification of large amounts of mutant protein needed for kinetic experiments at physiological pH, where the requirements for recombinant enzyme increases 20–100-fold for most mutants (Di Mauro et al., 2002; Wennberg et al., 2002). The kinetic parameters determined for released soluble-

Mammalian Alkaline Phosphatases: From Biology to Applications in Medicine and Biotechnology. J. L. Millán
Copyright © 2006 WILEY-VCH Verlag GmbH & Co. KGaA, Weinheim
ISBN: 3-527-31079-7

epitope tagged (*set*) and membrane-bound enzymes have proven comparable, so that there are no major concerns with respect to the kinetic characteristics of the resulting mutants. However, Magnusson et al. (2002) found that the binding of MAb 333 to TNAP was affected by the presence of the attached FLAG tag sequence in *set*TNAP. It should be stated that we opted for a FLAG tag and not for the more widely used His-tag system out of concern that the high affinity of this extraneous stretch of His residues would interfere with proper binding and saturation of the intrinsic Zn1 and Zn2 metal sites in the active-site pocket in some of the TNAP or PLAP mutant enzymes. This strategy using transient expression of *set*PLAP and *set*TNAP has allowed a large number of mutagenesis studies and remain our favorite conditions for *in vitro* expression to date (Kozlenkov et al., 2002, 2004).

However, although mammalian expression systems are suitable for research purposes, they are costly to scale up for production of recombinant proteins for biotechnological uses. Fortunately, progress has also been made expressing mammalian APs in other systems. TNAP and PLAP have been expressed successfully in insect cells. A soluble form of TNAP was purified to apparent homogeneity from the culture media of *Spodoptera frugiperda* Sf-9 insect cell line following infection with a recombinant *Autographa californica* multiple nuclear polyhedrovirus containing the TNAP cDNA gene under control of the polyhedrin promoter (Oda et al., 1999). To facilitate purification, an oligonucleotide consisting of six tandem codons for His and a stop codon was engineered into the TNAP cDNA. The enzyme was used for the production of antibodies specific for human TNAP (Oda et al., 1999). Using a similar system, a secreted form of human PLAP (SEAP) was also produced (Zhang et al., 2001). An affinity chromatographic column prepared by linking 4-aminobenzylphosphonic acid to histidylepoxy-Sepharose was used to isolate the SEAP from the cell supernatant following removal of cells and virus and 10-fold concentration through ultrafiltration. Marked alterations to *N*-linked glycosylation of human SEAP were observed with different baculovirus species, insect cell lines and cell culture media (Joshi et al., 2000). When a recombinant *Autographa californica* nucleopolyhedrovirus (AcMNPV) was used to produce SEAP in *Trichoplusia ni* (Tn-4h) cells cultured in serum-free medium, structural analyses indicated <1% hybrid and no complex oligosaccharides attached to SEAP, a typical result with the baculovirus expression vector system. However, when fetal bovine serum was added to the culture medium, 48 ± 4% of the oligosaccharides were hybrid or complex (but asialylated) glycans. When a recombinant *T. ni* nucleopolyhedrovirus (TnSNPV) was similarly used to express SEAP in Tn-4h cells cultured in serum-containing medium, only 24 ± 3% of the glycans contained terminal *N*-acetylglucosamine and/or galactose residues. In contrast, SEAP produced in Sf-9 cells grown in serum-containing medium with AcMNPV contained <1% hybrid oligosaccharides and no complex oligosaccharides. The results illustrate that the type of baculovirus, type of host cell and the growth medium all have a strong influence on the glycosylation pathway in insect cells, resulting in significant alterations in structures and relative abundance of *N*-linked glycoforms. In a subsequent report, SEAP was expressed in Sf-9 and also

Trichoplusia ni BTI-Tn-5B1–4 insect cell lines. The recombinant SEAP expression level was 7.0 U mL^{-1} in Tn-5B1–4, higher than the 4.1 U mL^{-1} produced by the Sf-9 cells. Kinetic analysis showed that V_{max} and K_m of the recombinant SEAP were ~10-fold lower than the values of the wt PLAP. Glycan analysis showed the presence of oligomannose-type *N*-linked glycans, i.e. Man(2–8)GlcNAc2 and Fuc-Man(3 or 4)GlcNAc2, in recombinant soluble PLAP produced in the Sf-9 and Tn-5B1–4 cell lines. The proportions of these oligosaccharide structures were different in the two cell lines. Man4GlcNAc2 and FucMan4GlcNAc2 were the major SEAP *N*-glycans produced in Sf-9 cells, whereas Man2GlcNAc2 was the major SEAP *N*-glycan produced in Tn-5B1–4 cells (Zhang et al., 2002a). Joosten and Shuler (2003) studied the effects of culture conditions on the degree of sialylation of SEAP expressed in insect cells. They found that when cultured in a high aspect ratio vessel or in tissue culture flasks, baculovirus-infected Tn-4s cells produced high levels of SEAP but under those conditions SEAP possessed only high-mannose, paucimannosidic and hybrid structures. In spinner flasks, lower SEAP yields were obtained but in such cultures, sialylation of SEAP could be achieved. Several spinner-flask culture conditions were tested and resulted in different SEAP specific yields and levels of sialylation. The highest level of sialylation (9%) was obtained in the culture with the lowest agitation rate and lowest yield (1.2 U/10^6 cells), suggesting a limiting capacity of the Tn-4s cells to process glycoproteins to sialylation. High specific yield, low passage number Tn5B1–4 cells did not produce SEAP with complex glycosylation when cultured in a low agitation rate spinner flask. In contrast to this level of sialylation, intermediate complex forms with terminal galactose or *N*-acetylglucosamine were found in low proportions (<3 and <1%, respectively) (Joosten and Shuler, 2003). Hence the lack of complex glycosylation has limited the use of the insect cell baculovirus expression vector system, despite its high productivity and versatility. Palomares et al. (2003) explored the capability of two novel cell lines, one from *Pseudaletia unipuncta* (A7S) and the other from *Danaus plexippus* (DpN1), to produce and glycosylate SEAP. Both the A7S and the DpN1 cells produced lower concentrations of SEAP than the Tn5B1–4 cells. Less than 5% of the glycans attached to SEAP produced by the Tn5B1–4 cells had complex forms. Glycans attached to SEAP from A7S cells contained 4% hybrid and 8% complex forms. Galactosylated biantennary structures were identified. Glycans attached to SEAP produced by the DpN1 cell line had 6% hybrid and 26% complex forms. Of the complex forms in SEAP from DpN1, 13% were identified as sialylated glycans. Even though neither novel cell line produced as much recombinant protein as the Tn5B1–4 cells, the glycosylation of SEAP expressed by both cell lines was more complete.

Recently, Estrada-Mondaca et al. (2005) assessed the effects of supplementation with mannosamine, *N*-acetylmannosamine and cytidine on the glycosylation of recombinant SEAP produced by suspension cultures of *Trichoplusia ni* (cabbage looper) BTI-Tn5B1–4 cells. Addition of mannosamine in the range 5–20 mM resulted in a 10-fold increase of the terminal GlcNAc associated with the recombinant protein produced after baculovirus infection. Such an increase yielded a maximum of 12.5% hybrid glycans having terminal GlcNAc with respect to total

N-linked glycans. In contrast, no changes in the glycan composition associated with SEAP were observed on supplementation with up to 20 mM N-acetylmannosamine or up to 1.5 mM cytidine.

A third system that has been used successfully is the methylotrophic yeast *Pichia pastoris*. A soluble form of human PLAP was expressed and the expression product was purified and characterized (Heimo et al., 1998). Yeast-derived PLAP was secreted into the medium to the level of 2 mg L^{-1}. The yeast recombinant PLAP displayed kinetic properties similar to those reported earlier for the membrane-bound PLAP, had specific activity of 774 U mg^{-1} and appeared in two subunit sizes, ca 62 and 65 kDa. This difference was due to heterogeneous N-glycosylation. Purified yeast-derived PLAP appeared as multiple forms in isoelectric focusing in the p*I* range 4.2–5.2. *Picchia pastoris* was also the system chosen by Roche Diagnostics to bring to the market the first recombinant mammalian AP, i.e. bIAP II, which our laboratory cloned and initially expressed using the Chinese hamster ovary expression system (Manes et al., 1998). The yeast-expressed bIAP II has kinetic and stability properties comparable to those of the recombinant enzyme produced in mammalian cells. However, the carbohydrate content of the recombinant enzyme is higher (31.4 versus 5–7% for the native enzyme), since expression in yeast leads to higher glycosylation – by a factor of 5 – than that found in nature (Bretthauer and Castellino, 1999).

Even plants have been used to express PLAP. Kormarnytsky et al. (2000) generated transgenic tobacco (*Nicotiana tabacum* L. cv Wisconsin) plants that secrete human PLAP through the leaf intercellular space into tobacco guttation fluid. Production rates of 1.1 µg g^{-1} of leaf dry weight per day were achieved for PLAP with this protein comprising almost 3% of total soluble protein in the guttation fluid. Guttation fluid can be collected throughout a plant's life, providing a continuous and nondestructive system for recombinant protein production, thus increasing yield, abolishing extraction and simplifying the downstream processing of the recombinant protein.

Surprisingly, many attempts to express mammalian APs in *Escherichia coli* have not yielded satisfactory results, although those data, including our own, have remained unpublished. However, Beck and Burtscher (1994) succeeded in expressing a shortened form of the human PLAP, lacking the 29 C-terminal amino acids that constitute the GPI anchoring signal in *E. coli* at about 5% of total protein. Most of the enzyme was present in an insoluble form, however, but soluble enzyme was detected in Western blots and activity tests. The protein was located in the periplasm since its signal peptide was cleaved off. Co-expression of potential folding aids like peptidylprolyl *cis–trans* isomerase or disulfide isomerase did not lead to an increase in soluble enzyme, either when overexpressed separately or from an operon (Beck and Burtscher, 1994). In a subsequent report, Beck et al. (1994) tested the expression of PLAP in a mutant strain of *E. coli* deficient in *dipZ*, a gene coding for a protein involved in cytochrome *c* biogenesis and the isogenic wt strain. The yield of soluble and active PLAP was significantly reduced in the mutant but could be fully recovered by expression of *dipZ* subcloned in a vector with low copy number.

11.2
APs as *In Vitro* and *In Vivo* Reporters

Three papers published in 1988 described the use of APs as reporter enzymes for a wide variety of *in vitro* applications. Yoon et al. (1988) constructed a plasmid containing the rat *ALPL* cDNA under the control of the simian virus 40 (SV40) early promoter and used it to transfect Chinese hamster ovary, SV40-transformed African Green Monkey kidney 7 and rat osteosarcoma 25/1 mammalian cells. AP activity in these cells, measured 3 days later, was 40–400-fold above background. When AP and chloramphenicol acetyltransferase (CAT) plasmids were cotransfected, the detection of AP activity was at least as sensitive as the detection of CAT activity using a radioactive substrate. Moreover, since TNAP was expressed as a membrane-bound ectoenzyme the transfected cells could be visualized by histochemical staining. Henthorn et al. (1988b) constructed the plasmid pSV2Apap containing a 4130 bp *Eco*RI to *Sca*I restriction fragment of the human *ALPP* gene inserted immediately downstream of a SV40 early promoter and upstream of the SV40 small tumor antigen intron and SV40 early poly(A) signal. Upon transfection into a variety of different cell types, PLAP activity could readily be detected by using whole cell suspensions or cell lysates. AP activity could also be visualized directly in individual transfected cells by histochemical staining. Cotransfections of cells with pSV2Apap and a related plasmid carrying the bacterial CAT gene (pSV2Acat) indicated that transcription of these two genes was detected with roughly the same sensitivity. Berger et al. (1988) pioneered the use of secreted PLAP (SEAP) as a marker for *in vitro* and *in vivo* detection. In transient expression experiments using transfected mammalian cells, these authors demonstrated that SEAP yields results that are qualitatively and quantitatively similar, at both the mRNA and protein levels, to parallel results obtained using established reporter genes. However, SEAP offered significant advantages in terms of ease of assay and assay expense and also has the potential for quantitative assay at levels as low as $0.2\,\text{pg}\,\text{mL}^{-1}$ of culture medium. These attributes suggested that SEAP could have general utility in experiments that rely on the accurate measurement of reporter gene expression levels as indeed has been the case as indicated by the widespread use of this approach.

Another application of this principle, published by Tate et al. (1990), involved using the plasmid pBC12/PLAP489 described previously (Berger et al., 1988) to transcribe SEAP mRNA *in vitro* and using the mRNA as an internal, coinjected standard to monitor translation of mRNAs in *Xenopus laevis* oocytes. The injection of as little as 1 ng of SEAP mRNA/oocyte resulted in the secretion of measurable amounts of SEAP activity in the medium. The procedure to express and detect SEAP activity has been described in detail (Cullen and Malim, 1992) for use primarily in the identification and functional dissection of the cis-acting sequences and trans-acting factors that regulate eukaryotic gene expression *in vivo* and as an alternative to using chloramphenicol acetyltransferase (CAT) that required radioactive substrates for detection. The stated advantages of using SEAP include: very high stability, efficient secretion by all cells tested and the availability of a simple, inexpensive and highly quantitative assay that does not require any unusual equipment or reagents. Furthermore, PLAP is only found in higher primates, including humans, and can be readily distinguished from the more prevalent TNAP and IAP isozymes based on its resistance to inactivation by incubation at 65 °C and to the drugs levamisole and L-hArg so that these other AP isozymes do not interfere with measurement of SEAP activity even if they are present (Cullen and Malim, 1992; Cullen, 2000). However, the SEAP assay initially reported did have one disadvantage relative to CAT, i.e. lower sensitivity. The assay described by Cullen and Malim (1992) could not reliably quantitate levels of SEAP enzyme expression that fell significantly below 50 pg of protein per milliliter. That level of sensitivity was 10–50-fold lower than the sensitivity of radioactivity-based assays for CAT expression. However, Cullen (2000) later described the use of chemiluminescent assays for SEAP based on the use of the phosphatase substrate disodium 3-(4-methoxyspiro{1,2-dioxetane-3,2′(5′-chloro)tricyclo[3.3.1.1.]decan}-4-yl)phenyl phosphate. The chemiluminescent assay for SEAP has been reported to be linear over the range 0.1 pg–1 ng of SEAP per sample and is at least as sensitive as the commercially available assays for luciferase. Flanagan and Leder (1990) used SEAP as a tag to facilitate identification of the ligand for the *kit* cell surface receptor. The fusion of the extracellular domain of *kit* to the N-terminus of the mature SEAP protein resulted in the synthesis of a secreted, soluble affinity reagent, bearing a fully active SEAP enzyme tag that could be tracked easily and sensitively. Using this reagent, these authors were able to demonstrate specific binding of *kit* to certain cell lines by *in situ* staining for SEAP activity. This approach was subsequently generalized and described in detail to use SEAP fusions of ligands or receptors as *in situ* probes for staining of cells, tissues and embryos (Flanagan et al., 2000). In another application, SEAP fusion proteins are described for use in the molecular characterization and cloning of receptors and their ligands (Flanagan and Cheng, 2000). This receptor-AP staining method is especially useful in situations where a reliable MAb is not available or if an orphan receptor is the focus of study. The technique permits localization of both receptors and ligands and is readily quantifiable for cell-surface binding assays (Brennan and Fabes, 2003). Yet another application is the use of PLAP as reporter in signal peptide sequence traps to clone secreted and transmembrane growth factors and signaling molecules. By fusing a cDNA library upstream and in frame with the PLAP reporter and assaying for PLAP activity at the cell surface, one can identify cDNA clones encoding functional signal peptides. Chen and Leder (1999) tested this peptide signal trap principle using mouse prostate and human prostatic carcinoma and identified several secreted and transmembrane proteins.

Fields-Berry et al. (1992) created a retrovirus vector encoding PLAP and found that it was at least as useful as a lacZ-encoding retrovirus with respect to high viral titer, stability of expression and identification of infected cells *in vivo*. Moreover, it was found to be neutral with respect to postnatal rodent retinal development and offered superior staining characteristics relative to lacZ. In another report, two histochemical marker genes, the *Drosophila* alcohol dehydrogenase (ADH) and human PLAP, were cloned into the recombinant retroviral vectors pLJ and pgag β-actin (Schreiber et al., 1993). The resulting vectors were transfected into retroviral producer cell lines, psi CRE and psi CRIP, and stable recombinant retrovirus producers were isolated. Recombinant virus was harvested and used to transduce genes into several cell lines, singly or in conjunction with lacZ (*E. coli* β-galactosidase)-containing retrovirus. Cell lines were then stained using standard histochemical methods for recombinant gene expression. The authors found that multiple gene products could be identified in the same cell populations and, in the case of PLAP and β-galactosidase, in the same cells. Means et al. (1997) used SEAP to design cell lines that permit the rapid, sensitive and quantitative assay of human immunodeficiency virus type 1 and simian immunodeficiency virus infectivity. SEAP has also been used to test the efficacy of new transfection procedures (Durocher et al., 2002) and as a marker to manipulate cell growth to optimize expression of recombinant proteins (Carvalhal et al., 2003). Norton et al. (2005) recently used SEAP as marker in co-transfection experiments with the middle glycoprotein of the hepatitis B virus with the goal of identifying new more effective inhibitors of glucosidase I and II, as potential anti-viral drugs. The authors found that SEAP gene product was itself sensitive to glucosidase inhibition and could be used as a surrogate marker.

PLAP has also been used as a reporter gene *in vivo*. These applications have favored the use of native GPI-anchored PLAP rather than SEAP. A construct was made linking the PLAP structural gene to an enhancer-promoter element from the human β-actin gene (DePrimo et al., 1996). This gene was inserted into the mouse genome by transfection of embryonic stem (ES) cells and by microinjection of fertilized eggs. Histochemical staining showed that the transgene was uniformly expressed in all stable ES cell lines and in all tissues examined from adult animals from five different lines of transgenic mice. Nontransgenic cells did not stain. These results suggested that the human PLAP gene could be useful in studies requiring phenotypic marking of cells in tissues of mice (DePrimo et al., 1996). Li et al. (1997) reported the development of convenient dicistronic transgenic markers for the rapid and efficient simultaneous analysis of transgene activity *in vivo* in transgenic mice. The lacZ gene and the human PLAP gene were fused to the internal ribosome entry sequence (IRES) from the encephalomyocarditis virus, which directs efficient mRNA cap-independent entry of the translation apparatus in mammalian cells. The IRES permits efficient translation of either lacZ or PLAP when placed anywhere within transgene exonic sequences, including both 5′ and 3′ untranslated regions. In addition, the production of constructs for transgenic analysis of DNA regulatory elements is greatly facilitated with IRES-lacZ or IRES-hpAP, since the IRES relieves the need for complicated in-frame transgene protein fusions to produce a functional β-gal or PLAP protein. Zsengeller et al. (1999) used a recombinant amphotropic retrovirus that expresses human PLAP and instilled it intratracheally into the mice to measure the transduction efficiency of mature pulmonary epithelium with or without pretreatment with keratinocyte growth factor. In another application, De Primo et al. (1998) used human PLAP transgenes to detect somatic mutations in mice *in situ*. The PLAP gene sequence was modified such that it could no longer produce functional PLAP enzyme. Mutant PLAP genes were placed in the mouse genome and populations of cells carrying these mutant PLAP genes were studied to determine the fraction of cells that would acquire PLAP activity. Spontaneous and induced reversion of mutant PLAP genes was studied in cultured cells and in the tissues of transgenic mice (DePrimo et al., 1998). Through construction of transgenic mice, ubiquitously expressing human PLAP, Skynner et al. (1999) demonstrated the suitability of PLAP as a reporter gene for use in conjunction with or as an alternative to lacZ. Their findings demonstrated that over-expression of PLAP has no adverse effects on mouse development or viability, despite a widespread pattern of expression. In another study, Hiramatsu et al. (2005) compared the usefulness of SEAP versus luciferase as secreted reporter molecules *in vivo* and found SEAP to be a much better serum reporter than luciferase. Another elegant use of PLAP as an *in vivo* marker involves the generation of an Flp recombinase (FRT) indicator strain that can be used to mark with PLAP activity most cells in the developing and adult mouse. This indicator strain can delineate the morphology of Flp-expressing cells and their descendant lineages. Awatramani et al. (2001) targeted the broadly expressed ROSA26 (R26) locus with an FRT-disrupted PLAP gene. They refer to this Flp indicator as R26:FRAP (FRT-disrupted PLAP). The authors created a construct containing an FRT-disrupted PLAP transgene into the previously described R26 vector and incorporated features into this construct that ensured that PLAP served as a faithful indicator of Flp activity and is not expressed in the absence of Flp-mediated excision of the FRT-flanked stop cassette. Consequently, any transcriptional read-through followed by internal translational initiation (downstream of the engineered stop codons) would generate a truncated, inactive enzyme. The R26:FRAP indicator strain is an important addition to the versatile Flp-FRT system (Awatramani et al., 2001).

Developing new anticancer therapeutic regimens requires the measurement of tumor cell growth in response to treatment. This is often accomplished by injecting immunocompromised mice with cells from cancer tissue or cell lines. After treating the animals, tumor weight or volume is measured. Such methods are complicated by inaccuracies in measuring tumor mass and often animals must be killed to measure tumor burden. To facilitate these studies, SEAP has been used *in vivo* as a surrogate marker to monitor *in vivo* tumor growth and anticancer drug efficacy in cancer xenografts. Bao et al. (2000) used an SEAP expression construct under control of the cytomegalovirus (CMV) promoter to stably transfect the A2780 cell line that was subsequently implanted into immunocompromised mice. Upon detection of subcutaneous tumors ,the serum SEAP activity correlated well with tumor volume and the plasma SEAP level was reduced after xenografted mice were treated with paclitaxel compared with untreated mice in both subcutaneous and intraperitoneal tumor models (Bao et al., 2000). These data suggest that the plasma SEAP activity can be used as an alternative to survival or tumor measurement in evaluating anticancer agents for efficacy, especially in the case of minimal or inaccessible disease. In another similar study, Nilsson et al. (2002) transfected OCC1 ovarian carcinoma (OC) cells with pCMV-SEAP. The OCC1-SEAP cells were maintained *in vitro* to monitor the relationship between cell number and SEAP production and were injected *in vivo* to determine whether SEAP levels in blood corresponded to tumor burden. Both subcutaneous and intraperitoneal tumor volumes correlated well with plasma SEAP levels. Experiments were performed to determine whether measuring SEAP levels could be used to monitor OCC1 cell response to platinum-containing chemotherapeutic drugs. OCC1-SEAP cells cultured *in vitro* were treated with the platinum-containing drug carboplatin. Carboplatin treatment decreased both cell proliferation and SEAP levels in culture medium. The constitutive rate of SEAP secretion per cell (ng SEAP per µg DNA) was not altered by carboplatin treatment. Therefore, changes in SEAP level reflect changes in OCC1 tumor cell number and not changes in regulation of SEAP secretion due to platinum-containing chemotherapeutic drug treatment. The results of these studies indicate that SEAP may be used as an *in vivo* reporter gene in a mouse model to monitor tumor growth and response to therapeutics (Nilsson et al., 2002).

The improvement of gene therapy vectors would benefit from the availability of a reporter gene that can be used for long-term studies in immunocompetent laboratory animals. Immunodeficient animals have been used in order to study long-term transgene expression. However, immunodeficient animals are not appropriate for assessing the impact of the immune response elicited by the gene transfer vector on transgene expression. Human PLAP and its secreted form, SEAP, are very immunogenic and therefore only show transient expression in immunocompetent animals since the hosts mount an immune response that eventually suppresses expression of this marker. Wang et al. (2001) described a variation of the principle of SEAP that uses the murine ortholog secreted EAP (MUSEAP), instead of human PLAP, as a reporter in order to bypass the immunogenicity problem in immunocompetent mice. Given that EAP is expressed as a membrane-bound enzyme at the two-cell to blastocyst stage of preimplantation development and re-expressed in adult animals in trace amounts in the thymus, intestine and testis (Section 2.2), mice would be expected to be immunotolerant to MUSEAP. Indeed, a comparison between injected and electrotransferred plasmid DNA encoding SEAP and MUSEAP indicated that whereas SEAP activity reached undetectable levels 4 weeks after gene transfer, the MUSEAP activity was stable up to 6 months in C57BL6 or 1 year in Balb/c mice after gene transfer. The mice of each group (injected with saline, SEAP- or MUSEAP-encoding plasmids) remained equally healthy during the time frame of the experiment; the expression of MUSEAP did not elicit any obvious adverse affects. The authors demonstrated by gene transfer in skeletal muscle of immunocompetent mice that MUSEAP is efficiently secreted and detected in the bloodstream and that injection of an increasing dose of DNA leads to a dose-dependent increase in plasma MUSEAP activity. They also showed that the expression of MUSEAP under the control of a constitutive promoter is stable and that the activity of MUSEAP in the bloodstream reflects the changes in the transcription rate of its gene. These properties make MUSEAP an ideal reporter gene that can be used for somatic gene transfer into immunocompetent mice in order to study the impact of the gene transfer vector, of metabolic, developmental or environmental factors on long-term gene expression (Wang et al., 2001). Recently, Maelandsmo et al. (2005) explored the use of MUSEAP as a reporter gene in the context of an early region 1 (E1)-deleted adenovirus (Ad) vector. The level of transgene expression from Ad-MUSEAP was similar to that observed for an Ad vector encoding the human SEAP gene. In agreement with the data from Wang et al. (2001), Maelandsmo et al. (2005) also found that after intravenous administration in mice, Ad-MUSEAP continued to express at high levels for the duration of the experiment (1 month), whereas expression from Ad-SEAP declined to background levels over the course of the experiment. The authors found that whereas cytotoxic T-lymphocytes were not detected against either the murine or the human enzyme, antibodies were readily detected against human SEAP but not MUSEAP.

11.3
APs as Molecular Biology and Diagnostic Reagents

Undoubtedly the most common biotechnological use of APs is in enzyme-linked immunosorbent assays (ELISAs), where AP, usually bIAP, is used as the detection molecule conjugated to antibodies (O'Sullivan and Marks, 1981). The most common method of conjugation is that employing glutaraldehyde, a homobifunctional aldehyde that reacts with amino acid residues in proteins to form a Schiff base. Conjugation with glutaraldehyde can be performed in either a one- or two-step procedure (Wisdom, 2004). However, other techniques, using heterobifunctional cross-linking reagents, have also been developed to provide greater control of the coupling reaction. For example, Husain and Bieniarz (1994) described a procedure for the specific labeling of the Fc region of immunoglobulins with calf IAP. The procedure involved the production of thiol-modified antibodies using cystamine (typically containing three or four thiol groups in the Fc region) and conjugation to maleimide-functionalized calf IAP to form a conjugate that retains superior antigen-binding activity (Husain and Bieniarz, 1994). Once the antibody conjugate binds to its target antigen, the binding is visualized by developing AP activity using a wide variety of substrates, through either absorbance, fluorescence or luminescence methods. Turner (1986) described the use of conjugated antibodies in conjunction with histochemical staining using nitrocellulose acetate as support. The author optimized this technique using three different substrates, β-naphthyl phosphate/Fast Blue, 5-bromo-4-chloro-3-indolyl phosphate and 4-methylumbelliferyl phosphate. This principle has also been adapted for use in immunoblots (Ey and Ashman, 1986). Other proteins, besides antibodies, have also been used for conjugation with APs. An example is the use of streptavidin conjugated to bIAP for the detection of biotinylated proteins or DNA (Binder et al., 1995). Even insoluble peptides have been coupled to APs in the presence of urea (Gerritse et al., 1991) for use in ELISA, *in situ* detection of antibody-forming cells and for receptor–ligand studies.

APs can also be conjugated to DNA. In particular, short synthetic oligonucleotides have been covalently cross-linked to calf IAP using the homobifunctional reagent disuccinimidyl suberate and used as hybridization probes (Jablonski et al., 1986). The oligomers, 21–26 bases in length, were complementary to unique sequences found in herpes simplex virus, hepatitis B virus, *Campylobacter jejuni* and enterotoxic *E. coli*. Each oligomer contained a single modified base with a 12-atom linker arm terminating in a reactive primary amine. Cross-linking through this amine resulted in oligomer–enzyme conjugates composed of one oligomer per enzyme molecule that had full AP activity and could hybridize to target DNA fixed to nitrocellulose within 15 min. The hybrids were detected directly with a dye precipitation assay at a sensitivity of 10^6 molecules (2×10^{-18} mol) of target DNA in a development time of 4 h. The enzyme had no apparent effect on selectivity or kinetics of oligonucleotide hybridization and the conjugates could be hybridized and melted off in a conventional manner. Farmar and Castaneda (1991) described a simple, low-cost protocol giving good yields of oligonucleotide–AP conjugates on a 7 or 35 nmol scale of oligomer. The cross-linking agent was disuccinimidyl suberate and *n*-butanol was used to remove excess disuccinimidyl suberate and side products away from the disuccinimidyl suberate–oligomer adduct before AP was added directly to the dried adduct. The crude conjugate was purified in one step using a DEAE HPLC column and an NaCl gradient. These conjugates were used to detect 0.4 pg of a hepatitis B virus sequence using a chemiluminescent assay. A variation of this approach involved labeling DNA with a bifunctional reagent consisting of a DNA-linking group and a competitive inhibitor AP, rather than AP itself (Davini et al., 1992). The nucleic acids labeled in such a way are able themselves to bind to AP, whose activity is restored in the presence of a chromogenic substrate. Applications of this principle to dot-blot procedures yielded a detection sensitivity of 25 pg. Benzinger et al. (1995) evaluated the specificity of AP-conjugated oligonucleotide probes for forensic DNA analysis. They found that although the AP detection method is slightly less sensitive than ^{32}P detection, the AP-conjugated oligonucleotide probes tested have specificity comparable to and are appropriate and suitable substitutes for ^{32}P-labeled plasmid inserts. A method using AP to label hepatitis B virus DNA as probe has been studied and used in clinical experiments to detect the virus DNA in hepatitis serum. AP, coupled with polyethylenimine and P-benzoquine as cross-linking reagent was used to label hepatitis B virus DNA. A comparison of the enzyme-labeled viral DNA probe and the ^{32}P-labeled probe gave 95.7% concordant results (Tu et al., 2004). Other small molecules can also be conjugated to APs. For example, AP–thyroxine conjugates have been used to develop an enzyme immunoassay for thyroxine (Eruk et al., 1984). Calf IAP was conjugated to thyroxine using cyanuric chloride, glutaraldehyde and 1-cyclohexyl-3-(2-morpholinoethyl)carbodiimide as coupling reagents. The conjugates produced were studied for incorporation of thyroxine, retention of enzyme activity and immunoreactivity in a thyroxine radioimmunoassay system against a range of thyroxine antisera.

In Chapter 5, we described the inhibition properties of AP, including competitive and uncompetitive inhibition. These properties have also been exploited to develop assays methods for a variety of substances. For example, Crans et al. (1990) developed a kinetic method for the determination of free vanadium(IV) and (V) at trace level concentrations using bIAP as sensing molecule. The method involves measuring the rate of bIAP-catalyzed hydrolysis of pNPP with (V_i) and without (V_0) a competitive inhibitor in the assay. Michaelis–Menten kinetics for a competitive inhibitor was used to express the relationship between V_0/V_i and the inhibitor concentration. Measuring both V_0 and V_i thus yields a V_0/V_i ratio that allows calculation of the competitive inhibitor concentration. Determination of free vanadium in complex fluids can be accomplished by comparing the ratio of rates of pNPP hydrolysis with and without a sequestering agent with the ratios of rates measured on addition of a known vanadium concentration. Free vanadium(V) can conveniently be measured from 10^{-7} to 10^{-5} M and free vanadium(IV) can be measured at 10^{-8} M and above. The error limits on the vanadium determinations range from ±3 to ±12% of the concentration depending on the assay conditions. In another example, the property of APs of being inhibited uncompetitively by theophylline has also been adapted to developing assays for theophylline. Theophylline is an effective bronchodilatator used in the treatment of asthma that requires frequent control because of its narrow therapeutic index. Methods for theophylline clinical monitoring have been devised based on the uncompetitive inhibition of AP in the presence of appropriate substrates. Vinet and Zizian (1979) described an original procedure for the determination of theophylline in serum. The drug was extracted from 0.4 mL of serum at pH 7.4 with chloroform–2-propanol (20:1, v/v) and back-extracted into sodium hydroxide (1 mmol L^{-1}). The inhibition of beef-liver AP by theophylline in this alkaline phase was measured at 25 °C, with pNPP as substrate, in 2-amino-2-methyl-1-propanol buffer, pH 9.4. The reciprocal of enzyme activity and theophylline concentration were linearly related in the range 2–60 mg L^{-1}. The maximum interference to be expected from 3-methylxanthine would increase the apparent theophylline concentration by no more than 1 mg L^{-1}. The method was found to be accurate, free from interference by other xanthines and often-coadministered drugs and the results correlated well with those of the immunoenzymic assay. Major advantages are reagent stability, low cost and simplicity of instrumentation. This principle was subsequently adapted to automated analysis (Ayers et al., 1989), amperometric detection (Gil et al., 1990) and electrochemical detection (Palmer et al., 1992). Other assays have involved the use of fluorogenic substrates (Jourquin and Kauffmann, 1998) and capillary electrophoresis (Whisnant et al., 2000), and now the detection limit for theophylline with this technique has been improved to levels of 3 µM, with 8.6 amol of bIAP required for each assay.

APs can also be used as part of an enzymatic cascade to increase the sensitivity of assay systems. For example, Obzansky et al. (1991) developed a highly sensitive flavin adenine dinucleotide-3′-phosphate (FADP)-based enzyme amplification cascade for determining AP activity. The cascade detects AP via the dephosphorylation of the novel substrate FADP to produce the cofactor FAD, which binds stoichiometrically to inactive apo D-amino acid oxidase (D-AAO). The resulting active holo D-AAO oxidizes D-proline to produce hydrogen peroxide, which is quantified by the horseradish peroxidase-mediated conversion of 3,5-dichloro-2-hydroxybenzenesulfonic acid and 4-aminoantipyrine to a colored product. The FADP-based enzyme amplification cascade has been used in a novel releasable linker immunoassay (RELIA) to quantify thyrotropin (TSH). In the assay, TSH is first captured onto antibody-coated chromium dioxide particles. After formation of an antibody–TSH sandwich with a dethiobiotinylated second antibody, the complex is reacted with a streptavidin–AP conjugate. Biotin is then used to release the conjugate into solution and AP is quantified in an automated version of the FADP-based amplification cascade on the ACA discrete clinical analyzer (Du Pont). The sensitivity of the colorimetric RELIA assay for TSH ($< 0.1\,mIU\,L^{-1}$) is comparable to that of fluorimetric assays. In another application of this principle, APs hydrolyze riboflavin 4'-phosphate to produce riboflavin. This is converted to riboflavin 5'-phosphate, using riboflavin kinase, which reconstitutes apoglycolate oxidase to give hologlycolate oxidase. That enzyme catalyzes the oxidation of glycolate with simultaneous production of hydrogen peroxide, which is detected via the formation of a colored product through the action of peroxidase. This system can be used to measure riboflavin and riboflavin 5'-phosphate (Harbron et al., 1991b).

Morimoto and Inouye (1997) prepared bispecific F(ab')$_2$ fragments recognizing both human thyroid-stimulating hormone (TSH) and AP disulfide bond exchange between F(ab')$_2$ fragments of IgG1 monoclonal antibodies (MAbs) against TSH and AP. AP was polymerized by glutaraldehyde and an ELISA for TSH was developed by using the AP polymers and bispecific F(ab')$_2$ fragments against TSH and AP. In this assay, the preparation of covalently linked enzyme–MAb conjugates was not needed and the interaction of MAb with nonspecific proteins was greatly reduced by the use of F(ab')$_2$ fragments. The sensitivity for TSH was shown to increase in proportion to the degree of polymerization of AP and the lower detection limit obtained with the AP trimer was $0.5\,\mu U\,mL^{-1}$. The sensitivity was 30 times higher than that of a conventional ELISA using covalently linked enzyme–MAb conjugates. The use of bispecific F(ab')$_2$ permits the use of monomers and polymers of the signal enzyme and, thereby, regulates the sensitivity of the assay system.

Bieber et al. (2004) developed metal-affinity pipettes and bioreactive AP probes as tools for the characterization of phosphorylated proteins and peptides. They used bIAP as bioreactive probe covalently bound to the mass spectrometry (MS) target in combination with affinity capture and matrix-assisted laser desorption/ionization time-of-flight (MALDI-TOF) MS to study the hydrolysis of several phosphoproteins found in human saliva. Human salivary proteins were extracted from diluted human saliva with immobilized metal-affinity pipettes and phosphoproteins were eluted directly from the affinity pipettes to the bioreactive probe with dilute ammonium hydroxide, which provided conditions appropriate for hydrolysis by AP covalently bound to the probe surface. The authors found that the combination of metal-affinity pipette extraction, AP-bioreactive probes and MALDI-TOF MS was an effective way to find and characterize phosphoproteins, known and unknown, in complex mixtures.

12
Use of APs in Prodrug Converting Strategies

Many anticancer drugs have limited bioavailability as a result of low chemical stability, limited absorption or rapid breakdown *in vivo*. To overcome these problems, investigators have designed prodrugs that can be activated *in situ* by the action of an enzyme. An excellent and comprehensive review of this subject was given by Rooseboom et al. (2004). In order to achieve a high concentration of anticancer drugs and to decrease undesirable side-effects, the prodrugs should be either activated specifically by the tumor cells or targeted specifically to the tumor by a variety of strategies, including antibody-directed enzyme prodrug therapy (ADEPT), gene-directed enzyme prodrug therapy (GDEPT), virus-directed enzyme prodrug therapy (VDEPT) and the use of homing peptides specific for the tumor vasculature (Ruoslahti, 2002). Many enzymes participate in activating prodrugs, including APs. However, the wide distribution of AP isozymes *in vivo* represents a significant problem that requires the use of directed or homing strategies. At least six prodrugs can be activated by APs, i.e. amifostine, 3-AP phosphate, etoposide phosphate, mitomycin C phosphate, paclitaxol and phenol mustard phosphate (Fig. 48).

Amifostine, formerly known as WR-2721, is an organic thiophosphate, full name *S*-2-(3-aminopropylamino)ethylphosphorothioic acid, that was developed to protect normal tissues selectively against the toxicities of chemotherapy and radiation. Amifostine is a prodrug that is dephosphorylated at the tissue site to its active chemoprotective thiol metabolite WR-1065 by APs (Calabro-Jones et al., 1985). Differences in the AP concentrations of normal versus tumor tissues can result in greater conversion of amifostine in normal tissues. Once inside the cell, the free thiol provides an alternative target to DNA and RNA for the reactive molecules of alkylating or platinum agents and acts as a potent scavenger of the oxygen free radicals induced by ionizing radiation and chemotherapy (Yuhas, 1980; Orditura et al., 1999). Preclinical animal studies demonstrated that the administration of amifostine protected against a variety of chemotherapy-related toxicities including cisplatin-induced nephrotoxicity, cisplatin-induced neurotoxicity, cyclophosphamide- and bleomycin-induced pulmonary toxicity and the cytotoxicities (including cardiotoxicity) induced by doxorubicin and related chemotherapeutic agents. Amifostine was shown to protect a variety of animal species from lethal doses of radiation. Thus, amifostine is at present broadly used as supportive treat-

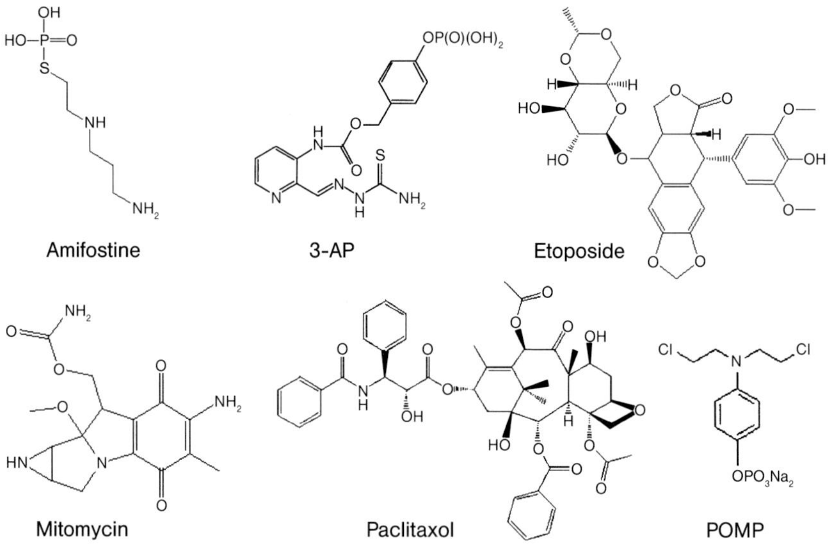

Fig. 48 Chemical structures of compounds that depend on AP activity for *in vivo* activation. Shown are the structures of the drugs amifostine, etoposide, mitomycin and paclitaxol and of the prodrugs 3-AP and phenol mustard phosphate (POMP).

ment during chemotherapy, in lymphomas and solid tumors (Orditura et al., 1999; Santini and Giles, 1999; Santini, 2001). A report by Giatromanolaki et al. (2002) suggested that down-regulation of IAP in the tumor vasculature and stroma provides a strong basis for explaining amifostine selectivity and the abundance of IAP expression in normal tissues, stromal and vascular, ensures an intense hydrolysis of WR-2721 and rapid intracellular accumulation of WR-1065.

The drug 3-AP, 3-aminopyridine–2-carboxaldehyde thiosemicarbazone, is an inhibitor of ribonucleotide reductase that plays a crucial role in the synthesis of DNA through the reductive conversion of ribonucleotides to deoxyribonucleotides. In order to increase the bioavailability of 3-AP, Li et al. (1998) developed a phosphate prodrug (3-AP phosphate). The prodrug showed improved solubility, was chemically more stable and was rapidly converted to 3-AP by bovine IAP and rat liver AP. When evaluated against the murine M-109 lung carcinoma and the B16-F10 murine melanoma xenograft models, the phosphate prodrug displayed improved efficacy and safety profiles compared with those found with the parent drug and demonstrated impressive antitumor effect using a once-a-day dosing regimen (Li et al., 1998, 2001).

Etoposide (VP-16) belongs to the general class of chemotherapy drugs known as plant alkaloids and is a semisynthetic derivative of podophyllotoxin, i.e. 4′-demethylepipodophyllotoxin 9-[4,6-O-(R)-ethylidene-β-D-glucopyranoside]. Etoposide acts by inhibiting human DNA topoisomerase II and is used to treat small cell lung, testicular and other cancers. Mitomycin, also known as mitomycin C and with the chemical structure 7-amino-9a-methoxymitosane, is an antibiotic isolated from *Streptomyces caespitosus* that has been shown to have antitumor activity. In one study, calf IAP was covalently linked to the two antitumor MAbs, L6 (anticarcinoma) and 1F5 (anti-B lymphoma), forming conjugates that could bind to antigen-positive tumor cells (Senter et al., 1989). The conjugates were capable of converting a relatively noncytotoxic prodrug, etoposide phosphate (EP), into etoposide, a drug with significant antitumor activity. *In vitro* studies with a human colon carcinoma cell line, H3347, demonstrated that whereas EP was less toxic than etoposide by a factor of >100, it was equally toxic when the cells were pretreated with L6–AP, a conjugate that bound to the surface of H3347 cells. The L6–AP conjugate localized in H3347 tumor xenografts in nude mice and histological evaluation indicated that the targeted AP was distributed throughout the tumor mass. A strong antitumor response was observed in H3347-bearing mice that were treated with L6–AP followed 18–24 h later by EP. This response, which included the rejection of established tumors, was superior to that of EP ($P < 0.005$) or etoposide ($P < 0.001$) given alone. The 1F5–AP conjugate did not bind to H3347 cells and did not enhance the toxicity of EP on these cells *in vitro*. In addition, 1F5–AP did not localize to H3347 tumors in nude mice and did not demonstrate enhanced antitumor activity in combination with the prodrug. In another report from this group, these same conjugates were able to convert the prodrugs mitomycin phosphate (MOP) and EP into an active mitomycin C derivative, mitomycin alcohol and etoposide, respectively (Senter et al., 1989). MOP and EP were less toxic to cultured cells from the H2981 lung adenocarcinoma than their respective hydrolysis products, mitomycin alcohol and etoposide, by a factor of >100, and they were also less toxic in mice. Pretreatment of H2981 cells with L6–AP greatly enhanced the cytotoxic effects of MOP and EP, whereas 1F5–AP caused no such enhancement. A strong antitumor response was observed in H2981-bearing mice that were treated with L6–AP followed 24 h later by either MOP or a combination of MOP and EP. This response was superior to that of MOP or combinations of MOP and EP given alone (Senter et al., 1989). Similarly, Haisma et al. (1992) used the anti-carcinoembryonic antigen MAb BW431/26 conjugated to AP and EP as a prodrug. Quantitative hydrolysis of EP to etoposide occurred within 10 min in the presence of AP. MAb BW431/26 and AP were conjugated using a thioether bond and the AP conjugate retained 93% of its calculated activity. SW1398 colon cancer cells were used to analyze the cytotoxicity of etoposide and EP. Etoposide (IC_{50} 22 µM) was 100 times more toxic than EP (20% growth inhibition at 200 µM) (Haisma et al., 1992). Sahin et al. (1990) developed a BsMAb by somatic hybridization of the two mouse hybridoma cell lines HRS-3 and AP-1, which produce MAbs with reactivity against the Hodgkin's- and Reed–Sternberg cell-associated CD30 antigen and AP, respectively. The whole immunoglobulin molecules and F(ab')$_2$ fragments of the BsMAb were equally effective in converting the relatively noncytotoxic prodrug, mitomycin phosphate (MOP), into mitomycin alcohol, which was 100 times more toxic to the Hodgkin's- and Reed–Sternberg cell line L540 (CD30 positive) than MOP. The cytotoxic activity of MOP was unaffected when the cells were pretreated with either the BsMAb or the enzyme alone. In order to sensitize tumor cells to etoposide, Kim et al. (2003) also considered a prodrug-converting system using membrane-bound IAP as the prodrug-activating enzyme to convert EP into etoposide. They used the retroviral vector for transducing IAP gene into SNU638 gastric cancer cells and EP was prepared by phosphorylation of etoposide. The approach greatly enhanced the cytotoxic effect in proportion to the concentration of EP, whereas control cells did not cause any cytotoxic effects after EP treatment (Kim et al., 2003).

Wallace and Senter (1991) also used the prodrug p-[N,N-bis(2-chloroethyl)amino]phenyl phosphate (phenol mustard phosphate, POMP), which had been shown to be 16 times less toxic to mice than its dephosphorylated from p-[N,N-bis(2-chloroethyl)amino]phenol (phenol mustard), to develop an ADEPT strategy using the L6–AP conjugate described above. POMP was prepared by phosphorylation of phenol mustard with phosphoryl chloride. The authors tested POMP first in vitro against H2981 human lung and H3396 human breast carcinoma cells. Pretreatment of the H2981 cells with L6–AP antibody conjugate enhanced the cytotoxic effects of POMP as a result of dephosphorylation of POMP. Subsequently, the L6–AP conjugate was administered in vivo to nude mice that had s.c. H2981 tumors that were ~75 mm^3 in volume. The L6–AP conjugate was administered i.v. 48 h prior to treatment with POMP (i.p.). The antitumor effects of the conjugate was compared with the prodrug without conjugate and to phenol mustard that was administered i.p. at the maximum tolerated dose. Neither soluble phenol mustard nor POMP had much of an effect on tumor growth, but the antitumor activity of POMP was increased significantly in animals that received the L6–AP conjugate 48 h before prodrug administration, and this level of activity was greater than with the drugs alone or a combination of 1F5–AP (nonbinding antibody–conjugate control) with POMP.

Paclitaxel is a taxoid drug extracted from the bark of the Pacific yew. It works against a wide variety of cancers, including breast, lung, head and neck, bladder and ovarian carcinoma, by interfering with mitosis as it binds to microtubules and inhibits their depolymerization (molecular disassembly) into tubulin, blocking a cell's ability to break down the mitotic spindle during mitosis. Some paclitaxel (taxol) phosphate derivatives, i.e. BMY46366, BMY-46489, BMS180661 and BMS180820, were used to determine the ability of bovine AP to convert these water-soluble potential prodrugs to tubulin-polymerizing metabolites (i.e. paclitaxel) (Mamber et al., 1995). HPLC/MS of BMS180661 treated with bovine AP confirmed the production of paclitaxel from the prodrug. In contrast, 2′- and 7-phosphate analogs BMY46366 and BMY46489 treated with AP were not active in tubulin assays. None of the paclitaxel phosphate prodrugs polymerized tubulin in the absence of metabolic activation. These results demonstrated that certain paclitaxel phosphate prodrugs could be metabolized by AP to yield effective tubulin polymerization (Mamber et al., 1995).

13
APs as Therapeutic Agents

13.1
In the Treatment of Hypophosphatasia

As described in detail in Section 7.1, TNAP plays a crucial role in controlling the concentration of the mineralization inhibitor PP_i and thus in regulating proper bone mineralization. A multitude of TNAP mutations lead to suboptimal activity of TNAP that in turn lead to hypophosphatasia. To date, however, although we have begun to understand the mechanisms that lead to the skeletal defects of this disease, there are no established treatments for this condition. However, some experimental trials with enzyme replacement therapy and cell therapy have been conducted. New research data have also opened up the possibility of using gene therapy and also inhibitors of PP_i production and transport as rational therapeutic approaches.

Whyte et al. (1982b) pioneered the testing of enzyme replacement therapy for hypophosphatasia on a severely affected 6-month-old girl. They administered repeated intravenous infusions of TNAP-rich plasma obtained by plasmapheresis from two men with Paget's bone disease. Circulating Paget's TNAP activity was found to have a half-life of 2 days, similar to that reported in adults, which did not change during a 5-week period of six AP infusions. Normalization of the patient's serum AP activity was followed by better control of her hypercalcemia and hypercalciuria. Sequential radiographic studies revealed an arrest of worsening rickets with slight remineralization of metaphyses, although urinary excretion of the AP substrates PEA and PP_i was unaltered by the therapy. This approach was used again on three additional patients with infantile hypophosphatasia (Whyte et al., 1984). The patients received weekly intravenous infusions of TNAP-rich plasma from patients with Paget's bone disease. Despite partial or complete correction of the deficiency of circulating TNAP activity, the authors observed no radiographic evidence for arrest of progressive osteopenia or improvement in rachitic defects in any of the patients. The authors concluded that the failure of infants with hypophosphatasia to show significant healing of rickets on correction of circulating TNAP activity supported the hypothesis that TNAP functions *in situ* during normal skeletal mineralization. In a similar study, Weninger et al. (1989) attempted enzyme replacement therapy for a severely affected premature boy (birth weight 2380 g at 36 weeks) with hypophosphatasia by infusions of purified human liver

TNAP. Treatment (1.2 IU kg^{-1} min^{-1}) started at age 3 weeks and was repeated at weekly intervals until age 10 weeks, when the child died. Samples of TNAP were diluted with 10 mL of physiological saline and infused over 30 min via an umbilical arterial catheter. No toxic or allergic side-effects were observed. Serum TNAP activity increased from 3 IU L^{-1} before treatment to a maximum level of 195 IU L^{-1} with a half-life between 37 and 62 h. Urinary excretion of PEA decreased during therapy. Calcium, phosphorus, parathyroid hormone and 1,25(OH)$_2$D3 levels remained within normal range. Sequential radiographic studies showed no improvement of bone mineralization. Bone morphology was studied by light and electron microscopy before treatment and post mortem. The borderline between mineralized and unmineralized matrix was more distinct after treatment and at the electron microscope level initial spots of mineralization were more frequent between the collagen fibrils compared with the biopsy specimen before treatment. In contrast to previous studies, however, only woven and bundle bone structures were studied from the tibial crest, where the lack of osteoblast-like cells upon the newly formed osteoid matrix was prominent (Weninger et al., 1989).

Of interest in this regard is also the report by Whyte et al. (1995) of the correction of substrate accumulation in carriers of hypophosphatasia during pregnancy. Given that three phospho compounds, PEA, PP$_i$ and PLP, accumulate in serum in hypophosphatasia patients, the authors tested whether the rise in serum PLAP associated with pregnancy would correct the disease. Blood or urine concentrations of PEA, PP$_i$ and PLP diminished substantially during that time. After childbirth, maternal circulating levels of PLAP decreased and PEA, PP$_i$ and PLP levels increased abruptly. The authors concluded that PLAP, like TNAP, is physiologically active towards PEA, PP$_i$ and PLP in humans. Results by Wennberg et al. (2002) on the properties of PLAP and GCAP to metabolize these physiological substrates also validate this conclusion, which may be of relevance if PLAP or GCAP were to be used in enzyme replacement approaches for hypophosphatasia. One of the most common problems with gene therapy aiming at replacing an absent or mutated gene is that recipients often develop an immune response to the therapeutic gene product, which is perceived by the body as a foreign antigen since it is either missing or mutated in the inherited disease. Hence, rather than attempting to introduce the wt TNAP gene in hypophosphatasia patients, PLAP but particularly the GCAP gene represent viable alternatives, since all men and women have low but detectable levels of GCAP in the circulation as a result of basal expression by type II pneumocytes (Section 2.2). Wennberg et al. (2002) found that the PLAP allozymes have K_ms in the range 1.4–1.5 mM for PP$_i$ or three times higher than the corresponding value for TNAP (0.48 mM). However, the turnover numbers of the PLAP allozymes for PP$_i$ (304, 404 and 244 s^{-1} for PLAP S, F and D, respectively) are 3.5–5.7-fold higher than for TNAP (70.4 s^{-1}). Thus the k_{cat}/K_m ratio is 203, 279 and 174 s^{-1} mM^{-1} for the PLAP S, F and D, respectively, compared with a k_{cat}/K_m of 147 s^{-1} mM^{-1} for TNAP. The fact that PLAP allozymes and in particular the D allozyme (owing to its similarity to GCAP), are able to hydrolyze PP$_i$ with specificity constants (k_{cat}/K_m) comparable to those of TNAP, will allow the use of PLAP and/or GCAP to develop gene therapy approaches to treating hypophosphatasia.

Cell therapy has been attempted as an alternative to enzyme-replacement therapy. Preliminary studies had indicated successful grafting of wt bone marrow cells in $Akp2^{-/-}$ mice (Fedde et al., 1996). Subsequently, Whyte et al. (2003) pioneered the use of bone marrow cell transplantation on an 8-month-old girl who seemed certain to die from the infantile form of hypophosphatasia. After cytoreduction, she was given T-cell-depleted, haplo-identical marrow from her healthy sister. Three months later, she was clinically improved, with considerable healing of rickets and generalized skeletal remineralization. However, 6 months post-transplantation, worsening skeletal disease recurred, with partial return of host hematopoiesis. At the age of 21 months, without additional chemotherapy or immunosuppressive treatment, she received a boost of donor marrow cells expanded *ex vivo* to enrich for stromal cells. Significant, prolonged clinical and radiographic improvement followed soon after. Nevertheless, biochemical features of hypophosphatasia have remained unchanged to date. Skeletal biopsy specimens were not performed. Now, at 6 years of age, she is intelligent and ambulatory but remains small. Among several hypotheses for this patient's survival and progress, the most plausible seems to be the transient and long-term engraftment of sufficient numbers of donor marrow mesenchymal cells, forming functional osteoblasts and perhaps chondrocytes, to ameliorate her skeletal disease (Whyte et al., 2003).

Given that TNAP can be found anchored to cell membranes (Section 4.3.2) and also circulating in serum in an anchor-depleted form (Section 9.1.1), it is conceivable that the circulating form of TNAP could also affect bone mineralization. Murshed et al. (2005) generated transgenic mice over-expressing a TNAP cDNA under control of the apolipoprotein E (*ApoE*) promoter and a liver-specific enhancer. These regulatory elements are active only after birth. Two lines of *ApoE*–TNAP mice that had more than a 10-fold increase in their TNAP serum level were used for subsequent experiments. The fact that *ApoE*–TNAP sera from both the lines released P_i from β-glycerophosphate at a much higher rate than wt serum confirmed that the TNAP cDNA transcribed by the transgene was biologically active in each of the transgenic lines used. *ApoE*–TNAP mice had no metabolic abnormalities and no histological evidence of ectopic ECM mineralization and also had normal serum P_i levels. The authors asked whether this increase in TNAP activity could affect the severity of the hyperosteoidosis characterizing rickets and osteomalacia by transferring the *ApoE*–TNAP transgene on to the *Akp2* null genetic background. [*ApoE*–TNAP; $Akp2^{\beta geo/\beta geo}$] mice had a normal life span and none of the neurological manifestation observed in $Akp2^{\beta geo/\beta geo}$ mice. Primary osteoblasts isolated from these mice did not stain for TNAP and did not mineralize in the presence of β-glycerophosphate. When bones of 1-month-old [*ApoE*-TNAP; $Akp2^{\beta geo/\beta geo}$] mice were analyzed by histology, a complete rescue of the hyperosteoidosis characterizing $Akp2^{\beta geo/\beta geo}$ mice was observed (Murshed et al., 2005). These results indicate that although it does not always affect bone mineralization in wt mice, high TNAP expression in liver and/or high levels of circulating TNAP can rescue a hyperosteoidosis. Indeed, the rescue of the hyperosteoidosis by circulating TNAP suggest that it is explained in part by the high collagen content in the bone ECM, although other molecular events may contribute to it. Hence the mechanism that explains the correction of the skeletal defect still needs to be elucidated. These data also appear at variance with the failure of past attempts to correct hyperosteoidosis in humans by administration of TNAP (Whyte et al., 1982, 1984; Weninger et al., 1989). However, it is possible that continuous delivery of high doses of TNAP is needed to achieve correction and that this continued supply was afforded by the use of this genetic approach but not by the injections of Paget's sera.

It was shown nearly 20 years ago that normalizing calcium and P_i concentration corrects the hyperosteoidosis of rickets patients (Balsan et al., 1986). The fact that extracellular calcium concentrations are more tightly regulated than extracellular P_i concentrations suggests that of these two ions, P_i may be the critical element in the induction of mineral crystals in a given ECM. Van den Bos et al. (1995) used rats that had been implanted with collagen slices complexed with bIAP to induce *de novo* mineralization and found a positive correlation between the degree of mineralization and serum P_i. Murshed et al. (2004) reported that correcting low serum P_i concentrations corrects a bone mineralization defect in hypophosphatasemic *Hyp* mice. It is unclear whether or not P_i supplementation may have a beneficial role in cases of hypophosphatasia. In fact, Wenkert et al. (2002) reported very preliminary observations that dietary restriction, rather than supplementation, of P_i was followed by improvement in the serum and urine parameters in one affected child, although no information was provided on any changes in the skeletal phenotype. However, the mineralizing role of P_i is antagonized by PP_i that prevents ectopic ECM mineralization in wt animals and the presence of PP_i in almost every ECM explains why raising the extracellular P_i concentration does not result in pathological ECM mineralization in wt mice. Thus, as discussed in Section 8.1.4.2, what triggers bone ECM mineralization is the ratio of P_i to PP_i, a ratio determined to a large extent by TNAP function. The notion that the P_i to PP_i ratio is important for inducing bone mineralization is in agreement with the observation that $Akp2^{-/-}$ mice that have abnormally high extracellular PP_i levels have hyperosteoidosis, whereas [$Akp2^{-/-}$; $Enpp1^{-/-}$] and [$Akp2^{-/-}$; ank/ank] mice have normal mineralization of the skull and vertebrae and normal extracellular PP_i concentration (Hessle et al., 2002; Harmey et al., 2004). Therefore, targeting the function of either NPP1 or ANK appears to be a viable therapeutic approach to treating the mineralization abnormalities of hypophosphatasia. Whereas the function of ANK can be targeted via administration of probenecid, there are currently no NPP1-specific inhibitors that can be used to target NPP1 function.

13.2
In the Treatment of CPPD Disease

As described in Section 7.1, TNAP has a very specific function regulating the size of the pool of extracellular PP_i and thus controlling proper mineralization. This pyrophosphatase activity of TNAP has been proposed to be of use in the therapeutic dissolution of calcium pyrophosphate dihydrate crystals (CPPD) such as are present in calcium arthropathies, particularly chondrocalcinosis. Early studies by Xu et al. (1991) indicated that that yeast pyrophosphatase effectively dissolved CPPD crystals in solutions. Maximum enzymatic dissolution of CPPD crystals was achieved at neutral pH and when the enzyme had access to the crystal surface. Similarly, CPPD dissolution was also achievable via TNAP at a pH optimum of 7.4, which is the optimum pH for its pyrophosphatase activity. TNAP acted more effectively on CPPD crystals than on soluble pyrophosphate relative to yeast pyrophosphatase. The authors also suggested that chondrocyte TNAP may play an important role in the dissolution of CPPD crystals in cartilage (Xu et al., 1991). Shinozaki et al. (1995) studied the mechanism of TNAP interaction with CPPD crystals *in vitro*. TNAP was incubated with CPPD crystals in an *in vitro* model system. They found that TNAP preferentially binds to the small end faces of CPPD crystals. Etch pits indicative of dissolution were demonstrated coexistent with TNAP crystal binding and TNAP pyrophosphohydrolytic activity. They also concluded that TNAP binding to CPPD was mediated by a nonenzymatic mechanism and that the CPPD crystal dissolution rate was limited by the availability of surface area on the crystal faces most susceptible to TNAP binding. Subsequently, Shinozaki and Pritzker (1996) studied the effects of enzyme inhibitors such as bisphosphonates, orthovanadate, calcium, cadmium and ascorbic acid on the pyrophosphatase and phosphohydrolase activity of TNAP and compared these effects with those on CPPD crystal dissolution. The authors concluded that dissolution of CPPD crystals by TNAP depends on the binding of TNAP to the CPPD crystals and also the pyrophosphatase activity of the bound TNAP. The strong inhibitory effects of bisphosphonates on TNAP CPPD crystal dissolution compared with those on TNAP pyrophosphatase activity suggest that bisphosphonates inhibit crystal dissolution by their affinity for the CPPD crystal surface (Shinozaki and Pritzker, 1996). These studies open the door to evaluating the therapeutic effects of local administration of purified or recombinant APs in the treatment of CPPD disease.

13.3
Endotoxin Treatment

The high pH optimum of APs *in vitro* as measured with the usual test substrates greatly exceeds the physiological pH range as it occurs in biological tissues. Poelstra et al. (1997a) hypothesized that the relatively high pH optimum of APs *in vitro* is related to dissociation of acidic groups in the protein preparation, which leads to the formation of negatively charged groups in the vicinity of the active site of the enzyme and that these negatively charged groups may promote the activity of AP. They examined the possibility that endotoxin might be a natural substrate for APs because this phosphorylated substance is able to supply multiple negatively charged residues in the microenvironment of the enzyme at a physiological pH level. Phosphate groups in the endotoxin molecule are known to be essential for the biological activities of this bacterial product. Poelstra and colleagues were able to demonstrate that in the intestine and renal tissue, APs are endowed with endotoxin dephosphorylating activity at pH levels closer to the physiological range. This is also illustrated by experiments *in vivo* showing that the toxicity of endotoxin is significantly reduced after exposure to AP preparations, as tested by inducing a local intradermal inflammatory reaction in rats. Collectively their data suggested that the ubiquitous AP isozymes might accomplish protection against endotoxin, an equally ubiquitous product of Gram-negative bacteria that may cause lethal complications after an infection with these microorganisms (Poelstra et al., 1997a). Further studies using intestinal cryostat and endotoxin from *Escherichia coli* and *Salmonella minnesota* R595 as substrate indicated that human PLAP was also able to dephosphorylate both preparations at pH 7.5 (Poelstra et al., 1997b). As phosphate residues in the lipid A moiety determine the toxicity of the endotoxin, they also examined the effect of levamisole *in vivo* using a rat septicemia model. The results showed that inhibition of endogenous AP by levamisole significantly reduces the survival of rats injected intraperitoneally with *E. coli* bacteria, whereas this drug does not influence survival of rats receiving a sublethal dose of the Gram-positive bacterium *Staphylococcus aureus*. The authors suggested that in view of the endotoxin-dephosphorylating properties of AP demonstrated *in vitro*, AP has a crucial role in host defense. They also stated that the effects of levamisole during Gram-negative bacterial infections and the localization of AP as an ectoenzyme in most organs as well as the induction of enzyme activity during inflammatory reactions and cholestasis are in accord with such a protective role (Poelstra et al., 1997b). In agreement with this hypothesis, Xu et al. (2002) showed that in acute liver injury, TNAP might contribute to protection from injury by a mechanism involving neutralization of endotoxin.

To test whether lipopolysaccharide (LPS) dephosphorylation could be used for intervention during sepsis, Bentala et al. (2002) investigated the effects of *Salmonella minnesota* R595 LPS and its dephosphorylated counterpart monophosphoryl lipid A (MPLA) on macrophage activation *in vivo* and *in vitro*. Exposure of RAW264.7 cells to LPS induced high levels of tumor necrosis factor-alpha (TNFα) and nitric oxide, whereas MPLA elicited no response. LPS *in vivo* induced a significant rise in TNFα levels in mice and an enhanced inflammatory cell influx in the lung, whereas MPLA did not. Having shown the relevance of this particular phosphate group of LPS, these authors subsequently explored the LPS-dephosphorylating ability of human PLAP in different tissues and the effect of PLAP administration in mice challenged with LPS. Histochemical data showed that PLAP dephosphorylated native LPS in all tissues examined, whereas MPLA was not dephosphorylated. When mice received PLAP immediately after the LPS challenge, the survival rate was 100%, over 57% in the control group. The authors concluded that the enzymatic removal of phosphate groups from LPS by PLAP represents a crucial detoxification reaction, which may provide a new strategy to treat LPS-induced diseases such as sepsis (Bentala et al., 2002). Since human PLAP is able to dephosphorylate LPS both *in vitro* and *in vivo*, Beumer et al. (2003) investigated whether AP derived from calf intestine (calf IAP) was also able to detoxify LPS. In mice administered calf IAP, 80% of the animals survived a lethal *E. coli* infection. In piglets, prior to LPS detoxification, the pharmacokinetic behavior of calf IAP was studied. IAP clearance was shown to be dose independent and showed a biphasic pattern with an initial t of 3–5 min and a second-phase t of 2–3 h. Although calf IAP cleared much faster than human PLAP, it attenuated LPS-mediated effects on hematology and TNFa responses at doses up to 10 µg kg^{-1} in piglets. LPS-induced hematological changes were antagonized and the TNFa response was reduced by up to 98%. Daily i.v. bolus administration of 4000 units of calf IAP, the highest dose used in the LPS intervention studies, in piglets for 28 days was tolerated without any sign of toxicity. Therefore, calf IAP potentially encompasses a novel therapeutic agent in the treatment of LPS-mediated diseases (Beumer et al., 2003).

Efforts were therefore made to establish optimal formulations of APs for *in vivo* administration. Ford and Dawson (1993) reported that lactose and trehalose maintained AP activity after freeze-drying and that preparations containing trehalose retained activity even when the material was subjected to temperatures of up to 45 °C for up to 84 days. Eriksson et al. (2003) investigated the formulation of sugar glass stabilized AP from bovine intestine (bIAP) into tablets. It was found that inulin was far superior to trehalose as a stabilizer of bIAP in tablets. The poor stabilizing capacities of trehalose after compaction are explained by crystallization of trehalose induced by the compaction process and moisture in the material. The results clearly show that inulin is an excellent stabilizer for bIAP. The tabletting properties are adequate, showing sufficient tablet strengths, low friability and good stability of inulin glass upon exposure to high relative humidity. Subsequently, Eriksson et al. (2003) performed final proof-of-concept studies in rats. This acid-labile calf IAP is potentially useful in the treatment of sepsis, a serious condition during which endotoxins can migrate into the bloodstream. The calf IAP was freeze-dried with inulin and subsequently compacted into round biconvex tablets with a diameter of 4 mm and a weight of 25–30 mg per tablet. The tablets were coated with an enteric coating in order to ensure their survival in the stomach. *In vitro* evaluation of tablets containing bIAP was the first step in the development. It was found that tablets without enteric coating dissolved rapidly in 0.10 M HCl with total loss of enzymatic activity of the AP. Tablets that were coated were stable for at least 2 h in 0.10 M HCl, but dissolved rapidly when the pH was increased to 6.8. Furthermore, it was shown that the enzymatic activity of the released bIAP was fully preserved. The *in vivo* test clearly showed that the oral administration of enteric-coated tablets resulted in the release of enzymatically active calf IAP in the intestinal lumen of rats. The location of the enhanced enzymatic activity of AP in the intestines varied with the time that had passed between the administration of the tablets and the killing of the rats. Also, the level of enzymatic activity increased with increase in the number of tablets that were administered (Eriksson et al., 2003). Based on the data mentioned above, human clinical trials have been initiated.

AP can be considered as a host defense molecule since this enzyme is able to detoxify bacterial endotoxin at physiological pH. The question emerged of whether this anti-endotoxin principle is inducible in the glomerulus and, if so, which glomerular cells might be involved in the expression of AP after stimulation with pro-inflammatory agents. Therefore, kidneys of rats treated with LPS, *E. coli* or nontoxic monophosphoryl lipid A (MPLA) were examined for AP activity 6 or 24 h after challenge. In addition cultures of endothelial cells or mesangial cells were evaluated for AP activity after stimulation with LPS, TNFα or IL-6 and mRNA for AP was studied in TNFα-stimulated and control mesangial cells. The results showed significant up-regulation of glomerular AP in LPS- or *E. coli*-injected rats compared with rats injected with MPLA. Endothelial and mesangial cells *in vitro* showed significant up-regulation of AP activity following stimulation with LPS, TNFα or IL-6, whereas increased mRNA for AP was observed in mesangial cells after TNFα stimulation compared with nonstimulated control cells. Since it appeared that hydrolysis occurred when endotoxin was used as a substrate in the histochemical staining, the authors concluded that inducible glomerular AP might reflect a local endotoxin detoxifying principle of the kidney (Kapojos et al., 2003). Recently, Van Veen et al. (2005) applied a single-dose intravenous administration of bIAP (0.15 IU g^{-1}) in a murine cecal ligation and puncture model of polymicrobial sepsis. Mice treated with bIAP showed decreased transaminase activity in plasma and decreased myeloperoxidase activity in the lung, indicating reduced associated hepatocellular and pulmonary damage, but survival was not significantly altered by bIAP in this single-dose regimen. In polymicrobial secondary peritonitis, both prophylactic and early bIAP treatment attenuated the inflammatory response both locally and systemically and reduced associated liver and lung damage.

13.4
TNAP as Therapeutic Target for the Management of Ectopic Calcification

Vascular calcification refers to the deposition of hydroxyapatite in cardiovascular tissues such as arteries and heart valves and is a significant risk factor in the pathogenesis of cardiovascular disease, being associated with myocardial infarction and coronary death. The mechanisms that lead to pathological vascular calcification parallel those of normal bone formation and, at the cellular level, the adult artery wall contains mesenchymal progenitor cells with myogenic and chondrogenic potential (Tintut et al., 2003). MVs, the membrane limited chondroblast- and osteoblast-derived structures in which the process of mineralization takes place (Sections 7.1.1 and 8.1.4), have also been documented in calcified atherosclerotic lesions (Kim, 1976; Tanimura et al., 1986a,b; Hsu et al., 2000) and at the molecular level, several proteins that play roles in the regulation of bone formation, such as TNAP, OPN, osteocalcin, osteoprotegerin, RANKL, BMP2 and BMP 4 and matrix Gla protein (MGP) have been detected in calcified arteries. The linkage between cardiovascular disease and bone formation has been further strengthened by observations in genetically modified mice. For example, MGP-deficient mice ($Mgp^{-/-}$) display an osteopenic bone phenotype and an arterial calcification (Speer et al., 2002). Mutations affecting the osteoclastic lineage, such OPG knockout mice, which have an osteoporotic phenotype, are also associated with arterial calcification (Bucay et al., 1998). Also, OPN, a mineralization inhibitor, has dual roles in bone and heart (Steitz et al., 2002); OPN is expressed in osteoblasts and in activated inflammatory cells in injured arteries. OPN appears to play a protective role against arterial calcification, as OPN null mice are compromised in responding to cardiovascular challenge (Myers et al., 2003). Similarly, PP_i, a potent inhibitor of mineralization, appears to be another protective factor since mice lacking NPP1, a major generator of PP_i, spontaneously develop articular cartilage, perispinal and aortic calcification at a young age (Okawa et al., 1998; Johnson et al., 2003). These mice share similar phenotypic features with a human disease, idiopathic infantile arterial calcification (IIAC). In IIAC, a deficiency in NPP1-mediated production of PP_i has been postulated to cause arterial calcification and periarticular calcifications of the large joints (Rutsch et al., 2001, 2003). Similarly, another mouse model in which PP_i levels are depressed, owing to defective transport function of the transmembrane protein ANK, display soft tissue ossification similar to NPP1 mice (Ho et al., 2000).

Recent studies, detailed in Section 8.1.4.2, have provided compelling proof that a major role for TNAP in bone tissue is to hydrolyze PP_i to maintain a proper concentration of this mineralization inhibitor ensuring normal bone mineralization (Johnson et al., 2000, 2003; Hessle et al., 2002). Normalization of PP_i levels in NPP1 null and ANK-deficient mice results in a correction of soft-tissue ossification abnormalities. Crossbreeding either the $Enpp1^{-/-}$ or the ank/ank mice to $Akp2^{-/-}$ mice normalizes PP_i levels. Importantly, these studies have indicated that TNAP may be a useful therapeutic target for the treatment of diseases such as ankylosis and osteoarthritis and also arterial calcification. There is substantial evidence pointing to the presence of AP-rich vesicles at sites of mineralization in human arteries. It has been demonstrated that increased levels of TNAP accelerate calcification in bovine vascular smooth muscle cells (VSMCs) and moreover, levamisole, a TNAP inhibitor, blocks bovine VSMC calcification in a dose-dependent manner (Shioi et al., 1995). Macrophages may induce a calcifying phenotype in human VSMCs by activating TNAP in the presence of IFNγ and $1,25(OH)_2D3$ (Shioi et al., 2002). The presence of TNAP-enriched MVs in human atherosclerotic lesions also suggests an active role in the promotion of vascular calcification (Tanimura et al., 1986a,b; Hui et al., 1997a; Hui and Tenenbaum, 1998; Hsu and Camacho, 1999). Recently, Mathieu et al. (2005) showed that calcification of human valve interstitial cells is dependent on AP activity. Hence there is ample evidence to explore the therapeutic potential of inhibiting TNAP activity at the site of arterial calcifications as a means of increasing the local concentration of PP_i, which by itself should antagonize the deposition of hydroxyapatite while simultaneously upregulating OPN expression by VSMCs and thus further contributing to reducing ectopic hydroxyapatite deposition.

14
APs in the Food Industry

Pasteurization of raw milk was introduced to extend product shelf-life and destroy pathogens. Measurements of residual AP activity have been used as an indicator of proper pasteurization in dairy products for more than 65 years. The standard for fluid milk products established by the Food and Drug Administration and cited in the 1995 Pasteurized Milk Ordinance (PMO) is <350 mU L^{-1}. In a study by Angelino et al. (1999), milk containing three levels of milk fat [skim (0.5%), low fat (2.0%) and whole (3.25%)], were heat-treated at five temperatures (59, 61, 63, 65 and 67 °C) using a laboratory scale, batch pasteurization method. Heated milk samples were removed at 5 min intervals, immediately cooled and then assayed using the quantitative fluorimetric method and the qualitative Scharer rapid test. Mean AP activity values as measured with the Fluorophos method decreased in all milk preparations as the time of sampling and temperature of heating increased. Samples representing the three fat levels and heat treated at 63 °C for 30 min, the minimum time/temperature allowed by the 1995 PMO, had AP activity values <100 mU L^{-1} (Angelino et al., 1999). A detailed kinetic study of AP, lactoperoxidase and β-lactoglobulin was carried out in the context of identifying intrinsic time/temperature indicators for controlling the heat processing of milk. The heat inactivation or denaturation of AP, lactoperoxidase and β-lactoglobulin under isothermal conditions was found to follow first-order kinetics (Claeys et al., 2001). Further studies indicated that whereas AP activity appeared lower in skimmed milk compared with semi-skimmed or whole milk, the kinetics of inactivation were comparable and fat content did not seem to affect the AP test result substantially for pasteurized milk (Claeys et al., 2002). Currently, adequate pasteurization of milk products is regarded as confirmed in samples that contain a residual bovine AP activity of <500 mU L^{-1}. This is equivalent to the statutory acceptable level of 4 µg phenol L^{-1} required by the EC analytical method (Allen et al., 2004). Anecdotally, the *Penicillium roqueforti* mold, used to produce blue cheeses, exhibits AP activity. Therefore, blue mold-ripened cheeses made from properly pasteurized milk test positive for AP activity. Because microbial phosphatases are considered to be more resistant to heat than is milk phosphatase, a statutory control test recommends the repasteurization of cheese at 66 C for 30 min to inactivate selectively the native milk enzyme. However, because of the thermal stability of *Penicillium roqueforti* phosphatase, this control test leads to confusion of the fungal enzyme with native milk AP and does not confirm whether the milk used to make cheese has been pasteurized (Rosenthal et al., 1996). Vega-Warner et al. (2000) produced polyclonal antibodies against the TNAP isoform found in bovine milk in order to develop a competitive indirect ELISA in order to determine if an ELISA can be used to verify pasteurization of fluid milk. However, such an approach still remains to be validated (Vega-Warner et al., 2000). Recently, Harding and Garry (2005) coordinated a collaborative international study to evaluate six different fluid dairy products at lower AP levels than previously verified using the Fluorophos test system. Thirteen laboratories participated to evaluate the fluorimetric test at 20, 40, 100, 350 and 500 mU L^{-1} and extend the scope of the method to include milk from not only cows but also goats and sheep. Initially, the statutory level of AP measured fluorimetrically was set to equivalent levels of colorimetric test standards (500 mU L^{-1}). The European Union recently announced its intention of lowering the legal limit from 500 to 350 mU L^{-1} and, in addition, setting a target value of 100 mU L^{-1}, which if exceeded would trigger an investigation into the pasteurizer plant performance. At 500 mU L^{-1} of AP, this trial generated relative standard deviation of repeatability values of 6.48, 5.69 and 1.74% and relative standard deviation of reproducibility values of 14.66, 13.30 and 5.33% for all cow's, sheep's and goat's milk samples, respectively. The authors concluded that the data from this study were comparable to those from previous studies and indicated that the Fluorophos test system method for measuring AP activity in milk from cows, sheep and goats was suitable not only at the current European statutory level of 500 mU L^{-1} but also at much lower levels.

15
Veterinary Uses of AP Determinations

Most applications in veterinary medicine parallel those in human medicine. However, the literature is much more restricted in this area. A few examples will illustrate the usefulness, or lack thereof, of monitoring AP activity in a few animal species other than rodents.

Two assay techniques (one based on wheat-germ lectin precipitation followed by a simple enzymatic reaction, the second on a specific enzyme-linked immunoassay) were used to measure serum levels of bone AP in 35 dogs of different ages. The correlation between bone AP and total AP activities was poor ($r = 0.20$ for enzymatic bone AP, $r = 0.31$ for immunoreactive bone AP), indicating that total AP should be considered unreliable as an indicator of bone AP activity in canine serum. The immunoassay demonstrated acceptable (13%) cross-reactivity with the liver isoform of TNAP. The commercial immunoassay kit was considered simple and fast whenever the activity of bone AP is the focus of interest although the wheat-germ lectin/enzymatic technique was preferred in situations where the activities of other AP isoforms are required (Allen et al., 2000). Itoh et al. (2002) studied the serum AP isozymes in normal dogs using a commercially available polyacrylamide gel disk electrophoresis kit (PAG/disk kit). Serum samples taken from the dogs were incubated with neuraminidase, after which most showed AP isozymes as two characteristic stained bands, one, the most anodic, was liver TNAP and the other was bone TNAP and both were corticosteroid-induced AP. As expected, the percentage of bone AP was highest in young dogs (age <1 year, 64.7%) and this value decreased with age. In contrast, the percentage of liver AP in young dogs (22.2%) was much lower than that in middle-aged dogs (ages 1–7 years, 59.3%) and old dogs (ages >7 years, 50.4%). However, Wiedeman et al. (2005) reported that corticosteroid-induced AP was not a useful prognostic indicator for response rate and remission duration in dogs with lymphoma. Barger et al. (2005) recently used AP staining to differentiate canine osteosarcoma from other vimentin-positive tumors. The authors evaluated 61 vimentin-positive neoplasms with a confirmed diagnosis based on histopathology. Tumors that expressed vimentin and were also positive for AP included 33 osteosarcomas, one multilobular tumor of bone, one amelanotic melanoma and one chondrosarcoma. The authors concluded that AP appears to be a highly sensitive and fairly specific marker in the diagnosis of osteosarcomas. In one rare case of a dysgerminoma in a horse (an Arabian filly), the tumor cells stained strongly for AP (Chandra et al., 1998).

Hoffmann et al. (1983) produced an antiserum directed against equine IAP and developed a sensitive and quantitative assay which was used to measure the half-life of intravenously injected IAP and to determine IAP levels in normal horse sera, in sera from horses with lesions not involving the gastrointestinal tract and sera from horses with lesions involving the gastrointestinal tract. The results indicated that IAP was not likely to appear in equine serum even when gastrointestinal disease might be present and, therefore, appeared to be of no diagnostic value. Also of interest is a report indicating that AP concentrations in semen can be used as an early indicator of unilateral or bilateral lack of patency of the epididymal and deferent ducts in the dog (Stornelli et al., 2003). Also, AP activity can be useful as an inexpensive, simple clinical assay for differentiating ejaculatory failure or duct blockages from azoospermia and oligospermia (Turner and McDonnell, 2003).

16
Methodologies

The following sections illustrate the diversity of assays and methodologies available for measuring AP expression. However, they are not intended to enumerate every method or approach that has been published or used, nor do they recount many of the established methods already covered by McComb et al. (1979).

16.1
Amperometric, Spectrophotometric and Potentiometric Assays

Spectrophotometric methods using *p*-nitrophenylphosphate (pNPP) as substrate for APs is by far the most commonly used method to detect AP catalytic activity (McComb et al., 1979). APs convert pNPP to *p*-nitrophenol that is highly colored with an absorption maximum at ~405 nm and a molar absorption coefficient of ~20 000 $M^{-1} cm^{-1}$. The spectrophotometric determination of pNPP gives a signal-to-noise ratio of ~200. Thompson et al. (1991) argued that the use of the hydrogenated form of pNPP, i.e. *p*-aminophenylphosphate (pAPP) could be determined by electrochemistry, being oxidized at a glassy carbon electrode to a quinone imine with high sensitivity, which should allow for a signal-to-noise ratio of ~1000 or five times better than with spectrophotometric methods. Tang et al. (1988) described the first use of pAPP as an alternative substrate for APs suitable for electrochemical detection. In that first paper, AP activity was determined using pAPP as the enzyme substrate and the enzyme-generated *p*-aminophenol was detected amperometrically. The oxidation potential obtained for the detection of *p*-aminophenol was lower than that for phenol and the detection limit for *p*-aminophenol was 0.20 pmol. Thompson et al. (1991) later compared the amperometric and spectrophotometric methods for APs. The Michaelis constant for pAPP in 0.10 M Tris buffer, pH 9.0, was 56 µM, whereas it was 82 µM for pNPP. The amperometric method had a detection limit of 7 nM for the product of the enzyme reaction, which was almost 20 times better than the spectrophotometric method. Similarly, with a 15-min reaction at room temperature and in a reaction volume of 1.1 mL, 0.05 µg L^{-1} AP could be detected by electrochemistry, almost an order of magnitude better than by absorption spectrophotometry, so amperometric detection appeared ideally suited for small-volume and trace immunoassays. Another

Mammalian Alkaline Phosphatases: From Biology to Applications in Medicine and Biotechnology. J. L. Millán
Copyright © 2006 WILEY-VCH Verlag GmbH & Co. KGaA, Weinheim
ISBN: 3-527-31079-7

approach involved the indirect determination of AP based on the amperometric detection of indigo carmine at a screen-printed electrode in a flow system. Indirect amperometric measurements of AP activity in solution were easily carried out using 3-indoxyl phosphate substrate. The AP-catalyzed hydrolysis gives rise to indigo product that is insoluble in aqueous solutions but easily converted into its soluble parent compound, indigo carmine, by addition of fuming sulfuric acid to the reaction medium. Using this approach, the authors achieved a linear range of more than one order of magnitude and a limit of detection of $1\,U\,L^{-1}$ AP, for an enzymatic reaction time of 60 min (Diaz-Gonzalez et al., 2002). Serra et al. (2005) recently reported the use of an amperometric graphite–Teflon composite tyrosinase biosensor for the rapid monitoring of AP, with no need for an incubation step and using phenylphosphate as the substrate. Phenol generated by the action of AP was monitored at the tyrosinase composite electrode through the electrochemical reduction of the o-quinone produced to catechol, which produces a cycle between the tyrosinase substrate and the electroactive product, giving rise to the amplification of the biosensor response and to the sensitive detection of AP. The current was measured at –0.10 V 5 min after the addition of AP. A linear calibration plot was obtained for AP between 2.0×10^{-13} and 2.5×10^{-11}, with a detection limit of 6.7×10^{-14} M. Venetz et al. (1990) devised a kinetic assay for APs based on the hydrolytic cleavage of the P–F bond in monofluorophosphate and fluoride ion-selective electrode. AP catalyzes the hydrolytic cleavage of the P–F bond in monofluorophosphate with the subsequent release of fluoride ions. A kinetic potentiometric method was described in which a fluoride ion-selective electrode is used for the sensitive and selective measurement of the released F^- for the determination of AP activity. The reaction showed a well-defined correlation with the hydrolysis of the P–O bond in pNPP.

Harbron et al. (1991a) described a sensitive luminometric method to detect AP activity based on the hydrolysis of riboflavin phosphates (5′FMN or 4′FMN) to produce riboflavin, which in turn is converted to 5′FMN using riboflavin kinase and then assayed using the bacterial bioluminescent system from *Vibrio harveyi* or *Vibrio fischeri*. The most sensitive assay was obtained using 4′FMN, which can measure <20 amol (10^{-18} mol) after a 1-h incubation. Obzansky et al. (1991) developed a highly sensitive flavin adenine dinucleotide-3′-phosphate (FADP)-based enzyme amplification cascade that detects AP via the dephosphorylation of the substrate FADP to produce the cofactor FAD, which binds stoichiometrically to inactive apo D-amino acid oxidase (D-AAO). The resulting active holo D-AAO oxidizes D-proline to produce hydrogen peroxide, which is quantified by the horseradish peroxidase-mediated conversion of 3,5-dichloro-2-hydroxybenzenesulfonic acid and 4-aminoantipyrine to a colored product. The same group described an amplified colorimetric assay for APs that uses riboflavin 4′-phosphate (Harbron et al., 1991b). AP hydrolyzes riboflavin 4′-phosphate to produce riboflavin, which is converted to riboflavin 5′-phosphate, using riboflavin kinase, which reconstitutes apoglycolate oxidase to give hologlycolate oxidase. This enzyme catalyzes the oxidation of glycolate with simultaneous production of hydrogen peroxide that is detected via the formation of a colored product through the action of peroxidase.

The system allows the detection of 4 amol after a 2-h incubation. Subsequently, Harbron et al. (1992) designed a simple, robust and extremely sensitive colorimetric assay for AP to be used as a detection system in diagnostic assays employing antibodies or gene probes. The method was based on the principle of prosthetogenesis, according to which a purpose-designed substrate (a prosthetogen) for a primary analyte-linked enzyme label is hydrolyzed to produce a prosthetic group for a detector enzyme system. The prosthetogen employed was a derivative of FAD that is phosphorylated at the 3′-position of the ribose ring (FADP), the label enzyme was AP and the detector was a D-amino acid oxidase–horseradish peroxidase coupled system. Essentially each turnover of every molecule of AP produces a molecule of D-amino acid oxidase for detection. Thus enormous amplification of the initial signal is achieved in short time periods because of the relatively high turnover number of AP for FADP. The system could be formatted as a stable, preformed, freeze-dried preparation containing all analytical components, which can be reconstituted simply by the addition of buffer solution. This methodology was able to quantify <0.1 amol of AP in 30 min at 25 C using microtiter plates. Subsequently, Fisher et al. (1995) combined this amplification principle with a luminescent end-point using the luminol–peroxidase system to produce an enzyme-amplified chemiluminescent assay based also on the principle of prosthetogenesis. When the assay was used to detect AP in solution, the detection limit was 0.4 amol in a 5-min assay. The inter-assay variance ranged from 4 to 20% and 7 to 19% across the dynamic range of the assay for a chemiluminescent assay and an enhanced chemiluminescent assay, respectively, employing two different preparations of luminol.

As discussed by Hallaway and O'Kane (2000), the most popular chemiluminescence assays utilize substituted dioxetane phosphate substrates. However, several problems exist with these assays that must be overcome before they can be used in the clinical motoring of AP activity. First, they are very sensitive, so patient samples usually need to be diluted prior to assaying, which may result in loss of linearity as serum components are also diluted. Second, the time to achieve a stable glow emission with dioxetane phosphates is longer than with colorimetric procedures, which increases the turnround time for assay results. Third, polymeric enhancers utilized to increase signal response in one commercial substrate cause precipitation of the serum matrix and increase the time to a stable glow emission. Hallaway and O'Kane (2000) found that incorporating a fluorescein-labeled serum as a combined sample matrix diluent and fluorescence enhancer to replace the polymeric enhancers can circumvent these problems. They optimized a method using 25 mM 1,2-dioxetane phosphate as substrate and FITC-labeled serum. Blum et al. (2001) also developed a method of detecting AP using the chemiluminescent substrate disodium 3-(4-methoxyspiro{1,2-dioxetane-3,2′-(5′-chloro)tricyclo[3.3.1.1(3,7)]decan}-4-yl)phenyl phosphate (CSPD) for enhanced AP sensitivity and a simplified assay but for use in examining AP activity in cellular assays of osteodifferentiation. They found that the chemiluminescent detection system was four orders of magnitude more sensitive than the standard colorimetric method in this system. Kokado et al. (2002) developed a chemiluminescent assay of AP using dihydroxyacetone phosphate (DHAP) or its ketal (DHAP-ketal). The substrates were transformed to dihydroxyacetone (DHA) after being hydrolyzed by AP, which reacts with lucigenin and produces strong chemiluminescence. Under the optimum assay conditions, the detection limits were 3.8×10^{-19} and 1.5×10^{-18} mol of AP, respectively. The mechanism of the chemiluminescence response was speculated to be as follows: the O_2^- generated by the reaction of DHA and O_2 in alkaline solution reacts with lucigenin and then emits light. The assay was applied to the enzyme immunoassay of 17β-estradiol, using AP as a label enzyme. The measurable range of 17β-estradiol was 15–4000 pg mL^{-1} and the proposed method was four times more sensitive than the colorimetric assay for AP by using pNPP as substrate. In another sensitive assay, adenosine-3'-phosphate-5'-phosphosulfate was hydrolyzed by AP to produce adenosine-5'-phosphosulfate, which was then converted into ATP by ATPsulfurylase in the presence of pyrophosphate. The ATP produced was detected by the luciferin–luciferase reaction. The measurable range was 1 zmol–100 fmol per assay (Arakawa et al., 2003).

16.2
Using Inhibitors and Heat Inactivation

Methods to quantify individual AP isozymes based on their differential heat stability and sensitivity to inhibitors were described in the 1970s (McComb et al., 1979). Classic among those papers are that by Green et al. (1971), which combined heat inactivation with the use of L-Phe inhibition to distinguish liver AP and IAP from "other" AP components; methods using L-hArg and levamisole to inhibit TNAP activity (Lin and Fishman, 1972; Van Belle, 1976), methods using L-hArg, L-Phe, L-Leu, L-Phe-Gly-Gly and L-Leu-Gly-Gly to discriminate each of the isozymes (Mulivor et al., 1978b) and the simplified heat inactivation method of Moss and Whitby (1975) where the liver TNAP isoform was determined in serum AP from measurements of residual activity made after incubating the samples for 15 and 25 min at 56 °C.

Since then, improvements to those early methodologies have been reported. Farley et al. (1981) used those approaches to quantitative skeletal AP activity in human serum but used organ-derived internal standards of skeletal, intestinal and biliary APs to minimize between-assay variation. Another method was devised by Kuwana and Rosalki (1991) to measure plasma IAP by inhibiting the nonintestinal component utilizing L-p-bromotetramisole as inhibitor. The method correlated well with measurements by an immunocapture assay. If carried out in parallel with wheat-germ agglutinin (WGA) precipitation of bone TNAP, subtraction of IAP activity from that of nonbone AP in the supernatant could be used to measure the TNAP that originates from the liver in men and nonpregnant women.

16.3
Electrophoretic Methods

Griffiths and Black (1987) developed an isoelectric focusing procedure for resolving AP isozymes and isoforms in serum. They used a thin-layer agarose gel film containing synthetic carrier ampholytes and a separator to flatten the pH gradient in the region of the isozyme and isoform isoelectric points. Sharp, highly resolved zones of enzyme activity were obtained by limiting diffusion by rapidly coupling the released product, 1-naphthol, to a diazonium salt, which forms a colored precipitate at the site of activity. With this procedure, the authors were able to resolved and identify 12 zones of AP activity in the serum of healthy individuals within a wide age range. The combination of isoelectric focusing and cellulose acetate electrophoresis also provided additional information concerning the origin of IAP bands in patients' sera (Griffiths et al., 1992). Van Hoof et al. (1990) separated AP isozymes in 1383 sera of normal individuals (aged 4–65 years) by agarose electrophoresis with the Isopal system. As expected, the predominant isozyme in children was of bone origin and almost all (99%) of the children had low activities of a second bone fraction, "bone variant" ALP. The "bone variant" disappeared after age 17 in girls and after age 20 in boys. The highest bone ALP activity was reached at age 9–10 in girls and 13–14 in boys, followed by a gradual decline in girls and a steep decline in boys. During adulthood, activity of the bone fraction was constant and no significant differences were observed between sexes, either for bone or for liver AP activity. The latter remained almost unchanged throughout life. The authors observed no high molecular mass (high-M_r) AP activity in children, whereas sera from 60% of the adults contained low activities of high-M_r AP. IAP (soluble form) and "intestinal variant" AP (hydrophobic form) were frequently present, in 21 and 37% of all samples, respectively. No significant differences were observed between age groups and sexes for the IAP isoforms (Van Hoof et al., 1990). Overlapping migration of IAP and bone AP are often found using the usual 7.5% polyacrylamide gel electrophoresis (PAGE) conditions, particularly in nonsecretor subjects after a high-fat meal. Matsushita et al. (2000) demonstrated that using instead 6.0% PAGE in the presence of 1% Triton X-100 both IAP and bone AP in serum were clearly separated regardless of the ABO blood group and the secretor status of the subjects.

16.4
Lectin-based Assays

Rosalki and Foo (1984) described two methods for the separation and quantification of the bone- and liver-type TNAP isoforms in plasma. In one approach, they used wheat-germ agglutinin (WGA) to precipitate the bone isoform preferentially. The activity of the bone isoform was calculated from measuring the AP activity in the precipitate and that of the liver isoform by subtracting the activity of the bone isoform from total AP activity. The liver fraction would also contain biliary, IAP and PLAP if these were present in the original plasma, but correction for such activity could be readily made. In the second method, the samples were separated on cellulose acetate membranes that, before electrophoresis, had been soaked in buffer containing WGA. The bone isoform was retarded and clearly separated from the liver fraction. Enzyme activity was demonstrated by staining using an indigogenic AP substrate incorporated into the agar gel and the stained fractions quantified by densitometry. Later the authors concluded that this second procedure was especially suitable for use in the diagnostic laboratory (Rosalki and Foo, 1989). Schreiber and Whitta (1986) found that the bone-type TNAP–lectin complex had a lower electrophoretic mobility than did the native unmodified bone isoform and by incorporating the lectin into an agarose gel, they could completely separate the bone- and liver-type TNAP isoforms. With this electrophoretic method, the liver, biliary and bone isoforms of TNAP were clearly separated on agarose gels. Activity staining with an indigogenic dye substrate revealed that the liver TNAP isoform migrated nearest the anode, followed by the biliary and bone TNAP isoforms. The results were similar to those of electrophoresis on cellulose acetate. However, the lectin–agarose gels resolved better the liver and bone isoforms and heat treatment of samples was not required before electrophoresis. Behr and Barnet (1986) slightly modified the original method of Rosalki and Foo and reported that activities of bone-type TNAP nearly always correlated with the clinical diagnosis. Only patients with hepatitis often had pathological bone activities not in accord with the other findings. Sorensen (1988) suggested standardizing the procedure by using a WGA concentration that would precipitate half of the AP activity of serum pooled from an equal number of healthy women and men. With those modifications, Sorensen obtained results for the isozymes in healthy subjects that agreed better with those obtained by the heat-inactivation methods.

Gonchoroff et al. (1989) used WGA affinity chromatography as a tool to investigate the structure of APs and to obtain fractions enriched in either bone or liver TNAP activity. The liver and bone isoforms in serum samples were incompletely resolved except that the activity in the nonretained fraction always represented pure liver-type isoform and constituted a larger percentage of total activity in pooled sera with increased liver TNAP activity than in pooled sera with increased bone TNAP activity. In contrast, a more avidly retained TNAP activity, presumably with high glycosylation, was found in human serum with high activity of bone-type TNAP. Burlina et al. (1991) also evaluated a method for quantifying the bone TNAP isoform utilizing WGA to precipitate this fraction. In precision studies, CVs ranged from 3.2 to 11.4% (within-day) and from 3.7 to 11.5% (between-day). The assay procedure was linear over a wide range of activities and was easily adapted to automated kinetic measurements. Comparison of the precipitation method with an affinity electrophoretic method, which utilizes cellulose acetate as a support, demonstrated a satisfactory correlation coefficient ($r = 0.886$). The authors found the method suitable for routine determination of bone-type TNAP and for the screening of bone metastases. Because of its technical simplicity and satisfactory analytical performance, it could be used instead of the heat-inactivation procedure. Another method was described by Ramasamy (1991) for the separation of liver and bone TNAP isoforms in serum using WGA affinity electrophoresis in a polyacrylamide gel matrix. The electrophoretic mobilities of liver-type TNAP and IAP are essentially not affected by lectin, but the bone TNAP isoform is retarded and separated from the liver fraction. Affinity electrophoresis in polyacrylamide gel, combined with agarose gel electrophoresis and a solid-phase linked antibody precipitation procedure for IAP, allowed the various AP fractions, biliary, liver, bone and intestinal, to be quantified. Crofton (1992) developed a modified lectin affinity electrophoresis method suitable for simultaneous measurement of liver, bone and high-M_r isoforms of TNAP in children. Day et al. (1992b) evaluated the performance of two methods for the analysis of AP isozymes designed for use in the routine chemical pathology laboratory: pre-incubation with neuraminidase before agarose electrophoresis and selective precipitation of the bone-type TNAP with WGA. They found a good correlation between the neuraminidase and WGA electrophoretic methods. The WGA precipitation method showed negative interference in the measurement of bone-type TNAP activity in samples containing biliary TNAP. They concluded that the neuraminidase electrophoretic method was a satisfactory alternative to the WGA affinity electrophoretic method, although it was more expensive, and that the WGA precipitation method cannot be recommended for use with serum samples from patients with suspected liver disease.

16.5
High-performance Liquid Chromatographic Methods

As presented in the previous section, Gonchoroff et al. (1989) had reported that the bone and liver TNAP isoforms were incompletely resolved by low-performance WGA affinity chromatography. Expanding upon this work, this group subsequently reported the first use of high-performance affinity chromatography (HPAC) in the separation of bone- and liver-type TNAP isoforms with use of WGA conjugated to 7 μm diameter silica particles and an eluent containing N-acetyl-D-glucosamine (NAG) (Anderson et al., 1990). On-line spectrophotometric detection of AP involved pumping diethanolamine-buffered pNPP solution post-column. Bone- and liver-type TNAP could be separated into two peaks with only 10% overlap when an exponential gradient was used. A linear step gradient separated 80.9% of liver-type TNAP and 91.6% of bone-type TNAP in two distinct peaks. True bone- and liver-type TNAP peak areas for the linear step gradient were determined by using correction factors, because each peak contained a co-eluted portion of the other TNAP isoform. The detection limit improved 10-fold over those of other techniques for TNAP isozymes, owing to the relatively large sample that could be applied to the column. Gonchoroff et al. (1991) evaluated this newly described HPAC method for the separation of human bone- and liver-type TNAP. Results obtained with the HPAC method correlated significantly well with bone AP levels obtained on the same patients by solid-phase immunoassay using MAbs specific for TNAP. Discordant results between these two methods were obtained in patients with primary biliary cirrhosis, where bone AP activity was undetectable by solid phase immunoassay but a fraction indistinguishable from bone AP was detected by HPAC.

Magnusson et al. (1993) optimized the use of HPLC for the separation, analysis and quantitation of bone- and liver-type TNAP. Factors affecting separation (analytical column type, column temperature, mobile phase buffer, mobile phase salt, gradient elution, flow-rate) and reaction conditions (substrate type, substrate concentration, fluorescent substrates, substrate buffer type, substrate buffer concentration, substrate pH, substrate activators, substrate detergent, substrate flow-rate, reaction temperature, postcolumn reaction detection) were evaluated. At least six different AP isoform peaks could be separated and quantified by weak anion-exchange HPLC in serum from healthy individuals: three bone AP isoform peaks (B/I, B1 and B2) and three liver AP isoform peaks (L1, L2 and L3) (see Fig. 42, Section 9.1.1). In healthy adults, the three bone isoforms, B/I, B1 and B2, account on average for 4, 16 and 37%, respectively, of the total serum AP activity (Magnusson et al., 1999). In serum, the minor fraction B/I is not a pure bone isoform as it co-elutes with the IAP isozyme and is composed, on average, of 70% bone and 30% IAP.

Some biochemical differences between the bone AP isoforms have been reported (Magnusson et al., 2002c), however, these isoforms are similar with respect to freeze–thaw stability, instability during concentration by vacuum centrifugation, solubility, heat inactivation, concanavalin A precipitation, reaction kinetics and inhibition by L-phenylalanine, L-homoarginine and levamisole. The bone AP isoforms differ with respect to sensitivity to precipitation with WGA, i.e. B1 and B2 have more (or more reactive) sialic acid residues than B/I. The reported differences in the sialic acid contents result also in differences in molecular weights. Native gradient gel electrophoresis gave estimated molecular weights of 126, 136 and 141 kDa for the B/I, B1 and B2 isoforms, respectively. These differences could be explained by the differences in the content of sialic acid residues since desialylation of B1 and B2 with neuraminidase decreased the apparent molecular weight to 127 kDa for both B1 and B2, that is, to approximately that observed for B/I (126 kDa). No change in molecular weight was observed for the B/I isoform after treatment with neuraminidase. It has also been reported that the sialic acid residues present in bone AP affect the immunoreactivity of MAbs against bone AP (Magnusson et al., 2002c; Section 6.3). It is of interest that preparations of bone AP from human osteosarcoma (SaOS-2) cells are currently used as calibrators in these immunoassay kits. A higher proportion of the bone AP isoform B/I (39–80%), compared with human serum, has been reported in these calibrators (Fig. 49) (Magnusson et al., 2001). In serum from healthy subjects, B/I has a lower activity than either the B1 and B2 isoforms and represents, on average, only 4% of the total AP activity measured (Magnusson et al., 1999). Some differences have been reported between these bone AP immunoassays in groups of patients who express different ratios of the bone AP isoforms, which could be explained by differences between the antibodies used in these immunoassays.

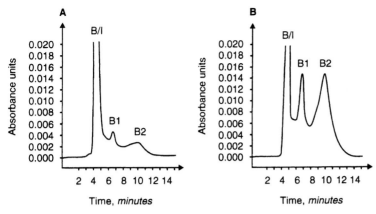

Fig. 49 Chromatographic profile of bone AP used as calibrators. (A) High bone AP calibrator of the Alkphase-B assay (140 U L^{-1}). Peaks, activities, retention times and peak heights, in order of elution, are as follows: B/I, 112 U L^{-1}, 4.83 min, 0.0662 AU (absorbance units); B1, 13 U L^{-1}, 6.77 min, 0.0040 AU; B2, 15 U L^{-1}, 9.92 min, 0.0022 AU.

(B) High BAP calibrator of the Tandem-R Ostase assay (120 µg L^{-1}). Peaks, activities, retention times and peak heights in order of elution are as follows: B/I, 47 µg L^{-1}, 4.87 min, 0.0505 AU; B1, 24 µg L^{-1}, 6.83 min, 0.0148 AU; B2, 49 µg L^{-1}, 9.82 min, 0.0149 AU. Taken from Magnusson et al. (2001) and reproduced with permission from Kidney International.

16.6
Specific Immunoassays with Polyclonal and MAbs

One of the first immunoassays for APs was a radioimmunoassay using polyclonal anti-PLAP antibodies for the quantitation of this isozyme during pregnancy (Holmgren et al., 1978). This was a double antibody solid-phase radioimmunoassay, highly sensitive with a minimum detectable dose of 9 ng of protein per milliliter. The assay was also specific for PLAP and was able to detect all common allozymes of PLAP equally well. With the described technique, the serum concentration of this enzyme during normal pregnancy was measured. A 25-fold increase from low levels during the first trimester up to 252 ± 70 ng mL^{-1} in gestation week 40 was observed. Subsequently, a highly sensitive ELISA also based on polyclonal antibodies was developed for the measurement of PLAP in serum and ascitis fluid (Millán and Stigbrand, 1981). The assay detected as little as 0.4 µg L^{-1}, significantly less than with a radioimmunoassay (RIA) performed with the same reagents (Nustad et al., 1984). It was highly specific for PLAP/GCAP and detected serum levels of PLAP/GCAP in an adult control population with an upper limit of normality of 1.85 µg L^{-1}. Concentrations of the analyte were increased in all pregnancy sera tested. Concentration and activity as measured by two different catalytic assays correlated well. Several MAb enzyme immunoassays (EIAs) were also developed at that time. McLaughlin et al. (1983) developed a solid-phase EIA using MAb H317 with specificity only for PLAP with a lower limit of detection of

0.1 U L^{-1} (~0.07 μg L^{-1}) PLAP. All plasma samples from healthy nonpregnant individuals had undetectable PLAP levels whereas, in pregnancy, a wide range (0–400 U L^{-1}) of PLAP levels was noted for 208 maternal plasma samples from 12 weeks gestation to term. Similarly, De Groote et al. (1983) used the PLAP-specific MAb E6 to develop and compare a solid-phase EIA with a solid-phase "sandwich" RIA involving immobilized polyclonal rabbit anti-PLAP in combination with iodinated E6. The EIA was based on polystyrene beads (5 mm in diameter) activated by glutaraldehyde and coated with the E6 MAb. The beads were incubated with 200 μL of sample overnight and after repeated washing the beads were transferred to a fresh tube for measurement of bound PLAP activity. For the solid-phase RIA design, the beads were coated with polyclonal rabbit anti-PLAP, followed by incubation with test samples, washed and then incubated again for 1 h at room temperature with 200 μL of ^{125}I-labeled E6 MAb followed by washing and measurement of the bound radioactivity. Both assays provided comparable specificity and detection limits. The EIA was preferred owing to its simplicity, although it required longer incubation times (around 1 h at 37 °C) to reach a sensitivity comparable to that of the RIA (De Groote, et al., 1983). Millán et al. (1985a) used the mouse H7 MAb to PLAP to develop an EIA for PLAP and GCAP. The antibody was bound to sheep anti-mouse IgG covalently coupled to tosylated shell-and-core light (1.07 g cm^{-3}) monodisperse polymer particles. Adding the H7-bound polymer particle suspension to a PLAP-containing sample gave maximal binding of the antigen within 10 min. PLAP and GCAP remained active and bound to the solid-phase throughout all assay manipulations and could therefore be saved for future testing. The assay was highly versatile and its sensitivity (routinely 0.05 pg L^{-1}) could be increased 1000-fold by adjusting the sample volume and incubation time (sample volume was irrelevant between 50 μL and 5 mL). Basal activities of serum GCAP in both men and women were detected (Millán et al., 1985a). Hirano et al. (1986) used the HPMS-1 MAb specific for PLAP to develop a sensitive monoclonal immunocatalytic assay (MICA) involving covalently binding the MAb to a paper disc. The minimum amount of PLAP detectable by this method was 0.0025 King–Armstrong units. Good correlation with the heat-treatment method was obtained. Kinoshita et al. (1990) established seven MAbs to PLAP and used one of them (MAb 7C6, of the IgG2a class) to also develop an EIA. At least two commercial EIAs for PLAP were developed and marketed at that time. Innogenetics (Ghent, Belgium) used the MAb 327 developed by De Broe and Pollet (1988), which reacts with both PLAP and GCAP, to develop the Innotest hPLAP kit. Sangtec Medical (Bromma, Sweden) used the C2 MAb developed by Millán and Stigbrand (1983) to detect PLAP specifically, as this antibody reacts poorly with most GCAP allozymes. Hendrix et al. (1990) also developed an EIA of human PLAP and GCAP in serum. Four MAbs raised against PLAP and recognizing different epitopes were selected to study the influence of the following variables on the accuracy of PLAP and GCAP measurement: phenotype, molecular form and glycation pattern of PLAP and GCAP; incubation temperature; and interferences by serum during immunobinding. Nine GCAP phenotypes were identified interacting with each antibody at a lower affinity than was seen for the more common PLAP phenotypes. Antibody affinity was higher for the free hydrophilic dimeric forms of PLAP and GCAP and was not influenced by the degree of glycation. In serum or tissue extracts, measurement of PLAP or GCAP was most accurate when immunoincubations were performed at 37 °C, with use of antibodies 327 and 7E8, respectively. In addition, correct measurements were achieved only when, during immunobinding, serum is incubated with an equal volume of deoxycholate (9 g L^{-1} final concentration) (Hendrix et al., 1990).

Specific immunoassays were also developed against the other human AP isozymes. Brock et al. (1984) used MAb AAP-1, specific for IAP to develop an immunoassay for use in amniotic fluid samples. Values in the immunoassay correlated closely with those obtained by direct determination of phenylalanine-inhibitable AP. Hirano et al. (1987b) developed specific immunoassays for human IAP and TNAP by use of MAbs 2HIMS-1 and HLMS-1 respectively, using the paper disc method used previously for the PLAP assay (Hirano, et al., 1986). In sera from 40 healthy individuals, the activity of TNAP was determined as 32 ± 12 IU L^{-1}. The activity of IAP was found to be 10-fold lower, 3.5 ± 6.3 IU L^{-1}, and of PLAP another 10-fold lower, 0.3 ± 0.2 IU L^{-1}, than that of TNAP. Several normal tissues contained all three isozymes, the intestinal mucosa, for example, which besides IAP also expresses trace amounts of PLAP and TNAP. Bailyes et al. (1988) used purified meconium IAP as antigen in the preparation of seven MAbs. Two of these antibodies were specific for IAP and reacted with different epitopes. Both bound adult IAP better than fetal IAP. One of these antibodies was used to establish a capture assay for IAP in human serum over the range 0.5–16 U L^{-1}. Reference ranges of serum IAP concentrations were established in relation to blood groups. Measurement of IAP in the serum of pregnant women showed no correlation with pre-term fetal passage of meconium. Deng and Parsons (1988) developed a solid-phase immunoassay for high molecular weight AP in human sera using a specific murine MAb HMAP-1 to human high-M_r AP. The antibody did not cross-react with the liver- and bone-type TNAP isoforms, IAP or PLAP isozymes and did not react with AP in bile, so it appears to react with the other components of this high molecular weight complex. Nevertheless, this antibody, bound to nitrocellulose membrane discs, permitted the development of a MAb immunocatalytic assay. The method was rapid and reproducible, giving good correlation with results from cellulose acetate membrane electrophoresis and having the added advantages of increased sensitivity and specificity. A large number of sera could then be assayed to evaluate the significance of serum high-M_r AP in relation to other AP activities and to the differential diagnosis of liver disease.

Panigrahi et al. (1994) developed a two-site IRMA, later named Tandem-R Ostase (Hybritech), using specific MAbs for measuring skeletal TNAP in human serum. Assay calibration was based on mass units (micrograms per liter) and was established with purified bone-type TNAP from a human osteosarcoma cell line, SAOS-2 (see Fig. 49, Section 16.3). Precision studies demonstrated intra- and inter-assay CVs of 3–5 and 5–7%, respectively. Relative reactivity studies showed that the assay has a seven-fold preference for detecting bone AP compared with the liver isoform in serum. The normal reference interval for 478 healthy adults was 5–22 µg L^{-1}. Method comparison studies showed good correlation between the bone AP assay and commercially available electrophoretic methods. Garnero and Delmas (1993) evaluated this IRMA on serum samples from patients with liver disease and patients with Paget's disease and determined a liver TNAP cross-reactivity of the IRMA of 16% that was confirmed by electrophoresis of the circulating AP isoforms. The authors concluded that this new IRMA for bone AP was reliable, had a low cross-reactivity with the liver isoform and appeared to be more sensitive than total AP activity for the clinical investigation of patients with osteoporosis and other metabolic bone diseases. Van Hoof et al. (1995) compared the agarose electrophoresis method (Isopal, Beckman) and the IRMA Ostase assay, in 293 patients. Overall correlation between the two methods was good ($r = 0.92$), except (a) for low values of bone TNAP and (b) in some samples with high total liver TNAP activity, both due to considerable cross-reactivity of the anti-bone TNAP antibodies of the Ostase kit with liver TNAP. This interference was not constant and was not evenly distributed across all concentrations of bone TNAP. Low bone AP determined with the IRMA (≤ 5 µg L^{-1}) was confirmed by electrophoresis (≤ 21 U L^{-1}), but bone TNAP activity determined by electrophoresis to be low (≤ 21 U L^{-1}) was not correlated with the IRMA results. The authors concluded that the IRMA for quantifying bone TNAP is acceptable as a screening method. However, when high values for bone TNAP are found with the Ostase method, confirmation by electrophoresis remains necessary to rule out cross-reactivity with high amounts of liver TNAP. For detecting low bone AP activities, electrophoresis remained the method of choice.

Gomez et al. (1995) described an immunocapture method in which serum samples were added to a microtiter plate coated with a MAb against bone-type TNAP and incubated for 3 h at room temperature. After the unbound materials had been washed off, the bound TNAP activity was measured by adding pNPP as substrate. The assay demonstrated no cross-reactivity to IAP or PLAP and only 3–8% cross-reactivity to liver-type TNAP. The assay detected increased bone-type TNAP in sera from patients with osteoporosis, Paget's disease, osteomalacia or primary hyperparathyroidism. This assay was later commercialized by Metra Biosystems as Alkphase-B and later as Metra® BAP, as part of Quidel Corporation. Bouman et al. (1996) compared the performance characteristics of the WGA precipitation assay and this immunocapture assay. The within- and between-run imprecision of the immunoassay (3.6–4.2 and 3.6–7.7%) was comparable to that of the WGA assay. The mean cross-reactivity with liver-type TNAP appeared to be 4% in the WGA assay and 11% in the immunocapture assay. Correlation studies of the WGA assay and the immunocapture results with total AP demonstrated $r = 0.98$ and 0.96, respectively (Bouman et al., 1996). Hata et al. (1996) also evaluated this immunocapture assay and found that the relative activity of the antibody was 100% with bone AP, 8.7% with liver AP and 0% with PLAP and IAP APs. Intra- and inter-assay coefficients of variation were <4%. The sensitivity of the assay was 0.7 U L^{-1} and linearity extended from 2 to 140 U L^{-1}. The recovery of bone-specific TNAP standard added to serum was 94–106%. The correlation coefficient between this method and a PAGE method was 0.94. The authors concluded that this new immunoassay of bone-specific AP would be useful for clinical investigation of patients with osteoporosis or other metabolic diseases of bone. Price et al. (1997) performed a detailed evaluation of two immunoassay principles, i.e. the immunometric that detects the bound TNAP isoform with a labeled second antibody and the immunocapture assay that simply measured the captured activity. The authors found a degree of imprecision that allowed the discrimination of changes within the reference range. The cross-reactivity of the liver isoform was found to be between 7.1 and 12.7% when two different methods of assessment were used. The comparison of results with an electrophoretic procedure showed that the immunocapture method recovered less of the bone isoform in samples from children than in samples from patients with Paget's disease; no such difference was found with the immunometric method. Hence the immunocapture antibody discriminated between different bone isoforms in children whereas the immunometric assay did not. The specification for the assay indicated cross-reactivity with rabbit, pig, dog, sheep, goat, cow, horse, cynomolgous macaque and human, but no recognition of rat or mouse TNAP. Brommage et al. (1999) used this assay to evaluate bone turnover following ovariectomy in macaques and Allen et al. (2000) used it to monitor bone formation in dogs. Broyles et al. (1998) examined the analytical and clinical performance characteristics of Tandem-MP Ostase, a new microplate immunoassay for bone-type TNAP in human sera. Bone-type TNAP was bound to streptavidin-coated microwells by a single biotinylated anti-bone TNAP MAb. Antigen was detected by the addition of pNPP. The assay was performed at room temperature in <90 min. Imprecision was 2.3–6.1% with a detection limit of 0.6 µg L^{-1}. Method comparison of bone TNAP measurements with the Tandem-MP Ostase assay and the mass-based Tandem-R Ostase assay revealed closely comparable results on all samples. The authors concluded that the Tandem-MP Ostase assay for serum bone TNAP is a rapid, simple, robust non-isotopic alternative to the Tandem-R Ostase immunoradiometric assay that provides an accurate and sensitive assessment of bone turnover.

16.7
mRNA-based Assays

Northern blot analysis represents the most commonly used procedure to monitor changes in mRNA expression in cell types or tissues and has been used in hundreds of papers to monitor changes in the mRNA level of a particular AP isozyme. However, other procedures have also been adapted to detecting each of the human and mouse AP mRNA levels simultaneously. Schär et al. (1997) developed an RT-PCR approach to distinguish between the RNA transcripts of all four human AP genes. Despite the high degree of sequence conservation, the authors devised 5' PCR primers residing in the 5' UTR of exon I that were unique for each isozyme mRNA whereas the 3' primer spanned the splicing site between exons II and III for each gene. This procedure had high specificity and high sensitivity and the authors used it to examine samples of TGCTs, where they found that expression of individual AP isozyme genes depends on the degree of differentiation of the tumors, but that simultaneous upregulation of all AP isozymes in all types of TGCTs does not occur (Section 10.2). In another study, Roelofs et al. (1999) used an RT-PCR approach in combination with a single nucleotide primer extension (SNuPE) assay to discriminate between PLAP and GCAP in TGCT samples. The first step involved RT-PCR amplification of both the PLAP and GCAP mRNAs using a common set of primers, one located in exon X and the other in exon XI. The cDNA template was then subjected to one cycle of amplification using either labeled dGTP or dATP using a forward SNuPE primer located in exon XI immediately upstream of codon 429, which differs between PLAP and GCAP (Section 4.2). Incorporation of dGTP denoted the presence of GCAP, while incorporation of dATP denoted the presence of PLAP. By use of this assay, Roelofs et al. (1999) were able to conclude that seminomas predominantly express GCAP whereas in embryonal carcinoma the ratio of PLAP and GCAP varies. Smans et al. (1999) used this same SNuPE assay to reveal changes in the expression of PLAP and GCAP in primary versus metastatic ovarian cancer samples (Section 10.3). Concerning the mouse genes, Hahnel et al. (1990) devised sets of specific primers for the *Akp2*, *Akp3* and *Akp5* mRNA and used them in RT-PCR experiments to identify their tissue-specific expression and in particular reported that both *Akp3* and *Akp5* are coexpressed in preimplantation mouse embryos (Section 2.2).

16.8
Histochemical and Immunohistochemical Detection

Ziomek et al. (1990) described a fluorescent histochemical technique for the detection of AP activities in cells. The technique utilizes standard azo dye chemistry with naphthol AS-MX phosphate as substrate and Fast Red TR as the diazonium salt. The reaction product is a highly fluorescent red precipitate. Preimplantation mouse embryos were used to establish optimal fixation and staining protocols and the specificity and sensitivity of the method. Fixation was in 4% paraformaldehyde for 1 h, as glutaraldehyde induced autofluorescence of the cells. Maximum staining was detected after 15–20 min in the stain solution. The stain solution itself proved to be nonfluorescent, thus allowing visual observation of the progress of the staining reaction by fluorescence microscopy in its presence. Similarly, Dikow et al. (1990) synthesized and purified menadiol diphosphate as a new substrate of APs. The menadiol released by AP action could be assayed by its reduction of tetrazolium salts or it could be coupled with diazonium salts. For both qualitative and semiquantitative histochemistry and immunohistochemistry, the best results were obtained by applying the method with nitro-blue tetrazolium to acetone–chloroform-pretreated cryostat sections. Tetranitro-blue tetrazolium, benzothiazolylphthalhydrazidyl tetrazolium and various diazonium salts were less suitable. Fast Blue BB and VB produced satisfactory results. The nitro-blue tetrazolium method with menadiol diphosphate was found to be superior to existing methods employing azo, azoindoxyl or tetrazolium salts and to metal precipitation methods. In qualitative histochemistry and immunohistochemistry, the NBT–menadiol diphosphate method resulted in higher quantities of precisely localized stain. Semiquantitative histochemistry with minimal incubation revealed more favorable kinetics for the menadiol diphosphate method, especially when using nitro-blue tetrazolium (Dikow et al., 1990). Two additional substrate chromogens for AP detection, synthesized for use in molecular biology research, salmon and magenta phosphate, were shown by Avivi et al. (1994) also to have advantageous characteristics for immunocytochemistry. Their relatively delicate pink- and magenta-colored products do not mask the colors produced by other staining procedures. In addition, the reaction products of these substrates are insoluble in water, ethanol and xylene, permitting the use of regressive hematoxylin staining procedures and coverslipping in permanent resin-based media. Most importantly, when these AP substrates were used in double-label immunocytochemistry in combination with horseradish peroxidase–diaminobenzidine (HRP–DAB) and counterstained with hematoxylin, all three colors could be easily distinguished.

Another high-resolution, fluorescence-based method for localization of endogenous AP activity utilizes ELF (enzyme-labeled fluorescence)-97 phosphate, which yields an intensely fluorescent yellow–green precipitate at the site of enzymatic activity. In fact, the ELF-97 AP substrate provides a bright, photostable, fluorescent signal amplification method for fluorescence *in situ* hybridization (FISH) (Paragas et al., 1997). Cox and Singer (1999) compared the staining on cryosections from a variety of AP-positive tissues using ELF-97 phosphate, the Gomori method, BCIP–NBT and naphthol AS-MX phosphate coupled with Fast Blue BB (colored) and Fast Red TR (fluorescent) diazonium salts. Although each method localized endogenous AP to the same specific sample regions, they found that sections labeled using ELF-97 phosphate exhibited significantly better resolution. The enzymatic product remained highly localized to the site of enzymatic activity, whereas signals generated using the other methods diffused. The authors found that the ELF-97 precipitate was more photostable than the Fast Red TR azo dye adduct. Using ELF-97 phosphate in cultured cells, they were able to detect an intracellular activity that was only weakly labeled with the other methods, but co-localized with an antibody against AP, suggesting that the ELF-97 phosphate provided greater sensitivity. They also found that detecting endogenous AP with ELF-97 phosphate was compatible with the use of antibodies and lectins (Cox and Singer, 1999). Similarly, Halbhuber et al. (2002) reported a high-resolution fluorescence method to localize AP activity in cells and tissue sections based on a naphthol–AS azo coupling procedure (Jenfluor® AP), which produced amorphous photostable fluorescent final reaction products without any diffusion artifacts.

References

Abu-Hasan, N. S. and R. G. Sutcliffe (1984) Molecular heterogeneity of human placental alkaline phosphatase associated with microvillous membranes. *Prog. Clin. Biol. Res.* **166**: 117–126.

Albert, J. L., S. A. Sundstrom and C. R. Lyttle (1990) Estrogen regulation of placental alkaline phosphatase gene expression in a human endometrial adenocarcinoma cell line. *Cancer Res.* 50: 3306–3310.

Ali, S. Y., S. W. Sajdera and H. C. Anderson (1970) Isolation and characterization of calcifying matrix vesicles from epiphyseal cartilage. *Proc. Natl. Acad. Sci. USA* **67**: 1513–1520.

Alkhoury, F. M. D. Malo, M. Mozumder, G. Mostafa and R. A. Hodin (2005) Differential regulation of intestinal alkaline phosphatase gene expression by Cdx1 and Cdx2. *Am. J. Phys. – Gastrointest. Liver Physiol.* **289**: G285–G290.

Allen, G., F. J. Bolton, D. R. Wareing, J. K. Williamson and P. A. Wright (2004) Assessment of pasteurization of milk and cream produced by on-farm dairies using a fluorimetric method for alkaline phosphatase activity. *Commun. Dis. Public Health* **7**: 96–101.

Allen, L. C., M. J. Allen, G. J. Breur, W. E. Hoffmann and D. C. Richardson (2000) A comparison of two techniques for the determination of serum bone-specific alkaline phosphatase activity in dogs. *Res. Vet. Sci.* **68**: 231–235.

Alpers, D. H., K. DeSchryver-Kecskemeti, C. L. Goodwin, C. A. Tindira, H. Harter and E. Slatopolsky (1988) Intestinal alkaline phosphatase in patients with chronic renal failure. *Gastroenterology* **94**: 62–67.

Alpers, D. H., R. Eliakim and K. DeSchryver-Kecskemeti (1990) Secretion of hepatic and intestinal alkaline phosphatases: similarities and differences. *Clin. Chim. Acta* **186**: 211–223.

Alpers, D. H., A. Mahmood, M. Engle, F. Yamagishi and K. DeSchryver-Kecskemeti (1994) The secretion of intestinal alkaline phosphatase (IAP) from the enterocyte. *J. Gastroenterol.* 29, Suppl. **7**: 63–67.

Alpers, D. H., Y. Zhang and D. J. Ahnen (1995) Synthesis and parallel secretion of rat intestinal alkaline phosphatase and a surfactant-like particle protein. *Am. J. Physiol.* **268**: E1205–E1214.

Alvaro, D., A. Benedetti, L. Marucci, M. Delle Monache, R. Monterubbianesi, E. Di Cosimo, L. Perego, G. Macarri, S. Glaser, G. Le Sage and G. Alpini (2000) The function of alkaline phosphatase in the liver: regulation of intrahepatic biliary epithelium secretory activities in the rat. *Hepatology* **32**: 174–184.

Amante, A., P. Borgiani, A. Gimelfarb and F. Gloria-Bottini (1996) Interethnic variability in birth weight and genetic background: a study of placental alkaline phosphatase. *Am. J. Phys. Anthropol.* **101**: 449–453.

Amthauer, R., K. Kodukula and S. Udenfriend (1992) Placental alkaline phosphatase: a model for studying C-terminal processing of phosphatidylinositol-glycan-anchored membrane proteins. *Clin. Chem.* **38**: 2510–2516.

Anderson, D. J., E. L. Branum and J. F. O'Brien (1990) Liver- and bone-derived isoenzymes of alkaline phosphatase in serum as determined by high-performance affinity chromatography. *Clin. Chem.* **36**: 240–246.

Anderson, H. C. and J. R. Reynolds (1973) Pyrophosphate stimulation of calcium uptake into cultured embryonic bones. Fine structure of matrix vesicles and their role in calcification. *Dev. Biol.* **34**: 211–227.

Anderson, H. C., H. H. Hsu, D. C. Morris, K. N. Fedde and M. P. Whyte (1997) Matrix vesicles in osteomalacic hypophosphatasia bone contain apatite-like mineral crystals. *Am. J. Pathol.* **151**: 1555–1561.

Anderson, H. C., J. B. Sipe, L. Hessle, R. Dhanyamraju, E. Atti, N. P. Camacho and J. L. Millán (2004) Impaired calcification around matrix vesicles of growth plate and bone in alkaline phosphatase-deficient mice. *Am. J. Pathol.* **164**: 841–847.

Anderson, H. C., R. Garimella and S. E. Tague (2005a) The role of matrix vesicles in growth plate development and biomineralization. *Front. Biosci.* **10**: 822–837.

Anderson, H. C., D. Harmey, N. P. Camacho, R. Garimella, J. B. Sipe, S. Tague, X. Bi, K. Johnson, R. Terkeltaub and J. L. Millán (2005b) Sustained osteomalacia of long bones despite major improvement in other hypophosphatasia-related mineral deficits in TNAP/NPP1 double deficient mice. *Am. J. Pathol.* **166**: 1711–1720.

Angelino, P. D., G. L. Christen, M. P. Penfield and S. Beattie (1999) Residual alkaline phosphatase activity in pasteurized milk heated at various temperatures – measurement with the fluorophos and Scharer rapid phosphatase tests. *J. Food Prot.* **62**: 81–85.

Anh, D. J., H. P. Dimai, S. L. Hall and J. R. Farley (1998) Skeletal alkaline phosphatase activity is primarily released from human osteoblasts in an insoluble form and the net release is inhibited by calcium and skeletal growth factors. *Calcif. Tissue Int.* **62**: 332–340.

Anh, D. J., A. Eden and J. R. Farley (2001) Quantitation of soluble and skeletal alkaline phosphatase and insoluble alkaline phosphatase anchor-hydrolase activities in human serum. *Clin. Chim. Acta* **311**: 137–148.

Anstrom, J. A., W. R. Brown, D. M. Moody, C. R. Thore, V. R. Challa and S. M. Block (2002) Temporal expression pattern of cerebrovascular endothelial cell alkaline phosphatase during human gestation. *J. Neuropathol. Exp. Neurol.* **61**: 76–84.

Antonarakis, S. E. (1998) Recommendations for a nomenclature system for human gene mutations. Nomenclature Working Group. *Hum. Mutat.* **11**: 1–3.

Arakawa, H., M. Shiokawa, O. Imamura and M. Maeda (2003) Novel bioluminescent assay of alkaline phosphatase using adenosine-3′-phosphate-5′-phosphosulfate as substrate and the luciferin–luciferase reaction and its application. *Anal. Biochem.* **314**: 206–211.

Arklie, J., J. Trowsdale and W. F. Bodmer (1981) A monoclonal antibody to intestinal alkaline phosphatase made against D98/AH-2 (HeLa) cells. *Tissue Antigens* **17**: 303–312.

Armesto, J., E. Hannappel, K. Leopold, W. Fischer, R. Bublitz, L. Langer, G. A. Cumme and A. Horn (1996) Microheterogeneity of the hydrophobic and hydrophilic part of the glycosylphosphatidylinositol anchor of alkaline phosphatase from calf intestine. *Eur. J. Biochem.* **238**: 259–269.

Ashwell, G. and J.Harford (1982) Carbohydrate-specific receptors of the liver. *Annu. Rev. Biochem.* **51**: 531–554.

Athanassiou, A., D. Pectasides, K. Pateniotis, L. Tzimis, A. Lafi, A. Cross and A. Epenetos (1990) ^{131}I-labeled monoclonal antibody H17E2 in the detection of subcutaneous metastasis from gestational choriocarcinoma. *Anticancer Res.* **10**: 913–915.

Avivi, C., O. Rosen and R. S. Goldstein (1994) New chromogens for alkaline phosphatase histochemistry: salmon and magenta phosphate are useful for single- and double-label immunohistochemistry. *J. Histochem. Cytochem.* **42**: 551–554.

Awatramani, R., P. Soriano, J. J. Mai and S. Dymecki (2001) An Flp indicator mouse expressing alkaline phosphatase from the ROSA26 locus. *Nat. Genet.* **29**: 257–259.

Ayers, G. J., A. J. Baldwin, A. M. Fowler, J. H. Goudie and D. Burnett (1989) Theophylline assay on Kodal Ektachem DTSC-Performance and interference by structurally-related compounds and salicylate. *Ann. Clin. Biochem.* **26**: 268–273.

Bacci, G., P. Picci, S. Ferrari, M. Orlandi, P. Ruggieri, R. Casadei, A. Ferraro, R. Biagini and A. Battistini (1993) Prognostic significance of serum alkaline phosphatase measurements in patients with osteosarcoma treated with adjuvant or neoadjuvant chemotherapy. *Cancer* **71**: 1224–1230.

Bacci, G., A. Longhi, S. Ferrari, S. Lari, M. Manfrini, D. Donati, C. Forni and M. Versari (2002) Prognostic significance of serum alkaline phosphatase in osteosarcoma of the extremity treated with neoadjuvant chemotherapy: recent experience at Rizzoli Institute. *Oncol. Rep.* **9**: 171–175.

Bachra, B. N. and H. R. Fischer (1968) Mineral deposition in collagen *in vitro*. *Calcif. Tissue Res.* **2**: 343–352.

Bailey, D., A. Marks, M. Stratis and R. Baumal (1991) Immunohistochemical staining of germ cell tumors and intratubular malignant germ cells of the testis using antibody to placental alkaline phosphatase and a monoclonal anti-seminoma antibody. *Mod. Pathol.* **4**: 167–171.

Bailyes, E. M., R. N. Seabrook, J. Calvin, G. A. Maguire, C. P. Price, K. Siddle and J. P. Luzio (1987) The preparation of monoclonal antibodies to human bone and liver alkaline phosphatase and their use in immunoaffinity purification and in studying these enzymes when present in serum. *Biochem. J.* **244**: 725–733.

Bailyes, E. M., P. M. Seymour, I. Fulton, C. P. Price and J. P. Luzio (1988) A monoclonal antibody capture assay for intestinal alkaline phosphatase and the measurement of this isoenzyme in pregnancy. *Clin. Chim. Acta* **172**: 267–274.

Balsan, S., M. Garabedian, M. Larchet, A. M. Gorski, G. Cournot, C. Tau, A. Bourdeau, C. Silve and C. Ricour (1986) Long-term nocturnal calcium infusions can cure rickets and promote normal mineralization in hereditary resistance to 1,25-dihydroxyvitamin D. *J. Clin. Invest.* **77**: 1661–1667.

Bao, R., M. Selvakumaran and T. C. Hamilton (2000) Use of a surrogate marker (human secreted alkaline phosphatase) to monitor *in vivo* tumor growth and anticancer drug efficacy in ovarian cancer xenografts. *Gynecol. Oncol.* **78**: 373–379.

Barger, A., R. Graca, K. Bailey, J. Messick, L. P. de Lorimier, T. Fan and W. Hoffmann (2005) Use of alkaline phosphatase staining to differentiate canine osteosarcoma from other vimentin-positive tumors. *Vet. Pathol.* **42**: 161–165.

Barka, T., S. Henderson and H. M. van der Noen (2000) Passive immunotherapy of mice bearing Ehrlich ascites tumor expressing human, membrane-bound placental alkaline phosphatase. *Tumor Biol.* **21**: 145–152.

Beck, R. and H. Burtscher (1994) Expression of human placental alkaline phosphatase in *Escherichia coli*. *Protein Expr. Purif.* **5**: 192–197.

Beck, R., H. Crooke, M. Jarsch, J. Cole and H. Burtscher (1994) Mutation in dipZ leads to reduced production of active human placental alkaline phosphatase in *Escherichia coli*. *FEMS Microbiol. Lett.* **124**: 209–214.

Beckman, G. (1970) Placental alkaline phosphatase, relation between phenotype and enzyme activity. *Hum. Hered.* **20**: 74–80.

Beckman, G. and L. Beckman (1975) The placental alkaline phosphatase variant D and natural selection. Relationship to complications of pregnancy and to "the Regan isoenzyme". *Hereditas* **81**: 85–88.

Beckman, L. and G. Beckman (1968) A genetic variant of placental alkaline phosphatase with unusual electrophoretic properties. *Acta Genet. Stat. Med.* **18**: 543–552.

Beckman, L., G. Bjorling and C. Christodoulou (1966) Pregnancy enzymes and placental polymorphism. I. Alkaline phosphatse. *Acta Genet. Stat. Med.* **16**: 59–73.

Beckman, G., L. Beckman and S. S. Magnusson (1972a) Placental alkaline phosphatase phenotypes and prenatal selection. Evidence from studies of spontaneous and induced abortions. *Hum. Hered.* **22**: 473–480.

Beckman, L., G. Beckman and S. S. Magnusson (1972b). Relationship between placental alkaline phosphatase phenotypes and the frequency of spontaneous abortion in previous pregnancies. *Hum. Hered.* **22**: 15–17.

Beckman, G., L. Beckman, E. Lundgren, J. L. Millán and C. Sikstrom (1989) Correlation between *Rsa*I restriction fragment length polymorphism and electrophoretic types of human placental alkaline phosphatase. *Hum. Hered.* **39**: 41–45.

Beckman, G., L. Beckman, A. Kivela, J. L. Millán and C. Sikstrom (1991) A new *Pst*I restriction fragment length polymorphism (RFLP) of placental alkaline phosphatase. RFLP haplotypes and correlation with electrophoretic types. *Hum. Hered.* **41**: 122–128.

Beckman, G., L. Beckman, C. Sikstrom and J. L. Millán (1992) DNA polymorphism of alkaline phosphatase isozyme genes: linkage disequilibria between placental and germ-cell alkaline phosphatase alleles. *Am. J. Hum. Genet.* **51**: 1066–1070.

Beckman, G., L. Beckman, C. Wennberg, C. Sikstrom and J. L. Millán (1994) *Pst*I restriction fragment length polymorphism of the human intestinal alkaline phosphatase gene. *Hum. Hered.* **44**: 175–177.

Beertsen, W. and T. Van den Bos (1992) Alkaline phosphatase induces the mineralization of sheets of collagen implanted subcutaneously in the rat. *J. Clin. Invest.* **89**: 1974–1980.

Beertsen, W., T. Van denBos and V. Everts (1999) Root development in mice lacking functional tissue non-specific alkaline phosphatase gene: inhibition of acellular cementum formation. *J. Dent. Res.* **78**: 1221–1229.

Beever, J. E. and H. A. Lewin (1992) RFLP at the bovine liver, bone and kidney alkaline phosphatase (*ALPL*) locus. *Anim. Genet.* **23**: 577.

Behr, W. and J. Barnert (1986) Quantification of bone alkaline phosphatase in serum by precipitation with wheat-germ lectin: a simplified method and its clinical plausibility. *Clin. Chem.* **32**: 1960–1966.

Behrens, C. M., C. A. Enns and H. H. Sussman (1983) Characterization of human foetal intestinal alkaline phosphatase. Comparison with the isoenzymes from the adult intestine and human tumour cell lines. *Biochem. J.* **211**: 553–558.

Bellows, C. G., J. E. Aubin and J. N. Heersche (1991) Initiation and progression of mineralization of bone nodules formed *in vitro*: the role of alkaline phosphatase and organic phosphate. *Bone Miner.* **14**: 27–40.

Benham, F. J., M. S. Povey and H. Harris (1978) Placental-like alkaline phosphatase in malignant and benign ovarian tumors. *Clin. Chim. Acta* **86**: 201–215.

Benham, F. J., P. W. Andrews, B. B. Knowles, D. L. Bronson and H. Harris (1981a). Alkaline phosphatase isozymes as possible markers of differentiation in human testicular teratocarcinoma cell lines. *Dev. Biol.* **88**: 279–287.

Benham, F. J., G. Balaban, D. Boccelli and H. Harris (1981b). Modulation of the expression of alkaline phosphatase genes in a human malignant cell line passaged through nude mice. *Int. J. Cancer* **28**: 257–264.

Bentala, H., W. R. Verweij, A. Huizinga-Van der Vlag, A. M. van Loenen-Weemaes, D. K. Meijer and K. Poelstra (2002) Removal of phosphate from lipid A as a strategy to detoxify lipopolysaccharide. *Shock* **18**: 561–566.

Bentouati, L., M. S. Baboli, H. Hachem, M. Hamza, P. Canal and G. Soula (1990) Hyperphosphatasemia related to three intestinal alkaline phosphatase isoforms: biochemical study. *Clin. Chim. Acta* **193**: 93–102.

Benzinger, E. A., A. K. Riech, R. E. Shirley and K. R. Kucharik (1995) Evaluation of the specificity of alkaline phosphatase-conjugated oligonucleotide probes for forensic DNA analysis. *Appl. Theor. Electrophor.* **4**: 161–165.

Berger, J., E. Garattini, J. C. Hua and S. Udenfriend (1987a). Cloning and sequencing of human intestinal alkaline phosphatase cDNA. *Proc. Natl. Acad. Sci. USA* **84**: 695–698.

Berger, J., A. D. Howard, L. Gerber, B. R. Cullen and S. Udenfriend (1987b). Expression of active, membrane-bound human placental alkaline phosphatase by transfected simian cells. *Proc. Natl. Acad. Sci. USA* **84**: 4885–4889.

Berger, J., J. Hauber, R. Hauber, R. Geiger and B. R. Cullen (1988) Secreted placental alkaline phosphatase: a powerful new quantitative indicator of gene expression in eukaryotic cells. *Gene* **66**: 1–10.

Bernard, G. W. (1978) Ultrastructural localization of alkaline phosphatase in initial intramembranous osteogenesis. *Clin. Orthop. Relat. Res.* **135**: 218–225.

Besman, M. and J. E. Coleman (1985) Isozymes of bovine intestinal alkaline phosphatase. *J. Biol. Chem.* **260**: 11190–11193.

Best, R. G., R. E. Meyer and C. F. Shipley (1991) Maternal serum placental alkaline phosphatase as a marker for low birth weight: results of a pilot study. *South. Med. J.* **84**: 740–742.

Beumer, C., M. Wulferink, W. Raaben, D. Fiechter, R. Brands and W. Seinen (2003) Calf intestinal alkaline phosphatase, a novel therapeutic drug for lipopolysaccharide (LPS)-mediated diseases, attenuates LPS toxicity in mice and piglets. *J. Pharmacol. Exp. Ther.* **307**: 737–744.

Bhudhisawasdi, V., K. Muisuk, P. Areejitranusorn, C. Kularbkaew, T. Khampitak, O. T. Saeseow and S. Wongkham (2004) Clinical value of biliary alkaline phosphatase in non-jaundiced cholangiocarcinoma. *J. Cancer Res. Clin. Oncol.* **130**: 87–92.

Bieber, A. L., K. A. Tubbs and R. W. Nelson (2004) Metal ligand affinity pipettes and bioreactive alkaline phosphatase probes: tools for characterization of phosphorylated proteins and peptides. *Mol. Cell. Proteomics* **3**: 266–272.

Binder, T., T. Berg, W. Siegert and C. A. Schmidt (1995) PCR–SSCP: non-radioisotopic detection with biotinylated primers and streptavidin–alkaline phosphatase conjugate. *Biotechniques* **18**: 780–781.

Blum JS. R. H. Li, A. G. Mikos and M. A. Barry (2001) An optimized method for the chemiluminescent detection of alkaline phosphatase levels during osteodifferentiation by bone morphogenetic protein 2. *J. Cell. Biochem.* **80**: 532–537.

Bonucci, E., G. Silvestrini and P. Bianco (1992) Extracellular alkaline phosphatase activity in mineralizing matrices of cartilage and bone: ultrastructural localization using a cerium-based method. *Histochemistry* **97**: 323–327.

Borregaard, N., L. Christensen, O. W. Bejerrum, H. S. Birgens and I. Clemmensen (1990) Identification of a highly mobilizable subset of human neutrophil intracellular vesicles that contains tetranectin and latent alkaline phosphatase. *J. Clin. Invest.* **85**: 408–416.

Bortolato, M., F. Besson and B. Roux (1999) Role of metal ions on the secondary and quaternary structure of alkaline phosphatase from bovine intestinal mucosa. *Proteins* **37**: 310–318.

Bossi, M., M. F. Hoylaerts and J. L. Millán (1993) Modifications in a flexible surface loop modulate the isozyme-specific properties of mammalian alkaline phosphatases. *J. Biol. Chem.* **268**: 25409–25416.

Bouman, A. A., C. M. de Ridder, J. H. Nijhof, J. C. Netelenbos and H. A. Delemarre-v. d. Waal (1996) Immunoadsorption assay for bone alkaline phosphatase compared with wheat-germ agglutinin precipitation assay in serum from (pre)-pubertal girls. *Clin. Chem.* **42**: 1970–1974.

Boyer, S. H. (1961) Alkaline phosphatase in human sera and placentae. *Science* **134**: 1002–1004.

Breathnach, R. and J. R. Knowles (1977) Phosphoglycerate mutase from wheat-germ: studies with ^{18}O-labeled substrate, investigations of the phosphatase and phosphoryl transfer activities and evidence for a phosphoryl–enzyme intermediate. *Biochemistry* **16**: 3054–3060.

Brenna, O., M. Perrella, M. Pace and P. G. Pietta (1975) Affinity-chromatography purification of alkaline phosphatase from calf intestine. *Biochem. J.* **151**: 291–296.

Brennan, C. and J. Fabes (2003) Alkaline phosphatase fusion proteins as affinity probes for protein localization studies. *Sci. STKE* PL2.

Bretthauer, R. K. and F. J. Castellino (1999) Glycosylation of *Pichia pastoris*-derived proteins. *Biotechnol. Appl. Biochem.* **30** (Pt. 3): 193–200.

Brock, D. J. and L. Barron (1988) Measurement of placental alkaline phosphatase in maternal plasma as an indicator of subsequent low birthweight outcome. *Br. J. Obstet. Gynaecol.* **95**: 79–83.

Brock, D. J., L. Barron, D. Bedgood and V. Van Heyningen (1984) Prenatal diagnosis of cystic fibrosis using a monoclonal antibody specific for intestinal alkaline phosphatase. *Prenat. Diagn.* **4**: 421–426.

Brommage, R., C. Allison, R. Stavisky and J. Kaplan (1999) Measurement of serum bone-specific alkaline phosphatase activity in cynomolgus macaques. *J. Med. Primatol.* **28**: 329–333.

Brown, D. and G. L. Waneck (1992) Glycosylphosphatidylinositol-anchored membrane proteins. *J. Am. Soc. Nephrol.* **3**: 895–906.

Broyles, D. L., R. G. Nielsen, E. M. Bussett, W. D. Lu, I. A. Mizrahi, P. A. Nunnelly, T. A. Ngo, J. Noell, R. H. Christenson and B. C. Kress (1998) Analytical and clinical performance characteristics of Tandem-MP Ostase, a new immunoassay for serum bone alkaline phosphatase. *Clin. Chem.* **44**: 2139–2147.

Brun-Heath, I. A. Taillandier, J. L. Serre and E. Mornet (2005) Characterization of 11 novel mutations in the tissue non-specific alkaline phosphatase gene responsible for hypophosphatasia and genotype-phenotype correlations. *Mol. Genet. Metab.* **84**: 273–277.

Brunette, M. G. and V. W. Dennis (1982) Effects of L-bromotetramisole on phosphate transport by the proximal renal tubule: failure to demonstrate a direct involvement of alkaline phosphate. *Can. J. Physiol. Pharmacol.* **60**: 276–281.

Bublitz, R., J. Armesto, E. Hoffmann-Blume, M. Schulze, H. Rhode, A. Horn, S. Aulwurm, E. Hannappel and W. Fischer (1993) Heterogeneity of glycosylphosphatidylinositol-anchored alkaline phosphatase of calf intestine. *Eur. J. Biochem.* **217**: 199–207.

Bublitz, R., H. Hoppe, G. A. Cumme, M. Thiele, A. Attey and A. Horn (2001) Structural study on the carbohydrate moiety of calf intestinal alkaline phosphatase. *J. Mass Spectrom.* **36**: 960–972.

Bucay, N., I. Sarosi, C. R. Dunstan, S. Morony, J. Tarpley, C. Capparelli, S. Scully, H. L. Tan, W. Xu, D. L. Lacey, W. J. Boyle and W. S. Simonet (1998) Osteoprotegerin-deficient mice develop early onset osteoporosis and arterial calcification. *Genes Dev.* **12**: 1260–1268.

Burke, A. P. and F. K. Mostofi (1988a). Intratubular malignant germ cells in testicular biopsies: clinical course and identification by staining for placental alkaline phosphatase. *Mod. Pathol.* **1**: 475–479.

Burke, A. P. and F. K. Mostofi (1988b). Placental alkaline phosphatase immunohistochemistry of intratubular malignant germ cells and associated testicular germ cell tumors. *Hum. Pathol.* **19**: 663–670.

Burlina, A., M. Plebani, S. Secchiero, M. Zaninotto and L. Sciacovelli (1991) Precipitation method for separating and quantifying bone and liver alkaline phosphatase isoenzymes. *Clin. Biochem.* **24**: 417–423.

Butterworth, P. J. (1994) Time-dependent irreversible inhibition of bovine kidney alkaline phosphatase by oxidized adenosine. Use of this compound as a site-directed inhibitor for studying uncompetitive inhibition. *Cell. Biochem. Funct.* **12**: 263–266.

Buttery, J. E., C. R. Milner, P. Nenadovic and P. R. Pannall (1980) Detection of alkaline phosphatase/immunoglobulin complexes. *Clin. Chem.* **26**: 1620–1621.

Calabro-Jones, P. M., R. C. Fahey, G. D. Smoluk and J. F. Ward (1985) Alkaline phosphatase promotes radioprotection and accumulation of WR-1065 in V79-171 cells incubated in medium containing WR-2721. *Int. J. Radiat. Biol. Relat. Stud. Phys. Chem. Med.* **47**: 23–27.

Calhau, C., F. Martel, C. Hipolito-Reis and I. Azevedo (2000) Differences between duodenal and jejunal rat alkaline phosphatase. *Clin. Biochem.* **33**: 571–577.

Calvo, M.S., D.R. Eyre and C.M. Gundberg (1996) Molecular basis and clinical application of biological markers of bone turnover. *Endocrinol. Rev.* **17**: 333–368.

Camolezi, F. L., K. R. Daghastanli, P. P. Magalhaes, J. M. Pizauro and P. Ciancaglini (2002) Construction of an alkaline phosphatase-liposome system: a tool for biomineralization study. *Int. J. Biochem. Cell. Biol.* **34**: 1091–1101.

Carvalhal, A. V., S. S. Santos, J. Calado, M. Haury and M. J. Carrondo (2003) Cell growth arrest by nucleotides, nucleosides and bases as a tool for improved production of recombinant proteins. *Biotechnol. Prog.* **19**: 69–83.

Caswell, A. M., M. P. Whyte and R. G. Russell (1986) Normal activity of nucleoside triphosphate pyrophosphatase in alkaline phosphatase-deficient fibroblasts from patients with infantile hypophosphatasia. *J. Clin. Endocrinol. Metab.* **63**: 1237–1241.

Chakrabartty, A. and R. A. Stinson (1985a). Properties of membrane-bound and solubilized forms of alkaline phosphatase from human liver. *Biochim. Biophys. Acta* **839**: 174–180.

Chakrabartty, A. and R. A. Stinson (1985b). Tetrameric alkaline phosphatase in human liver plasma membranes. *Biochem. Biophys. Res. Commun.* **131**: 328–335.

Challa, V. R., D. M. Moody, W. R. Brown and D. Zagzag (2004) A morphologic study of the vasculature of malignant gliomas using thick celloidin sections and alkaline phosphatase stain. *Clin. Neuropathol.* **23**: 167–172.

Chandra, A. M., J. C. Woodard and A. M. Merritt (1998) Dysgerminoma in an Arabian filly. *Vet. Pathol.* **35**: 308–311.

Chang, G. G. and S. L. Shiao (1994) Possible kinetic mechanism of human placental alkaline phosphatase *in vivo* as implemented in reverse micelles. *Eur. J. Biochem.* **220**: 861–870.

Chang, T. C. and G. G. Chang (1984) Essential tyrosyl residues of human placental alkaline phosphatase. *Int. J. Biochem.* **16**: 1237–1243.

Chang, C. H. D. Angellis and W. H. Fishman (1980) Presence of the rare D-variant heat-stable, placental-type alkaline phosphatase in normal human testis. *Cancer Res.* **40**: 1506–1510.

Chang, G. G., S. C. Wang and F. Pan (1981) Periodate-oxidized AMP as a substrate, an inhibitor and an affinity label of human placental alkaline phosphatase. *Biochem. J.* **199**: 281–287.

Chang, G. G., M. S. Shiao, K. R. Lee and J. J. Wu (1990) Modification of human placental alkaline phosphatase by periodate-oxidized $1,N^6$-ethenoadenosine monophosphate. *Biochem. J.* **272**: 683–690, (1990)

Chang, W. S., K. R. Zachow and D. Bentley (1993) Expression of epithelial alkaline phosphatase in segmentally iterated bands during grasshopper limb morphogenesis. *Development* **118**: 651–663.

Chapin, R. E., J. L. Phelps, B. E. Miller and T. J. Gray (1987) Alkaline phosphatase histochemistry discriminates peritubular cells in primary rat testicular cell culture. *J. Androl.* **8**: 155–161.

Chen, H. and P. Leder (1999) A new signal sequence trap using alkaline phosphatase as a reporter. *Nucleic Acids Res.* **27**: 1219–1222.

Chen, R., E. I. Walter, G. Parker, J. P. Lapurga, J. L. Millán, Y. Ikehara, S. Udenfriend and M. E. Medof (1998) Mammalian glycophosphatidylinositol anchor transfer to proteins and post-transfer deacylation. *Proc. Natl. Acad. Sci. USA* **95**: 9512–9517.

Chen, S. S., K. K. Chen, A. T. Lin, Y. H. Chang, H. H. Wu, T. H. Hsu and L. S. Chang (1997) The significance of serum alkaline phosphatase bone isoenzyme in prostatic carcinoma with bony metastasis. *Br. J. Urol.* **79**: 217–220.

Chodirker, B. N., S. P. Coburn, L. E. Seargeant, M. P. Whyte and C. R. Greenberg (1990) Increased plasma pyridoxal-5′-phosphate levels before and after pyridoxine loading in carriers of perinatal/infantile hypophosphatasia. *J. Inher. Metab. Dis.* **13**: 891–896.

Chou, J. Y. and S. Takahashi (1987) Control of placental alkaline phosphatase gene expression in HeLa cells: induction of synthesis by prednisolone and sodium butyrate. *Biochemistry* **26**: 3596–3602.

Cimmino, M. A. and S. Accardo (1990) Changes in the isoenzyme pattern of alkaline phosphatase in patients with rheumatoid arthritis. *Clin. Chem.* **36**: 1376–1377.

Cimmino, M. A., L. Buffrini, G. Barisone, M. Bruzzone and S. Accardo (1990) Alkaline phosphatase activity in the serum of patients with rheumatoid arthritis. *Z. Rheumatol.* **49**: 143–146.

Cirera Nogueras, L., J. Vivancos Lleida, M. Salazar Badia, G. Ercilla Gonzalez, F. Ballesta Martinez, C. Martin Vega and J. Carbonell Abello (1982) Raised serum alkaline phosphatase activity in one family. *Arch. Intern. Med.* **142**: 188–189.

Claeys, W. L., L. R. Ludikhuyze, A. M. van Loey and M. E. Hendrickx (2001) Inactivation kinetics of alkaline phosphatase and lactoperoxidase and denaturation kinetics of beta-lactoglobulin in raw milk under isothermal and dynamic temperature conditions. *J. Dairy Res.* **68**: 95–107.

Claeys, W. L., A. M. Van Loey and M. E. Hendrickx (2002) Kinetics of alkaline phosphatase and lactoperoxidase inactivation and of beta-lactoglobulin denaturation in milk with different fat content. *J. Dairy Res.* **69**: 541–553.

Coburn, S.P. (1996) Modeling vitamin B_6 metabolism. *Adv. Food. Nutr. Res.* **40**: 107–132.

Coburn, S. P., J. D. Mahuren, M. Jain, Y. Zubovic and J. Wortsman (1998) Alkaline phosphatase (EC 3.1.3.1) in serum is inhibited by physiological concentrations of inorganic phosphate. *J. Clin. Endocrinol. Metab.* **83**: 3951–3957.

Coburn, S. P., A. Slominski, J. D. Mahuren, J. Wortsman, L. Hessle and J. L. Millán (2003) Cutaneous metabolism of vitamin B-6. *J. Invest. Dermatol.* **120**: 292–300.

Coleman, J. E. (1992) Structure and mechanism of alkaline phosphatase. *Annu. Rev. Biophys. Biomol. Struct.* **21**: 441–483.

Cooper, E. H., P. Whelan and D. Purves (1994) Bone alkaline phosphatase and prostate-specific antigen in the monitoring of prostate cancer. *Prostate* **25**: 236–242.

Cornish-Bowden, A. (2004) *Fundamentals of Enzyme Kinetics*, 3rd edn. London: Portland Press, Chapter 5, pp. 113–144.

Cox, W. G. and V. L. Singer (1999) A high-resolution, fluorescence-based method for localization of endogenous alkaline phosphatase activity. *J. Histochem. Cytochem.* **47**: 1443–1456.

Crans, D. C., M. S. Gottlieb, J. Tawara, R. L. Bunch and L. A. Theisen (1990) A kinetic method for determination of free vanadium(IV) and -(V) at trace level concentrations. *Anal. Biochem.* **188**: 53–64.

Croce, M. A., G. L. Kramer and D. L. Garbers (1979) Inhibition of alkaline phosphatase by substituted xanthines. *Biochem. Pharmacol.* **28**: 1227–1231.

Crofton, P. M. (1981) Site of alkaline phosphatase attachment in alkaline phosphatase–immunoglobulin G complexes. *Clin. Chim. Acta* **112**: 33–42.

Crofton, P. M. (1987) Properties of alkaline phosphatase isoenzymes in plasma of preterm and term neonates. *Clin. Chem.* **33**: 1778–1782.

Crofton, P. M. (1992) Wheat-germ lectin affinity electrophoresis for alkaline phosphatase isoforms in children: age-dependent reference ranges and changes in liver and bone disease. *Clin. Chem.* **38**: 663–670.

Crofton, P. M. and R. Hume (1987) Alkaline phosphatase isoenzymes in the plasma of preterm and term infants: serial measurements and clinical correlations. *Clin. Chem.* **33**: 1783–1787.

Crofton, P. M. and A. F. Smith (1981a). High-molecular-mass alkaline phosphatase in serum and bile: nature and relationship with lipoprotein-X. *Clin. Chem.* **27**: 867–874.

Crofton, P. M. and A. F. Smith (1981b). High-molecular-mass alkaline phosphatase in serum and bile: physical properties and relationship with other high-molecular-mass enzymes. *Clin. Chem.* **27**: 860–866.

Crofton, P. M., D. C. Kilpatrick and A. G. Leitch (1981) Complexes in serum between alkaline phosphatase and immunoglobulin G: immunological and clinical aspects. *Clin. Chim. Acta* **111**: 257–265.

Cullen, B. R. (2000) Utility of the secreted placental alkaline phosphatase reporter enzyme. *Methods Enzymol.* **326**: 159–164.

Cullen, B. R. and M. H. Malim (1992) Secreted placental alkaline phosphatase as a eukaryotic reporter gene. *Methods Enzymol.* **216**: 362–368.

Culvenor, J. G., A. W. Harris, T. E. Mandel, A. Whitelaw and E. Ferber (1981) Alkaline phosphatase in hematopoietic tumor cell lines of the mouse: high activity in cells of the B lymphoid lineage. *J. Immunol.* **126**: (1974)–(1977)

Curatolo, C., G. M. Ludovico, M. Correale, A. Pagliarulo, I. Abbate, E. Cirrillo Marucco and A. Barletta (1992) Advanced prostate cancer follow-up with prostate-specific antigen, prostatic acid phosphatase, osteocalcin and bone isoenzyme of alkaline phosphatase. *Eur. Urol.* **21** (Suppl. 1): 105–107.

Dadak, V., P. Janda and O. Janiczek (1999) Evidence of complex formation between alkaline phosphatase and a pro-apoptotic hemoprotein cytochrome *c*. *Gen. Physiol. Biophys.* **18**: 387–400.

Dadak, V., O. Janiczek and O. Vrana (2002) Cytochrome *c* forms complexes and is partly reduced at interaction with GPI-anchored alkaline phosphatase. *Biochim. Biophys. Acta* **1570**: 9–18.

Dairiki, K., S. Nakamura, S. Ikegami, M. Nakamura, T. Fujimori, N. Tamaoki and N. Tada (1989) Mouse Ly-31.1 is an alloantigenic determinant of alkaline phosphatase predominantly expressed in the kidney and bone. *Immunogenetics* **29**: 235–240.

Davies, J. O., E. R. Davies, K. Howe, P. Jackson, E. Pitcher, B. Randle, C. Sadowski, G. M. Stirrat and C. A. Sunderland (1985a). Practical applications of a monoclonal antibody (NDOG2) against placental alkaline phosphatase in ovarian cancer. *J. R. Soc. Med.* **78**: 899–905.

Davies, J. O., E. R. Davies, K. Howe, P. C. Jackson, E. M. Pitcher, C. S. Sadowski, G. M. Stirrat and C. A. Sunderland (1985b). Radionuclide imaging of ovarian tumours with ^{123}I-labeled monoclonal antibody (NDOG2) directed against placental alkaline phosphatase. *Br. J. Obstet. Gynaecol.* **92**: 277–286.

Davini, E., C. Di Leo, A. Rossodivita and P. Zappelli (1992) Alkaline phosphatase inhibitors as labels of DNA probes. *Genet. Anal. Tech. Appl.* **9**: 39–47.

Davitz, M. A., D. Hereld, S. Shak, J. Krakow, P. T. Englund and V. Nussenzweig (1987) A glycan–phosphatidylinositol-specific phospholipase D in human serum. *Science* **238**: 81–84.

Day, A. P., M. D. Feher, R. Chopra and P. D. Mayne (1992a). Triglyceride fatty acid chain length influences the post prandial rise in serum intestinal alkaline phosphatase activity. *Ann. Clin. Biochem.* **29**: 287–291.

Day, A. P., S. Saward, C. M. Royle and P. D. Mayne (1992b). Evaluation of two new methods for routine measurement of alkaline phosphatase isoenzymes. *J. Clin. Pathol.* **45**: 68–71.

De Bernard, B., P. Bianco, E. Bonucci, M. Costantini, G. C. Lunazzi, P. Martinuzzi, C. Modricky, L. Moro, E. Panfili and P. Pollesello (1986) Biochemical and immunohistochemical evidence that in cartilage an alkaline phosphatase is a Ca^{2+}-binding glycoprotein. *J. Cell. Biol.* **103**: 1615–1623.

De Broe, M. E. and D. E. Pollet (1988) Multicenter evaluation of human placental alkaline phosphatase as a possible tumor-associated antigen in serum. *Clin. Chem.* **34**: 1995–1999.

De Broe, M. E., F. Roels, E. J. Nouwen, L. Claeys and R. J. Wieme (1985) Liver plasma membrane: the source of high molecular weight alkaline phosphatase in human serum. *Hepatology* **5**: 118–128.

De Groote, G., P. De Waele, A. Van de Voorde, M. De Broe and W. Fiers (1983) Use of monoclonal antibodies to detect human placental alkaline phosphatase. *Clin. Chem.* **29**: 115–119.

Deftos, L. J., R. L. Wolfert and C. S. Hill (1991) Bone alkaline phosphatase in Paget's disease. *Horm. Metab. Res.* **23**: 559–561.

Dehghani, H., S. Narisawa, J. L. Millán and A. C. Hahnel (2000) Effects of disruption of the embryonic alkaline phosphatase gene on preimplantation development of the mouse. *Dev. Dyn.* **217**: 440–448.

Dekker, I., T. Rozeboom, J. Delemarre, A. Dam and J. W. Oosterhuis (1992) Placental-like alkaline phosphatase and DNA flow cytometry in spermatocytic seminoma. *Cancer* **69**: 993–996.

Demenis, M. A. and F. A. Leone (2000) Kinetic characteristics of ATP hydrolysis by a detergent-solubilized alkaline phosphatase from rat osseous plate. *IUBMB Life* **49**: 113–119.

Demers, L. M., L. Costa, V. M. Chinchilli, L. Gaydos, E. Curley and A. Lipton (1995) Biochemical markers of bone turnover in patients with metastatic bone disease. *Clin. Chem.* **41**: 1489–1494.

Deng, G., G. Liu, L. Hu, J. R. Gum, Jr and Y. S. Kim (1992a). Transcriptional regulation of the human placental-like alkaline phosphatase gene and mechanisms involved in its induction by sodium butyrate. *Cancer Res.* **52**: 3378–3383.

Deng, J. T. and P. G. Parsons (1988) Solid phase immunoassay for high molecular weight alkaline phosphatase in human sera using a specific monoclonal antibody. *Clin. Chim. Acta* **176**: 291–301.

Deng, J. T., M. F. Hoylaerts, V. O. Van Hoof and M. E. De Broe (1992b). Differential release of human intestinal alkaline phosphatase in duodenal fluid and serum. *Clin. Chem.* **38**: 2532–2538.

Deng, J. T., M. F. Hoylaerts, M. E. De Broe and V. O. van Hoof (1996) Hydrolysis of membrane-bound liver alkaline phosphatase by GPI-PLD requires bile salts. *Am. J. Physiol.* **271**: G655–G663.

DePrimo, S. E., P. J. Stambrook and J. R. Stringer (1996) Human placental alkaline phosphatase as a histochemical marker of gene expression in transgenic mice. *Transgenic Res.* **5**: 459–466.

DePrimo, S. E., J. Cao, M. N. Hersh and J. R. Stringer (1998) Use of human placental alkaline phosphatase transgenes to detect somatic mutation in mice in situ. *Methods* **16**: 49–61.

DeSchryver-Kecskemeti, K., R. Eliakim, S. Carroll, W. F. Stenson, M. A. Moxley and D. H. Alpers (1989) Intestinal surfactant-like material. A novel secretory product of the rat enterocyte. *J. Clin. Invest.* **84**: 1355–1361.

DeSchryver-Kecskemeti, K., R. Eliakim, K. Green and D. H. Alpers (1991) A novel intracellular pathway for rat intestinal digestive enzymes (alkaline phosphatase and sucrase) via a lamellar particle. *Lab. Invest.* **65**: 365–373.

Devgun, M. S., C. R. Paterson and B. T. Martin (1981) Seasonal changes in the activity of serum alkaline phosphatase. *Enzyme* **26**: 301–305.

Di Mauro, S., T. Manes, L. Hessle, A. Kozlenkov, J. M. Pizauro, M. F. Hoylaerts and J. L. Millán (2002) Kinetic characterization of hypophosphatasia mutations with physiological substrates. *J. Bone Miner. Res.* **17**: 1383–1391.

Diaz-Gonzalez, M., C. Fernandez-Sanchez and A. Costa-Garcia (2002) Indirect determination of alkaline phosphatase based on the amperometric detection of indigo carmine at a screen-printed electrode in a flow system. *Anal. Sci.* **18**: 1209–1213.

Dikow, A., R. Gossrau and H. G. Frank (1990) Menadiol diphosphate, a new substrate for non-specific alkaline phosphatase in histochemistry and immunohistochemistry. *Histochemistry* **94**: 217–223.

Doellgast, G. J. and K. Benirschke (1979) Placental alkaline phosphatase in Hominidae. *Nature* **280**: 601–602.

Doellgast, G. J. and W. H. Fishman (1976) L-Leucine, a specific inhibitor of a rare human placental alkaline phosphatase phenotype. *Nature* **259**: 49–51.

Doellgast, G. J. and W. H. Fishman (1977) Inhibition of human placental-type alkaline phosphatase variants by peptides containing L-leucine. *Clin. Chim. Acta* **75**: 449–454.

Doellgast, G. J. and H. D. Homesley (1984) Placental-type alkaline phosphatase in ovarian cancer fluids and tissues. *Obstet. Gynecol.* **63**: 324–329.

Doellgast, G. J. and S. C. Wei (1984) Immunochemical data suggesting a pattern for the evolution of human placental alkaline phosphatase. *Mol. Immunol.* **21**: 197–203.

Doellgast, G. J., M. Kennedy and L. J. Donald (1980) The '18' variant of human placental alkaline phosphatase is identical to the 'D-variant'. *Hum. Hered.* **30**: 18–20.

Doellgast, G. J., S. C. Wei, M. Kennedy, H. Stills and K. L. Benirschke (1981) Primate placental alkaline phosphatase. *FEBS Lett.* **135**: 61–64.

Domar, U., A. Danielsson, K. Hirano and T. Stigbrand (1988) Alkaline phosphatase isozymes in non-malignant intestinal and hepatic diseases. *Scand. J. Gastroenterol.* **23**: 793–800.

Domar, U., K. Hirano and T. Stigbrand (1991) Serum levels of human alkaline phosphatase isozymes in relation to blood groups. *Clin. Chim. Acta* **203**: 305–313.

Domar, U., F. Karpe, A. Hamsten, T. Stigbrand and T. Olivecrona (1993) Human intestinal alkaline phosphatase – release to the blood is linked to lipid absorption, but removal from the blood is not linked to lipoprotein clearance. *Eur. J. Clin. Invest.* **23**: 753–760.

Donald, L. J. and E. B. Robson (1974a). The genetics of placental alkaline phosphatase: a possible 'null' allele. *Ann. Hum. Genet.* **38**: 7–18.

Donald, L. J. and E. B. Robson (1974b). Rare variants of placental alkaline phosphatase. *Ann. Hum. Genet.* **37**: 303–313.

Dotti, G., E. Garattini, G. Borleri, K. Masuhara, O. Spinelli, T. Barbui and A. Rambaldi (1999) Leucocyte alkaline phosphatase identifies terminally differentiated normal neutrophils and its lack in chronic myelogenous leukaemia is not dependent on p210 tyrosine kinase activity. *Br. J. Haematol.* **105**: 163–172.

Durbin, H., E. M. Milligan, S. J. Mather, D. F. Tucker, R. Raymond and W. F. Bodmer (1988) Monoclonal antibodies to placental alkaline phosphatase: preclinical evaluation in a human xenograft tumour model of F(ab')2 and Fab fragments. *Int. J. Cancer Suppl.* **2**: 59–66.

Durocher, Y., S. Perret and A. Kamen (2002) High-level and high-throughput recombinant protein production by transient transfection of suspension-growing human 293-EBNA1 cells. *Nucleic Acids Res.* **30**: E9.

Eliakim, R., K. DeSchryver-Kecskemeti, L. Nogee, W. F. Stenson and D. H. Alpers (1989) Isolation and characterization of a small intestinal surfactant-like particle containing alkaline phosphatase and other digestive enzymes. *J. Biol. Chem.* **264**: 20614–20619.

Eliakim, R., M. J. Becich, K. Green and D. H. Alpers (1990a). Both tissue and serum phospholipases release rat intestinal alkaline phosphatase. *Am. J. Physiol.* **259**: G618–G625.

Eliakim, R., S. Seetharam, C. C. Tietze and D. H. Alpers (1990b). Differential regulation of mRNAs encoding for rat intestinal alkaline phosphatase. *Am. J. Physiol.* **259**: G93–G98.

Eliakim, R., A. Mahmood and D. H. Alpers (1991) Rat intestinal alkaline phosphatase secretion into lumen and serum is coordinately regulated. *Biochim. Biophys. Acta* **1091**: 1–8.

Eliakim, R., G. S. Goetz, S. Rubio, B. Chailley-Heu, J. S. Shao, R. Ducroc and D. H. Alpers (1997) Isolation and characterization of surfactant-like particles in rat and human colon. *Am. J. Physiol.* **272**: G425–G434.

Elsafi, M. E., J. T. Holmberg, B. Hultberg, I. Hagerstrand, A. Isaksson, B. Joelsson and K. Melen (1989) Alkaline phosphatase in cholestatic and cirrhotic rats. A biochemical and histochemical study. *Enzyme* **42**: 145–151.

Ey, P. L. and L. K. Ashman (1986) The use of alkaline phosphatase-conjugated anti-immunoglobulin with immunoblots for determining the specificity of monoclonal antibodies to protein mixtures. *Methods Enzymol.* 121: 497–509.

Endo, T., H. Ohbayashi, Y. Hayashi, Y. Ikehara, N. Kochibe and A. Kobata (1988) Structural study on the carbohydrate moiety of human placental alkaline phosphatase. *J. Biochem. (Tokyo)* **103**: 182–187.

Endo, T., K. Higashino, T. Hada, H. Imanishi, K. Muratani, N. Kochibe and A. Kobata (1990) Structures of the asparagine-linked oligosaccharides of an alkaline phosphatase, Kasahara isozyme, purified from FL amnion cells. *Cancer Res.* **50**: 1079–1084.

Engle, M. J. and D. H. Alpers (1992) The two mRNAs encoding rat intestinal alkaline phosphatase represent two distinct nucleotide sequences. *Clin. Chem.* **38**: 2506–2509.

Engle, M. J., M. L. Grove, M. J. Becich, A. Mahmood and D. H. Alpers (1995a). Appearance of surfactant-like particles in apical medium of Caco-2 cells may occur via tight junctions. *Am. J. Physiol.* **268**: C1401–1413.

Engle, M. J., A. Mahmood and D. H. Alpers (1995b). Two rat intestinal alkaline phosphatase isoforms with different C-terminal peptides are both membrane-bound by a glycan phosphatidylinositol linkage. *J. Biol. Chem.* **270**: 11935–11940.

Engle, M. J., A. Mahmood and D. H. Alpers (2001) Regulation of surfactant-like particle secretion by Caco-2 cells. *Biochim. Biophys. Acta* **1511**: 369–380.

Epenetos, A. A., P. Travers, K. C. Gatter, R. D. Oliver, D. Y. Mason and W. F. Bodmer (1984) An immunohistological study of testicular germ cell tumours using two different monoclonal antibodies against placental alkaline phosphatase. *Br. J. Cancer* **49**: 11–15.

Epenetos, A. A., A. J. Munro, D. F. Tucker, W. Gregory, W. Duncan, R. H. MacDougall, M. Faux, P. Travers and W. F. Bodmer (1985a). Monoclonal antibody assay of serum placental alkaline phosphatase in the monitoring of testicular tumours. *Br. J. Cancer* **51**: 641–644.

Epenetos, A. A., D. Snook, G. Hooker, R. Begent, H. Durbin, R. T. Oliver, W. F. Bodmer and J. P. Lavender (1985b). Indium-111 labeled monoclonal antibody to placental alkaline phosphatase in in the detection of neoplasms of testis, ovary and cervix. *Lancet* **2**: 350–353.

Epenetos, A. A., D. Carr, P. M. Johnson, W. F. Bodmer and J. P. Lavender (1986) Antibody-guided radiolocalization of tumours in patients with testicular or ovarian cancer using two radioiodinated monoclonal antibodies to placental alkaline phosphatase. *Br. J. Radiol.* **59**: 117–125.

Eriksson, H. J., W. R. Verweij, K. Poelstra, W. L. Hinrichs, G. J. de Jong, G. W. Somsen and H. W. Frijlink (2003) Investigations into the stabilization of drugs by sugar glasses: II. Delivery of an inulin-stabilised alkaline phosphatase in the intestinal lumen via the oral route. *Int. J. Pharm.* **257**: 273–281.

Eruk, A. A., R. J. Washington and I. Laing (1984) The preparation and characterization of alkaline phosphatase-thyroxine conjugates and their use in the enzyme immunoassay of thyroxine. *Ann. Clin. Biochem.* **21**: 434–443.

Escalante-Alcalde, D., F. Recillas-Targa, D. Hernandez-Garcia, S. Castro-Obregon, M. Terao, E. Garattini and L. Covarrubias (1996) Retinoic acid and methylation cis-regulatory elements control the mouse tissue non-specific alkaline phosphatase gene expression. *Mech. Dev.* **57**: 21–32.

Estrada-Mondaca, S., L. A. Delgado-Bustos and O. T. Ramirez (2005) Mannosamine supplementation extends the N-acetyl-glucosaminylation of recombinant human secreted alkaline phosphatase produced in *Trichoplusia ni* (cabbage looper) insect cell cultures. *Biotech. Appl. Biochem.* **42**: 25–34.

Fallon, M. D., M. P. Whyte and S. L. Teitelbaum (1980) Stereospecific inhibition of alkaline phosphatase by L-tetramisole prevents *in vitro* cartilage calcification. *Lab. Invest.* **43**: 489–494.

Fallon, M. D., S. L. Teitelbaum, R. S. Weinstein, S. Goldfischer, D. M. Brown and M. P. Whyte (1984) Hypophosphatasia: clinicopathologic comparison of the infantile, childhood and adult forms. *Medicine* **63**: 12–24.

Farley, J.R. (1995) Phosphate regulates the stability of skeletal alkaline phosphatase activity in human osteosarcoma (SaOS-2) cells without equivalent effects on the level of skeletal alkaline phosphatase immunoreactive protein. *Calcif. Tissue Int.* **57**: 371–378.

Farley, J. R. and D. J. Baylink (1986) Skeletal alkaline phosphatase activity as a bone formation index *in vitro*. *Metabolism* **35**: 563–571.

Farley, J. R. and P. Magnusson (2005) Effects of tunicamycin, mannosamine and other inhibitors of glycoprotein processing on skeletal alkaline phosphatase in human osteoblast-like cells. *Calcif. Tissue Int.* **76**: 63–74.

Farley, J. R. and B. Stilt-Coffing (2001) Apoptosis may determine the release of skeletal alkaline phosphatase activity from human osteoblast-line cells. *Calcif. Tissue Int.* **68**: 43–52.

Farley, J. R., J. L. Ivey and D. J. Baylink (1980) Human skeletal alkaline phosphatase. Kinetic studies including pH dependence and inhibition by theophylline. *J. Biol. Chem.* **255**: 4680–4686.

Farley, J. R., C. H. Chesnut, III and D. J. Baylink (1981) Improved method for quantitative determination in serum of alkaline phosphatase of skeletal origin. *Clin. Chem.* **27**: 2002–2007.

Farley, J. R., N. M. Tarbaux, K. H. Lau and D. J. Baylink (1987) Monofluorophosphate is hydrolyzed by alkaline phosphatase and mimics the actions of NaF on skeletal tissues, *in vitro*. *Calcif. Tissue Int.* **40**: 35–42.

Farley, J. R., S. L. Hall, D. Ilacas, C. Orcutt, B. E. Miller, C. S. Hill and D. J. Baylink (1994) Quantification of skeletal alkaline phosphatase in osteoporotic serum by wheat-germ agglutinin precipitation, heat inactivation and a two-site immunoradiometric assay. *Clin. Chem.* **40**: 1749–1756.

Farmar, J. G. and M. Castaneda (1991) An improved preparation and purification of oligonucleotide–alkaline phosphatase conjugates. *Biotechniques* **11**: 588–589.

Fedde, K. N (1992) Human osteosarcoma cells spontaneously release matrix-vesicle-like structures with the capacity to mineralize. *Bone. Miner.* **17**: 145–151.

Fedde, K. N., M. P. Michel and M. P. Whyte (1993) Evidence against a role for alkaline phosphatase in the dephosphorylation of plasma membrane proteins: hypophosphatasia fibroblast study. *J. Cell. Biochem.* **53**: 43–50.

Fedde, K. N., L. Blair, F. Terzic, H. C. Anderson, S. Narisawa, J. L. Millán and M. P. Whyte (1996a). Amelioration of the skeletal disease in hypophosphatasia by bone-marrow transplantation using alkaline phosphatase knockout mice. *Am. J. Hum. Genet.* **59**: A15.

Fedde, K. N., M. P. Michell, P. S. Henthorn and M. P. Whyte (1996b). Aberrant properties of alkaline phosphatase in patient fibroblasts correlate with clinical expressivity in severe forms of hypophosphatasia. *J. Clin. Endocrinol. Metab.* **81**: 2587–2594.

Fedde, K. N., C. C. Lane and M. P. Whyte (1988) Alkaline phosphatase is an ectoenzyme that acts on micromolar concentrations of natural substrates at physiologic pH in human osteosarcoma (SAOS-2) cells. *Arch. Biochem. Biophys.* **264**: 400–409.

Fedde, K. N., L. Blair, J. Silverstein, S. P. Coburn, L. M. Ryan, R. S. Weinstein, K. Waymire, S. Narisawa, J. L. Millán, G. R. MacGregor and M. P. Whyte (1999) Alkaline phosphatase knock-out mice recapitulate the metabolic and skeletal defects of infantile hypophosphatasia. *J. Bone. Miner. Res.* **14**: 2015–2026.

Felix, R. and H. Fleisch (1974) The pyrophosphatase and (Ca^{2+}–Mg^{2+})-ATPase activity of purified calf bone alkaline phosphatase. *Biochim. Biophys. Acta* **350**: 84–94.

Fields-Berry, S. C., A. L. Halliday and C. L. Cepko (1992) A recombinant retrovirus encoding alkaline phosphatase confirms clonal boundary assignment in lineage analysis of murine retina. *Proc. Natl. Acad. Sci. USA* **89**: 693–697.

Fisher, M., S. Harbron and B. R. Rabin (1995) An amplified chemiluminescent assay for the detection of alkaline phosphatase. *Anal. Biochem.* **227**: 73–79.

Fishman, L., H. Miyayama, S. G. Driscoll and W. H. Fishman (1976) Developmental phase-specific alkaline phosphatase isoenzymes of human placenta and their occurrence in human cancer. *Cancer Res.* **36**: 2268–2273.

Fishman, W. H. (1974) Perspectives on alkaline phosphatase isoenzymes. *Am. J. Med.* **56**: 617–650.

Fishman, W. H. (1990) On the importance of being (stereo) specific. *Clin. Chim. Acta* **186**: 129–131.

Fishman, W. H. and H. G. Sie (1971) Organ-specific inhibition of human alkaline phosphatase isoenzymes of liver, bone, intestine and placenta; L-phenylalanine, L-tryptophan and L-homoarginine. *Enzymologia* **41**: 141–167.

Fishman, W. H., N. K. Ghosh, N. R. Inglis and S. Green (1968a). Quantitation of the placental isoenzyme of alkaline phosphatase in pregnancy sera. *Enzymologia* **34**: 317–321.

Fishman, W. H., N. R. Inglis, S. Green, C. L. Anstiss, N. K. Gosh, A. E. Reif, R. Rustigian, M. J. Krant and L. L. Stolbach (1968b). Immunology and biochemistry of Regan isoenzyme of alkaline phosphatase in human cancer. *Nature* **219**: 697–699.

Fisken, J., R. C. Leonard, G. Shaw, A. Bowman and J. E. Roulston (1989) Serum placental-like alkaline phosphatase (PLAP): a novel combined enzyme linked immunoassay for monitoring ovarian cancer. *J. Clin. Pathol.* **42**: 40–45.

Fjeld, J. G., O. S. Bruland, H. B. Benestad, L. Schjerven, T. Stigbrand and K. Nustad (1990) Radioimmunotargeting of human tumour cells in immunocompetent animals. *Br. J. Cancer.* **62**: 573–578.

Fjeld, J. G., H. B. Benestad, T. Stigbrand and K. Nustad (1992) *In vivo* measurement of the association constant of a radiolabeled monoclonal antibody in experimental immunotargeting. *Br. J. Cancer.* **66**: 74–78.

Flanagan, J. G. and P. Leder (1990) The kit ligand: a cell surface molecule altered in steel mutant fibroblasts. *Cell* **63**: 185–194.

Flanagan, J. G. and H. J. Cheng (2000) Alkaline phosphatase fusion proteins for molecular characterization and cloning of receptors and their ligands. *Methods Enzymol.* **327**: 198–210.

Flanagan, J. G., H. J. Cheng, D. A. Feldheim, M. Hattori, Q. Lu and P. Vanderhaeghen (2000) Alkaline phosphatase fusions of ligands or receptors as *in situ* probes for staining of cells, tissues and embryos. *Methods Enzymol.* **327**: 19–35.

Fleshood, H. L. and H. C. Pitot (1969) O-phosphorylethanolamine ammonia lyase, a new pyridoxal phosphate-dependent enzyme. *Biochem. Biophys. Res. Commun.* **36**: 110–118.

Fleshood, H. L. and H. C. Pitot (1970) The metabolism of O-phosphorylethanolamine in animal tissues. II. Metabolic regulation of O-phosphorylethanolamine phospho-lyase *in vivo*. *Arch. Biochem. Biophys.* **141**: 423–429.

Fonta, C., L. Negyessy, L. Renaud and P. Barone (2004) Areal and subcellular localization of the ubiquitous alkaline phosphatase in the primate cerebral cortex: evidence for a role in neurotransmission. *Cereb. Cortex* **14**: 595–609.

Fonta, C., L. Negyessy, L. Renaud and P. Barone (2005) Postnatal development of alkaline phosphatase activity correlates with the maturation of neurotransmission in the cerebral cortex. *J. Comp. Neurol.* **486**: 179–196.

Ford, A. W. and P. J. Dawson (1993) The effect of carbohydrate additives in the freeze-drying of alkaline phosphatase. *J. Pharm. Pharmacol.* **45**: 86–93.

Franceschi, R. T. and J. Young (1990) Regulation of alkaline phosphatase by 1,25-dihydroxyvitamin D3 and ascorbic acid in bone-derived cells. *J. Bone. Miner. Res.* **5**: 1157–1167.

Franceschi, R. T., P. R. Romano and K. Y. Park (1988) Regulation of Type I collagen synthesis by 1,25-dihydroxyvitamin D3 in human osteosarcoma cells. *J. Biol. Chem.* **263**: 18938–18945.

Franceschi, R. T., B. S. Iyer and Y. Cui (1994) Effects of ascorbic acid on collagen matrix formation and osteoblast differentiation in murine MC3T3-E1 cells. *J. Bone. Miner. Res.* **9**: 843–854.

Fraser, D. (1957) Hypophosphatasia. *Am. J. Med.* **22**: 730–746.

Frederiks, W. M., C. J. Van Noorden, D. C. Aronson, F. Marx, K. S. Bosch, G. N. Jonges, I. M. Vogels and J. James (1990) Quantitative changes in acid phosphatase, alkaline phosphatase and 5′-nucleotidase activity in rat liver after experimentally induced cholestasis. *Liver* **10**: 158–166.

Freeman, J. L., P. W. Rhyne, D. O. Dean, G. A. Beard and G. J. Doellgast (2001) Isolation of antibody to human placental alkaline phosphatase (PLAP) from extracts of human placentae. *Am. J. Reprod. Immunol.* **46**: 149–160.

Friden, B. E., R. Makiya, B. M. Nilsson, S. Holm and T. I. Stigbrand (1994) The human placental immunoglobulin G receptor and immunoglobulin G transport. *Am. J. Obstet. Gynecol.* **171**: 258–263.

Fröhlander, N. and J. L. Millán (2001) Intron-size and exon polymorphism in the mouse tissue-nonspecific alkaline phosphatase gene. In: Xue, G., Y. Xue, Z. Xu, R. Holmes, G. L. Hammond and H. A. Lim (eds), *Gene Families, Studies of DNA, RNA, Enzymes and Proteins*. Singapore: World Scientific, pp. 233–242.

Fukui, K., T. Hada, H. Imanishi, W. Liu, A. Iwasaki, K. Hirano and K. Higashino (1997) The tumor-derived fetal-intestinal alkaline phosphatase cDNA is identical in sequence to the adult intestinal alkaline phosphatase isozyme gene. *Clin. Chim. Acta* **265**: 57–63.

Fukushi, M., N. Amizuka, K. Hoshi, H. Ozawa, H. Kumagai, S. Omura, Y. Misumi, Y. Ikehara and K. Oda (1998) Intracellular retention and degradation of tissue-nonspecific alkaline phosphatase with a Gly317→Asp substitution associated with lethal hypophosphatasia. *Biochem. Biophys. Res. Commun.* **246**: 613–618.

Fukushi-Irie, M., M. Ito, Y. Amaya, N. Amizuka, H. Ozawa, S. Omura, Y. Ikehara and K. Oda (2000) Possible interference between tissue-non-specific alkaline phosphatase with an Arg54→Cys substitution and acounterpart with an Asp277→Ala substitution found in a compound heterozygote associated with severe hypophosphatasia. *Biochem. J.* **348**: 633–642.

Fukushima, K., Y. Ikehara, M. Kanai, N. Kochibe, M. Kuroki and K. Yamashita (2003) A beta-N-acetylglucosaminyl phosphate diester residue is attached to the glycosylphosphatidylinositol anchor of human placental alkaline phosphatase: a target of the channel-forming toxin aerolysin. *J. Biol. Chem.* **278**: 36296–36303.

Gainer, A. L. and R. A. Stinson (1982) Evidence that alkaline phosphatase from human neutrophils is the same gene product as the liver/kidney/bone isoenzyme. *Clin. Chim. Acta* **123**: 11–17.

Gallo, R. L., R. A. Dorschner, S. Takashima, M. Klagsbrun, E. Eriksson and M. Bernfield (1997) Endothelial cell surface alkaline phosphatase activity is induced by IL-6 released during wound repair. *J. Invest. Dermatol.* **109**: 597–603.

Galperin, M. Y. and M. J. Jedrzejas (2001) Conserved core structure and active site residues in alkaline phosphatase superfamily enzymes. *Proteins* **45**: 318–324.

Galperin, M. Y., A. Bairoch and E. V. Koonin (1998) A superfamily of metalloenzymes unifies phosphopentomutase and cofactor-independent phosphoglycerate mutase with alkaline phosphatases and sulfatases. *Protein Sci.* **7**: 1829–1835.

Gandhy, M. and S. G. Damle (2003) Relation of salivary inorganic phosphorus and alkaline phosphatase to the dental caries status in children. *J. Indian Soc. Pedod. Prev. Dent.* **21**: 135–138.

Garattini, E. and M. Gianni (1996) Leukocyte alkaline phosphatase a specific marker for the post-mitotic neutrophilic granulocyte: regulation in acute promyelocytic leukemia. *Leuk. Lymphoma* **23**: 493–503.

Garattini, E., J. Margolis, E. Heimer, A. Felix and S. Udenfriend (1985) Human placental alkaline phosphatase in liver and intestine. *Proc. Natl. Acad. Sci. USA* **82**: 6080–6084.

Garattini, E., J. C. Hua, Y. C. Pan and S. Udenfriend (1986) Human liver alkaline phosphatase, purification and partial sequencing: homology with the placental isozyme. *Arch. Biochem. Biophys.* **245**: 331–337.

Garattini, E., J. C. Hua and S. Udenfriend (1987) Cloning and sequencing of bovine kidney alkaline phosphatase cDNA. *Gene* **59**: 41–46.

Garcia-Rozas, C., A. Plaza, F. Diaz-Espada, M. Kreisler and C. Martinez-Alonso (1982) Alkaline phosphatase activity as a membrane marker for activated B cells. *J. Immunol.* **129**: 52–55.

Garnero, P. and P. D. Delmas (1993) Assessment of the serum levels of bone alkaline phosphatase with a new immunoradiometric assay in patients with metabolic bone disease. *J. Clin. Endocrinol. Metab.* **77**: 1046–1053.

Gazzarrini, C., N. Stagni, P. Pollesello, P. D'Andrea and B. De Bernard (1989) Possible mechanism of inhibition of cartilage alkaline phosphatase by insulin. *Acta Diabetol. Lat.* **26**: 321–327.

Gerbitz, K. D., H. J. Kolb and O. H. Wieland (1977) Human alkaline phosphatases. I. Purification and some structural properties of the enzyme from human liver. *Hoppe Seylers Z. Physiol. Chem.* **358**: 435–446.

Gerritse, K., M. Fasbender, W. Boersma and E. Claassen (1991) Conjugate formation in urea: coupling of insoluble peptides to alkaline phosphatase for ELISA and *in situ* detection of antibody-forming cells. *J. Histochem. Cytochem.* **39**: 987–992.

Ghosh, A. K. and J. I. Mullins (1995) cDNA encoding a functional feline liver/bone/kidney-type alkaline phosphatase. *Arch. Biochem. Biophys.* **322**: 240–249.

Gianni, M., M. Terao, S. Sozzani and E. Garattini (1993) Retinoic acid and cyclic AMP synergistically induce the expression of liver/bone/kidney-type alkaline phosphatase gene in L929 fibroblastic cells. *Biochem. J.* **296**: 67–77.

Giatromanolaki, A., E. Sivridis, E. Maltezos and M. I. Koukourakis (2002) Down-regulation of intestinal-type alkaline phosphatase in the tumor vasculature and stroma provides a strong basis for explaining amifostine selectivity. *Semin. Oncol.* **29**: 14–21.

Gijsbers, R., H. Ceulemans, W. Stalmans and M. Bollen (2001) Structural and catalytic similarities between nucleotide pyrophosphatases/phosphodiesterases and alkaline phosphatases. *J. Biol. Chem.* **276**: 1361–1368.

Gil, E. P., H. T. Tang, H. B. Halsall, W. R. Heineman and A. S. Misiego (1990) Competitive heterogeneous enzyme immunoassay for theophylline by flow-injection analysis with electrochemical detection of *p*-aminophenol. *Clin. Chem.* **36**: 662–665.

Ginsburg, M., M. H. Snow and A. McLaren (1990) Primordial germ cells in the mouse embryo during gastrulation. *Development* **110**: 521–528.

Giwercman, A., L. Cantell and A. Marks (1991) Placental-like alkaline phosphatase as a marker of carcinoma-*in-situ* of the testis. Comparison with monoclonal antibodies M2A and 43-9F. *Apmis* **99**: 586–594.

Glimcher, M. J (1998) The nature of mineral phase in bone, biological and clinical implications. In: Avioli, L. V. and S. M. Krane (eds), *Metabolic Bone Disease and Clinically Related Disorders*. London: Academic Press, pp. 23–46.

Goetz, G. S., J.-S. Shao and D. H. Alpers (1997) Purification and initial characterization of a surfactant-like particle (SLP) from human stomach. *Gastroenterology* **112**: A128.

Goetz, G. S., A. Mahmood, S. J. Hultgren, M. J. Engle, K. Dodson and D. H. Alpers (1999) Binding of pili from uropathogenic *Escherichia coli* to membranes secreted by human colonocytes and enterocytes. *Infect. Immun.* **67**: 6161–6163.

Gogolin, K. J., C. A. Slaughter and H. Harris (1981) Electrophoresis of enzyme–monoclonal antibody complexes: studies of human placental alkaline phosphatase polymorphism. *Proc. Natl. Acad. Sci. USA* **78**: 5061–5065.

Goldenberg, R. L., T. Tamura, M. DuBard, K. E. Johnston, R. L. Copper and Y. Neggers (1997) Plasma alkaline phosphatase and pregnancy outcome. *J. Matern. Fetal Med.* **6**: 140–145.

Goldsmith, J. D., B. Pawel, J. R. Goldblum, T. L. Pasha, S. Roberts, P. Nelson, J. S. Khurana, F. G. Barr and P. J. Zhang (2002) Detection and diagnostic utilization of placental alkaline phosphatase in muscular tissue and tumors with myogenic differentiation. *Am. J. Surg. Pathol.* **26**: 1627–1633.

Goldstein, D. J. and H. Harris (1979) Human placental alkaline phosphatase differs from that of other species. *Nature* **280**: 602–605.

Goldstein, D. J., L. Blasco and H. Harris (1980) Placental alkaline phosphatase in nonmalignant human cervix. *Proc. Natl. Acad. Sci. USA* **77**: 4226–4228.

Goldstein, D. J., C. Rogers and H. Harris (1982) A search for trace expression of placental-like alkaline phosphatase in non-malignant human tissues: demonstration of its occurrence in lung, cervix, testis and thymus. *Clin. Chim. Acta* **125**: 63–75.

Goldstein, D. J., R. Levy, P. L. Yu and H. Harris (1985) Genetic variability of alkaline phosphatase expression in inbred mouse tissues. *Biochem. Genet.* **23**: 155–167.

Gomez, B., Jr, S. Ardakani, J. Ju, D. Jenkins, M. J. Cerelli, G. Y. Daniloff and V. T. Kung (1995) Monoclonal antibody assay for measuring bone-specific alkaline phosphatase activity in serum. *Clin. Chem.* **41**: 1560–1566.

Gonchoroff, D. G., E. L. Branum and J. F. O'Brien (1989) Alkaline phosphatase isoenzymes of liver and bone origin are incompletely resolved by wheat-germ-lectin affinity chromatography. *Clin. Chem.* **35**: 29–32.

Gonchoroff, D. G., E. L. Branum, S. L. Cedel, B. L. Riggs and J. F. O'Brien (1991) Clinical evaluation of high-performance affinity chromatography for the separation of bone and liver alkaline phosphatase isoenzymes. *Clin. Chim. Acta* **199**: 43–50.

Goseki, M., S. Oida, A. Nifuji and S. Sasaki (1990) Properties of alkaline phosphatase of the human dental pulp. *J. Dent. Res.* **69**: 909–912.

Goseki-Sone, M., S. Oida, T. Iimura, A. Yamamoto, H. N. Matsumoto, N. Omi, K. Takeda, Y. Maruoka, I. Ezawa and S. Sasaki (1996) Expression of mRNA encoding intestinal type alkaline phosphatase in rat liver and its increase by fat-feeding. *Liver* **16**: 358–364.

Goseki-Sone, M., H. Orimo, T. Iimura, H. Miyazaki, K. Oda, H. Shibata, M. Yanagishita, Y. Takagi, H. Watanabe, T. Shimada and S. Oida (1998) Expression of the mutant (1735T-DEL) tissue-nonspecific alkaline phosphatase gene from hypophosphatasia patients. *J. Bone Miner. Res.* **13**: 1827–1834.

Goseki-Sone, M., H. Orimo, T. Iimura, Y. Takagi, H. Watanabe, K. Taketa, S. Sato, H. Mayanagi, T. Shimada and S. Oida (1998) Hypophosphatasia: identification of five novel missense mutations (G507A, G705A, A748G, T1155C, G1320A) in the tissue-nonspecific alkaline phosphatase gene among Japanese patients. *Hum. Mutat. (Suppl. 1)*: S263–S267.

Goseki-Sone M. N. Sogabe, M. Fukushi-Irie, L. Mizoi, H. Orimo, T. Suzuki, H. Nakamura, H. Orimo and T. Hosoi (2005) Functional analysis of the single nucleotide polymorphism (787T>C) in the tissue-nonspecific alkaline phosphatase gene associated with BMD. *J. Bone Miner. Res.* **20**: 773–782.

Green, S., C. L. Antiss and W. H. Fishman (1971) Automated differential isoenzyme analysis II: the fractionation of serum alkaline phosphatase into "liver", "intestinal" and "other" components. *Enzymologia* **41**: 9–26.

Greenberg, C. R., C. L. Taylor, J. C. Haworth, L. E. Seargeant, S. Philipps, B. Triggs-Raine and B. N. Chodirker (1993) A homoallelic Gly317→Asp mutation in *ALPL* causes the perinatal (lethal) form of hypophosphatasia in Canadian mennonites. *Genomics* **17**: 215–217.

Greene, P. J. and H. H. Sussman (1973) Structual comparison of ectopic and normal placental alkaline phosphatase. *Proc. Natl. Acad. Sci. USA* **70**: 2936–2940.

Greenspan, S.L., R. Dresner-Pollak, R.A. Parker, D. London and L. Ferguson (1997) Diurnal variation of bone mineral turnover in elderly men and women. *Calcif. Tissue Int.* **60**: 419–423.

Griffin, C. A., M. Smith, P. S. Henthorn, H. Harris, M. J. Weiss, M. Raducha and B. S. Emanuel (1987) Human placental and intestinal alkaline phosphatase genes map to 2q34–q37. *Am. J. Hum. Genet.* **41**: 1025–1034.

Griffiths, J. and J. Black (1987) Separation and identification of alkaline phosphatase isoenzymes and isoforms in serum of healthy persons by isoelectric focusing. *Clin. Chem.* **33**: 2171–2177.

Griffiths, J., A. Vernocchi and E. Simoni (1995) Transient hyperphosphatasemia of infancy and childhood. A study of serum alkaline phosphatase by electrofocusing techniques. *Arch. Pathol. Lab. Med.* **119**: 784–789.

Griffiths, W. C., P. D. Camara, M. Rosner, R. Lev and E. M. Brooks (1992) Prevalence and properties of the intestinal alkaline phosphatase identified in serum by cellulose acetate electrophoresis. *Clin. Chem.* **38**: 507–511.

Grim, M. and B. M. Carlson (1990) Alkaline phosphatase and dipeptidylpeptidase IV staining of tissue components of skeletal muscle: a comparative study. *J. Histochem. Cytochem.* **38**: (1907)–(1912)

Grøn, I. H (1978) Mammalian O-phosphorylethanolamine phospho-lyase activity and its inhibition. *Scand. J. Clin. Lab. Invest.* **38**: 107–112.

Grozdea, J., G. Bourrouillou, C. Mounie and P. Colombies (1980) Increase of neutrophil alkaline phosphatase in the parents of trisomy 21 children. Hematological and cytogenetic studies. *Acta Haematol.* **64**: 38–41.

Grozdea, J., A. Maret, H. Vergnes, G. Bourrouillou, J. Verdier, J. Martin, R. Salvayre and P. Colombies (1984) Cytochemical and biochemical studies on neutrophil alkaline phosphatase in parents of trisomy 21 children. *Hum. Genet.* **67**: 313–316.

Grozdea, J., A. Brisson-Lougarre, H. Vergnes, R. Bierme, S. Alie-Daram, J. Martin and A. Fournie (1988a). IgG/placental alkaline phosphatase complexes in Rh-incompatible pregnancy with fetal hydrops. *Vox Sang.* **55**: 42–43.

Grozdea, J., H. Vergnes, A. Brisson-Lougarre, G. Bourrouillou, J. Martin, C. Blum and P. Colombies (1988b). Heat resistance, immunological and quantitative changes of neutrophil alkaline phosphatase in trisomy 21 pregnancies. *Hum. Genet.* **78**: 240–243.

Grozdea, J., H. Vergnes, A. Brisson-Lougarre, R. Bierme, G. Bourrouillou, E. Duchayne, J. Martin and P. Colombies (1991) Difference in activity properties and subcellular distribution of neutrophil alkaline phosphatase between normal individuals and patients with trisomy 21. *Br. J. Haematol.* **77**: 282–286.

Gum, J. R., W. K. Kam, J. C. Byrd, J. W. Hicks, M. H. Sleisenger and Y. S. Kim (1987) Effects of sodium butyrate on human colonic adenocarcinoma cells. Induction of placental-like alkaline phosphatase. *J. Biol. Chem.* **262**: 1092–1097.

Gum, J. R., J. W. Hicks, T. L. Sack and Y. S. Kim (1990) Molecular cloning of complementary DNAs encoding alkaline phosphatase in human colon cancer cells. *Cancer Res.* **50**: 1085–1091.

Hada, T., K. Higashino, T. Okochi, Y. Yamamura, M. Matsuda, M. Osafune, T. Kotake and T. Sonoda (1979) Another Kasahara-variant alkaline phosphatase in renal cell carcinomas. *Gann.* **70**: 503–508.

Hadjis, N. S., J. I. Blenkharn, G. Hatzis, A. Adam, J. Beacham and L. H. Blumgart (1990) Patterns of serum alkaline phosphatase activity in unilateral hepatic duct obstruction: a clinical and experimental study. *Surgery* **107**: 193–200.

Hahnel, A. C. and G. A. Schultz (1990) Cloning and characterization of a cDNA encoding alkaline phosphatase in mouse embryonal carcinoma cells. *Clin. Chim. Acta* **186**: 171–174.

Hahnel, A. C., D. A. Rappolee, J. L. Millán, T. Manes, C. A. Ziomek, N. G. Theodosiou, Z. Werb, R. A. Pedersen and G. A. Schultz (1990) Two alkaline phosphatase genes are expressed during early development in the mouse embryo. *Development* **110**: 555–564.

Haisma, H. J., E. Boven, M. van Muijen, R. De Vries and H. M. Pinedo (1992) Analysis of a conjugate between anti-carcinoembryonic antigen monoclonal antibody and alkaline phosphatase for specific activation of the prodrug etoposide phosphate. *Cancer Immunol. Immunother.* **34**: 343–348.

Halbhuber, K. J., R. Krieg, P. Fischer, K. Konig, H. Nasse and W. Dietz (2002) Jenfluor ap – a novel fluorogenic substrate for *in situ* detection of alkaline phosphatase activity. *Cell. Mol. Biol. (Noisy-le-Grand)* **48**: OL343–OL358.

Hall, S. L., H. P. Dimai and J. R. Farley (1999) Effects of zinc on human skeletal alkaline phosphatase activity *in vitro*. *Calcif. Tissue Int.* **64**: 163–172.

Hallaway, B. J. and D. J. O'Kane (2000) Chemiluminescence assay of serum alkaline phosphatase and phosphoprotein phosphatases. *Methods Enzymol.* **305**: 391–401.

Hamilton, B. A., J. L. McPhee, K. Hawrylak and R. A. Stinson (1990) Alkaline phosphatase releasing activity in human tissues. *Clin. Chim. Acta* **186**: 249–254.

Hamilton-Dutoit, S. J., H. Lou and G. Pallesen (1990) The expression of placental alkaline phosphatase (PLAP) and PLAP-like enzymes in normal and neoplastic human tissues. An immunohistological survey using monoclonal antibodies. *APMIS* **98**: 797–811.

Handjiski, B. K., S. Eichmuller, U. Hofmann, B. M. Czarnetzki and R. Paus (1994) Alkaline phosphatase activity and localization during the murine hair cycle. *Br. J. Dermatol.* **131**: 303–310.

Harada, T., I. Koyama, A. Shimoi, D. H. Alpers and T. Komoda (2002) Identification of pulmonary surfactant that bears intestinal-type and tissue-nonspecific-type alkaline phosphatase in endotoxin-induced rat bronchoalveolar fluid. *Cell. Tissue Res.* **307**: 69–77.

Harada, T., I. Koyama, T. Kasahara, D. H. Alpers and T. Komoda (2003) Heat shock induces intestinal-type alkaline phosphatase in rat IEC-18 cells. *Am. J. Physiol. Gastrointest. Liver Physiol.* **284**: G255–262.

Harada, T., I. Koyama, T. Matsunaga, A. Kikuno, T. Kasahara, M. Hassimoto, D. H. Alpers and T. Komoda (2005) Characterization of structural and catalytic differences in rat intestinal alkaline phosphatase isozymes. *FEBS J.* **272**: 2477–2486.

Harbron, S., H. J. Eggelte, S. M. Benson and B. R. Rabin (1991a). Amplified luminometric assays of alkaline phosphatase using riboflavin phosphates. *J. Biolumin. Chemilumin.* **6**: 251–258.

Harbron, S., H. J. Eggelte and B. R. Rabin (1991b). Amplified colorimetric assay of alkaline phosphatase using riboflavin 4′-phosphate: a simple method for measuring riboflavin and riboflavin 5′-phosphate. *Anal. Biochem.* **198**: 47–51.

Harbron, S., H. J. Eggelte, M. Fisher and B. R. Rabin (1992) Amplified assay of alkaline phosphatase using flavin-adenine dinucleotide phosphate as substrate. *Anal. Biochem.* **206**: 119–124.

Harding, F. and E. Garry (2005) Collaborative evaluation of a fluorimetric method for measuring alkaline phosphatase activity in cow's, sheep's and goat's milk. *J. Food Prot.* **68**: 1047–1053.

Harmenberg, U., J. E. Frodin, E. Ljungdahl-Stahle, H. Mellstedt, B. Wahren and T. Stigbrand (1989) Significance of alkaline phosphatase isozymes in the monitoring of patients with colorectal carcinoma. *Tumor Biol.* **10**: 225–231.

Harmenberg, U., M. Koha, R. Makiya, K. Koshida, B. Brismar, T. Stigbrand and B. Wahren (1991) Identification and characterization of alkaline phosphatase isozymes in human colorectal adenocarcinomas. *Tumour Biol.* **12**: 237–248.

Harmey, D., L. Hessle, S. Narisawa, K. A. Johnson, R. Terkeltaub and J. L. Millán (2004) Concerted regulation of inorganic pyrophosphate and osteopontin by *Akp2*, *Enpp1* and *Ank*: an integrated model of the pathogenesis of mineralization disorders. *Am. J. Pathol.* **164**: 1199–1209.

Harris, H. (1980) Multilocus enzyme systems and the evolution of gene expression: the alkaline phosphatases as a model example. *Harvey Lect.* **76**: 95–124.

Harris, H. (1990) The human alkaline phosphatases: what we know and what we don't know. *Clin. Chim. Acta* **186**: 133–150.

Harris, M. I. and J. E. Coleman (1968) The biosynthesis of apo- and metalloalkaline phosphatases of *Escherichia coli*. *J. Biol. Chem.* **243**: 5063–5073.

Harrison, G., I. M. Shapiro and E. E. Golub (1995) The phosphatidylinositol–glycolipid anchor on alkaline phosphatase facilitates mineralization initiation *in vitro*. *J. Bone Miner. Res.* **10**: 568–573.

Hata, K., H. Tokuhiro, K. Nakatsuka, T. Miki, Y. Nishizawa, H. Morii and M. Miura (1996) Measurement of bone-specific alkaline phosphatase by an immunoselective enzyme assay method. *Ann. Clin. Biochem.* **33**: 127–131.

Hatta, M., H. Daitoku, H. Matsuzaki, Y. Deyama, Y. Yoshimura, K. Suzuki, A. Matsumoto and A. Fukamizu (2002) Regulation of alkaline phosphatase promoter activity by forkhead transcription factor FKHR. *Int. J. Mol. Med.* **9**: 147–152.

Hawrylak, K. and R. A. Stinson (1987) Tetrameric alkaline phosphatase from human liver is converted to dimers by phosphatidylinositol phospholipase C. *FEBS Lett.* **212**: 289–291.

Hawrylak, K. and R. A. Stinson (1988) The solubilization of tetrameric alkaline phosphatase from human liver and its conversion into various forms by phosphatidylinositol phospholipase C or proteolysis. *J. Biol. Chem.* **263**: 14368–14373.

Hawrylak, K., L. Kihn, D. Rutkowski and R. A. Stinson (1990) Purified tetrameric alkaline phosphatase: the effect of treatments with phosphatidylinositol phospholipase C and sodium dodecyl sulfate. *Clin. Chim. Acta* **186**: 197–201.

Heath, J. K (1978) Characterization of a xenogeneic antiserum raised against the fetal germ cells of the mouse: cross-reactivity with embryonal carcinoma cells. *Cell* **15**: 299–306.

Heath, J. K., L. J. Suva, K. Yoon, M. Kiledjian, T. J. Martin and G. A. Rodan (1992) Retinoic acid stimulates transcriptional activity from the alkaline phosphatase promoter in the immortalized rat calvarial cell line, RCT-1. *Mol. Endocrinol.* **6**: 636–646.

Heimo, H., K. Palmu and I. Suominen (1998) Human placental alkaline phosphatase: expression in *Pichia pastoris*, purification and characterization of the enzyme. *Protein Expr. Purif.* **12**: 85–92.

Hendrix, P. G., M. F. Hoylaerts, E. J. Nouwen and M. E. De Broe (1990) Enzyme immunoassay of human placental and germ-cell alkaline phosphatase in serum. *Clin. Chem.* **36**: 1793–1799.

Henthorn, P. S. and M. P. Whyte (1992) Missense mutations of the tissue-nonspecific alkaline phosphatase gene in hypophosphatasia. *Clin. Chem.* **38**: 2501–2505.

Henthorn, P. S. and M. P. Whyte (1995) Infantile hypophosphatasia: successful prenatal assessment by testing for tissue-non-specific alkaline phosphatase isoenzyme gene mutations. *Prenatal Diagn.* **15**: 1001–1006, (1995)

Henthorn, P. S., B. J. Knoll, M. Raducha, K. N. Rothblum, C. Slaughter, M. Weiss, M. A. Lafferty, T. Fischer and H. Harris (1986) Products of two common alleles at the locus for human placental alkaline phosphatase differ by seven amino acids. *Proc. Natl. Acad. Sci. USA* **83**: 5597–5601.

Henthorn, P. S., M. Raducha, Y. H. Edwards, M. J. Weiss, C. Slaughter, M. A. Lafferty and H. Harris (1987) Nucleotide and amino acid sequences of human intestinal alkaline phosphatase: close homology to placental alkaline phosphatase. *Proc. Natl. Acad. Sci. USA* **84**: 1234–1238.

Henthorn, P. S., M. Raducha, T. Kadesch, M. J. Weiss and H. Harris (1988a). Sequence and characterization of the human intestinal alkaline phosphatase gene. *J. Biol. Chem.* **263**: 12011–12019.

Henthorn, P., P. Zervos, M. Raducha, H. Harris and T. Kadesch (1988b). Expression of a human placental alkaline phosphatase gene in transfected cells: use as a reporter for studies of gene expression. *Proc. Natl. Acad. Sci. USA* **85**: 6342–6346.

Henthorn, P. S., M. Raducha, K. N. Fedde, M. A. Lafferty and M. P. Whyte (1992) Different missense mutations at the tissue-nonspecific alkaline phosphatase gene locus in autosomal recessively inherited forms of mild and severe hypophosphatasia. *Proc. Natl. Acad. Sci. USA* **89**: 9924–9928.

Herasse, M., M. Spentchian, A. Taillandier, K. Keppler-Noreuil, A. N. Fliorito, J. Bergoffen, R. Wallerstein, C. Muti, B. Simon-Bouy and E. Mornet (2003) Molecular study of three cases of odontohypophosphatasia resulting from heterozygosity for mutations in the tissue non-specific alkaline phosphatase gene. *J. Med. Genet.* **40**: 605–609.

Herz, F. (1984) Induction of alkaline phosphatase expression in cultured cancer cells. *Prog. Clin. Biol. Res.* **166**: 139–166.

Hessle, L., K. A. Johnson, H. C. Anderson, S. Narisawa, A. Sali, J. W. Goding, R. Terkeltaub and J. L. Millán (2002) Tissue-nonspecific alkaline phosphatase and plasma cell membrane glycoprotein-1 are central antagonistic regulators of bone mineralization. *Proc. Natl. Acad. Sci. USA* **99**: 9445–9449.

Higashino, K., M. Hashinotsume, K. Y. Kang, Y. Takahashi and Y. Yamamura (1972) Studies on a variant alkaline phosphatase in sera of patients with hepatocellular carcinoma. *Clin. Chim. Acta* **40**: 67–81.

Higashino, K., S. Kudo and Y. Yamamura (1974) Further investigation of a variant of the placental alkaline phosphatase in human hepatic carcinoma. *Cancer Res.* **34**: 3347–3351.

Higashino, K., S. Kudo, R. Otani and Y. Yamamura (1975a). A hepatoma-associated alkaline phosphatase, the Kasahara isozyme, compared with one of the isozymes of FL amnion cells. *Ann. N. Y. Acad. Sci.* **259**: 337–346.

Higashino, K., S. Kudo, Y. Yamamura, T. Honda and J. Sakurai (1975b). Possible identity between the hepatoma alkaline phosphatase and an isozyme of human amniotic membrane (FL cells) *Clin. Chim. Acta* **60**: 267–272.

Higashino, K., R. Otani, S. Kudo and Y. Yamamura (1977) A fetal intestinal-type alkaline phosphatase in hepatocellular carcinoma tissue. *Clin. Chem.* **23**: 1615–1623.

Higashino, K., K. Muratani, T. Hada, H. Imanishi, Y. Amuro, Y. Yamamoto, J. Furuyama, K. Hirano, Y. M. Hong, M. Shimokura, T. Hirano and T. Kishimoto (1990) Purification and some properties of the fast migrating alkaline phosphatase in FL-amnion cells (the Kasahara isoenzyme) and its cDNA cloning. *Clin. Chim. Acta* **186**: 151–164.

Hill, C. S. and R. L. Wolfert (1990) The preparation of monoclonal antibodies which react preferentially with human bone alkaline phosphatase and not liver alkaline phosphatase. *Clin. Chim. Acta* **186**: 315–320.

Hinnebusch, B. F., J. W. Henderson, A. Siddique, M. S. Malo, W. Zhang, M. A. Abedrapo and R. A. Hodin (2003) Transcriptional activation of the enterocyte differentiation marker intestinal alkaline phosphatase is associated with changes in the acetylation state of histone H3 at a specific site within its promoter region *in vitro*. *J. Gastroint. Surg.* **7**: 237–244; discussion, 244–235.

Hinnebusch, B. F., A. Siddique, J. W. Henderson, M. S. Malo, W. Zhang, C. P. Athaide, M. A. Abedrapo, X. Chen, V. W. Yang and R. A. Hodin (2004) Enterocyte differentiation marker intestinal alkaline phosphatase is a target gene of the gut-enriched Kruppel-like factor. *Am. J. Physiol. Gastrointest. Liver Physiol.* **286**: G23–G30.

Hiramatsu, N., A. Kasai, Y. Meng, K. Hayakawa, J. Yao and M. Kitamura (2005) Alkaline phosphatase vs luciferase as secreted reporter molecules *in vivo*. *Anal. Biochem.* **339**: 249–256.

Hirano, K., Y. Iiizumi, Y. Mori, K. Toyoshi, M. Sugiura and S. Iino (1985) Role of alkaline phosphatase in phosphate uptake into brush border membrane vesicles from human intestinal mucosa. *J. Biochem.* **97**: 1461–1466.

Hirano, K., Y. Iiizumi, Y. Hayashi, T. Tanaka, M. Sugiura, K. Hayashi, Z. D. Lu and S. Iino (1986) A highly sensitive assay method for human placental alkaline phosphatase involving a monoclonal antibody bound to a paper disk. *Anal. Biochem.* **154**: 624–631.

Hirano, K., U. M. Domar, H. Yamamoto, E. E. Brehmer-Andersson, B. E. Wahren and T. I. Stigbrand (1987a). Levels of alkaline phosphatase isozymes in human seminoma tissue. *Cancer Res.* **47**: 2543–2546.

Hirano, K., H. Matsumoto, T. Tanaka, Y. Hayashi, S. Iino, U. Domar and T. Stigbrand (1987b). Specific assays for human alkaline phosphatase isozymes. *Clin. Chim. Acta* **166**: 265–273.

Hirano, K., K. Kusano, Y. Matsumoto, T. Stigbrand, S. Iino and K. Hayashi (1989) Intestinal-like alkaline phosphatase expressed in normal human adult kidney. *Eur. J. Biochem.* **183**: 419–423.

Hirano, K., I. Koyama and T. Stigbrand (1990) Purification and partial characterization of the placental-like alkaline phosphatase in human lung tissue. *Clin. Chim. Acta* **186**: 265–273.

Ho, A. M., M. D. Johnson and D. M. Kingsley (2000) Role of the mouse *Ank* gene in control of tissue calcification and arthritis. *Science* **289**: 265–270.

Hodin, R. A., J. R. Graham, S. Meng and M. P. Upton (1994) Temporal pattern of rat small intestinal gene expression with refeeding. *Am. J. Physiol.* **266**: G83–89.

Hoffmann, W. E., J. L. Dorner and H. Morris (1983) Diagnostic value of intestinal alkaline phosphatase in horse serum. *Vet. Clin. Pathol.* **12**: 33–38.

Hofmann, M. C., W. Jeltsch, J. Brecher and H. Walt (1989) Alkaline phosphatase isozymes in human testicular germ cell tumors, their precancerous stage and three related cell lines. *Cancer Res.* **49**: 4696–4700.

Holmgren, P. A. and T. Stigbrand (1978) Catalytic properties and stability of three common variants of placental alkaline phosphatase. *Biochem. Genet.* **16**: 433–442.

Holmgren, P. A., T. Stigbrand, M. G. Damber and B. von Schoultz (1978) A double antibody solid phase radioimmunoassay for placental alkaline phosphatase. *Clin. Chim. Acta* **83**: 205–210.

Holmgren, P. A., T. Stigbrand, M. G. Damber and B. von Schoultz (1979) Serum levels of placental alkaline phosphatase in high-risk pregnancies. *Obstet. Gynecol.* **54**: 631–634.

Hooper, N. M (1997) Glycosyl-phosphatidyl-inositol anchored membrane enzymes. *Clin. Chim. Acta* **266**: 3–12.

Hording, U., K. Toftager-Larsen, A. Dreisler, B. Lund, S. Daugaard, F. Lundvall, J. Arends, P. Winkel and M. Rorth (1990) CA 125, placental alkaline phosphatase and tissue polypeptide antigen in the monitoring of ovarian carcinoma. A comparative study of three different tumor markers. *Gynecol. Obstet. Invest.* **30**: 178–183.

Horwich, A., D. F. Tucker and M. J. Peckham (1985) Placental alkaline phosphatase as a tumour marker in seminoma using the H17 E2 monoclonal antibody assay. *Br. J. Cancer* **51**: 625–629.

Hoshi, K., N. Amizuka, K. Oda, Y. Ikehara and H. Ozawa (1997) Immunolocalization of tissue non-specific alkaline phosphatase in mice. *Histochem. Cell. Biol.* **107**: 183–191.

Houston, B., A. J. Stewart and C. Farquharson (2004) PHOSPHO1 – a novel phosphatase specifically expressed at sites of mineralization in bone and cartilage. *Bone* **34**: 629–637.

Hoylaerts, M. F. and J. L. Millán (1991) Site-directed mutagenesis and epitope-mapped monoclonal antibodies define a catalytically important conformational difference between human placental and germ cell alkaline phosphatase. *Eur. J. Biochem.* **202**: 605–616.

Hoylaerts, M. F., T. Manes and J. L. Millán (1992a). Molecular mechanism of uncompetitive inhibition of human placental and germ-cell alkaline phosphatase. *Biochem. J.* **286**: 23–30.

Hoylaerts, M. F., T. Manes and J. L. Millán (1992b). Allelic amino acid substitutions affect the conformation and immunoreactivity of germ-cell alkaline phosphatase phenotypes. *Clin. Chem.* **38**: 2493–2500.

Hoylaerts, M. F., T. Manes and J. L. Millán (1997) Mammalian alkaline phosphatases are allosteric enzymes. *J. Biol. Chem.* **272**: 22781–22787.

Hoylaerts, M. F., L. Ding, S. Narisawa, S. Vankerckhoven and J. L. Millán (2005) Intramolecular active site transition in mammalian alkaline phosphatases depends on a structurally intact N-terminal α-helix. Submitted.

Hsu, H. H. and N. P. Camacho (1999) Isolation of calcifiable vesicles from human atherosclerotic aortas. *Atherosclerosis* **143**: 353–362.

Hsu, H. H., N. P. Camacho, F. Sun, O. Tawfik and H. Aono (2000) Isolation of calcifiable vesicles from aortas of rabbits fed with high cholesterol diets. *Atherosclerosis* **153**: 337–348.

Hu, J. C., R. Plaetke, E. Mornet, C. Zhang, X. Sun, H. F. Thomas and J. P. Simmer (2000) Characterization of a family with dominant hypophosphatasia. *Eur. J. Oral Sci.* **108**: 189–194.

Hua, J. C., E. Garattini, Y. C. Pan, J. D. Hulmes, M. Chang, L. Brink and S. Udenfriend (1985) Purification and partial sequencing of bovine liver alkaline phosphatase. *Arch. Biochem. Biophys.* **241**: 380–385.

Hua, J. C., J. Berger, Y. C. Pan, J. D. Hulmes and S. Udenfriend (1986) Partial sequencing of human adult, human fetal and bovine intestinal alkaline phosphatases: comparison with the human placental and liver isozymes. *Proc. Natl. Acad. Sci. USA* **83**: 2368–2372.

Huang, T. M., H. C. Hung, T. C. Chang and G. G. Chang (1998) Solvent kinetic isotope effects of human placental alkaline phosphatase in reverse micelles. *Biochem. J.* **330**: 267–275.

Hui, M. and H. C. Tenenbaum (1998) New face of an old enzyme: alkaline phosphatase may contribute to human tissue aging by inducing tissue hardening and calcification. *Anat. Rec.* **253**: 91–94.

Hui, M., S. Q. Li, D. Holmyard and P. Cheng (1997a). Stable transfection of nonosteogenic cell lines with tissue nonspecific alkaline phosphatase enhances mineral deposition both in the presence and absence of beta-glycerophosphate: possible role for alkaline phosphatase in pathological mineralization. *Calcif. Tissue Int.* **60**: 467–472.

Hui, M. Z., H. C. Tenenbaum and C. A. McCulloch (1997b). Collagen phagocytosis and apoptosis are induced by high level alkaline phosphatase expression in rat fibroblasts. *J. Cell. Physiol.* **172**: 323–333.

Hummer, C. and J. L. Millán (1991) Gly429 is the major determinant of uncompetitive inhibition of human germ cell alkaline phosphatase by L-leucine. *Biochem. J.* **274**: 91–95.

Hung, H. C. and G. G. Chang (1998) Biphasic denaturation of human placental alkaline phosphatase in guanidinium chloride. *Proteins* **33**: 49–61.

Hung, H. C. and G. G. Chang (2001a). Differentiation of the slow-binding mechanism for magnesium ion activation and zinc ion inhibition of human placental alkaline phosphatase. *Protein Sci.* **10**: 34–45.

Hung, H. C. and G. G. Chang (2001b). Multiple unfolding intermediates of human placental alkaline phosphatase in equilibrium urea denaturation. *Biophys. J.* **81**: 3456–3471.

Husain, M. and C. Bieniartz (1994) Fc site-specific labeling of immunoglobulins with calf intestinal alkaline phosphatase. *Bioconj. Chem.* **5**: 482–490.

Hustin, J., J. Collette and P. Franchimont (1987) Immunohistochemical demonstration of placental alkaline phosphatase in various states of testicular development and in germ cell tumours. *Int. J. Androl.* **10**: 29–35.

Hustin, J., Y. Gillerot, J. Collette and P. Franchimont (1990) Placental alkaline phosphatase in developing normal and abnormal gonads and in germ-cell tumours. *Virchows Arch. A. Pathol. Anat. Histopathol.* **417**: 67–72.

Ierardi, D. F., J. M. Pizauro and P. Ciancaglini (2002) Erythrocyte ghost cell–alkaline phosphatase: construction and characterization of a vesicular system for use in biomineralization studies. *Biochim. Biophys. Acta* **1567**: 183–192.

Imai, K., Y. Tomaru, T. Ohnuki, H. Yamanaka, H. Sakai, H. Kanetake, Y. Minami, K. Nomata and Y. Saito (1992) Significance of a new stratification of alkaline phosphatase and extent of disease in patients with prostate carcinoma with bone metastasis. *Cancer* **69**: 2983–2989.

Imanishi, H., T. Hada, K. Muratani, K. Hirano and K. Higashino (1990a). Expression of a hybrid form of alkaline phosphatase isoenzyme in a newly established cell line (HuG-1) from a gastric cancer patient. *Cancer Res.* **50**: 3408–3412.

Imanishi, H., T. Hada, K. Muratani, K. Hirano and K. Higashino (1990b). An alkaline phosphatase reacting with both monoclonal antibodies to intestinal and placental isoenzymes. *Clin. Chim. Acta* **186**: 309–314.

Inglis, N. R., S. Kirley, L. L. Stolbach and W. H. Fishman (1973) Phenotypes of the Regan isoenzyme and identity between the placental D-variant and the Nagao isoenzyme. *Cancer Res.* **33**: 1657–1661.

Iqbal, S. J., T. Davies, R. Cole, P. Whitaker and C. Chapman (2000) Neutrophil alkaline phosphatase (NAP) score in the diagnosis of hypophosphatasia. *Clin. Chim. Acta* **302**: 49–57.

Ishida, Y., K. Komaru, M. Ito, Y. Amaya, S. Kohno and K. Oda (2003) Tissue-nonspecific alkaline phosphatase with an Asp(289)→Val mutation fails to reach the cell surface and undergoes proteasome-mediated degradation. *J. Biochem.* **134**: 63–70.

Ito, M., N. Amizuka, H. Ozawa and K. Oda (2002) Retention at the cis-Golgi and delayed degradation of tissue-non-specific alkaline phosphatase with an Asn153→Asp substitution, a cause of perinatal hypophosphatasia. *Biochem. J.* **361**: 473–480.

Itoh, H., T. Kakuta, G. Genda, I. Sakonju and K. Takase (2002) Canine serum alkaline phosphatase isoenzymes detected by polyacrylamide gel disk electrophoresis. *J. Vet. Med. Sci.* **64**: 35–39.

Iyer, S. K., H. H. Daron and J. L. Aull (1988) Purification and properties of alkaline phosphatase from boar seminal plasma. *J. Reprod. Fertil.* **82**: 657–664.

Izquierdo, L., T. Lopez and P. Marticorena (1980) Cell membrane regions in preimplantation mouse embryos. *J. Embryol. Exp. Morphol.* **59**: 89–102.

Jablonski, E., E. W. Moomaw, R. H. Tullis and J. L. Ruth (1986) Preparation of oligodeoxynucleotide-alkaline phosphatase conjugates and their use as hybridization probes. *Nucleic Acids Res.* **14**: 6115–6128.

Jacobs, S. C. (1983) Spread of prostatic cancer to bone. *Urology* **21**: 337–344.

Jansonius, J. N. (1998) Structure, evolution and action of vitamin B_6-dependent enzymes. *Curr. Opin. Struct. Biol.* **8**: 759–69.

Jansson, U.H.G., B. Kristiansson, P. Magnusson, L.Larsson, K. Albertsson-Wikland and R. Bjarnason (2001) The decrease of IGF-I, IGF-binding protein-3 and bone alkaline phosphatase isoforms during gluten challenge correlates with small intestinal inflammation in children with coeliac disease. *Eur. J. Endocrinol.* **144**: 417–423.

Jemmerson, R. and T. Stigbrand (1984) Monoclonal antibodies block the trypsin cleavage site on human placental alkaline phosphatase. *FEBS. Lett.* **173**: 357–359.

Jemmerson, R., M. Takeya, N. Shah and W. H. Fishman (1984a). Tumor immunolocalization using monoclonal antibodies which bind placental alkaline phosphatase. *Prog. Clin. Biol. Res.* **166**: 245–256.

Jemmerson, R., N. Shah, M. Takeya and W. H. Fishman (1984b). Functional organization of the placental alkaline phosphatase polypeptide chain. *Prog. Clin. Biol. Res.* **166**: 105–115.

Jemmerson, R., F. G. Klier and W. H. Fishman (1985a). Clustered distribution of human placental alkaline phosphatase on the surface of both placental and cancer cells. Electron microscopic observations using gold-labeled antibodies. *J. Histochem. Cytochem.* **33**: 1227–1234.

Jemmerson, R., J. L. Millán, F. G. Klier and W. H. Fishman (1985b). Monoclonal antibodies block the bromelain-mediated release of human placental alkaline phosphatase from cultured cancer cells. *FEBS Lett.* **179**: 316–320.

Jemmerson, R., N. Shah and W. H. Fishman (1985c). Evidence for homology of normal and neoplastic human placental alkaline phosphatases as determined by monoclonal antibodies to the cancer-associated enzyme. *Cancer Res.* **45**: 3268–3273.

Jemmerson, R., N. Shah, M. Takeya and W. H. Fishman (1985d). Characterization of the placental alkaline phosphatase-like (Nagao) isozyme on the surface of A431 human epidermoid carcinoma cells. *Cancer Res.* **45**: 282–287.

Jeppsson, A., B. Wahren, E. Brehmer-Andersson, C. Silfversward, T. Stigbrand and J. L. Millán (1984a). Eutopic expression of placental-like alkaline phosphatase in testicular tumors. *Int. J. Cancer* **34**: 757–761.

Jeppsson, A., B. Wahren, J. L. Millán and T. Stigbrand (1984b). Tumour and cellular localization by use of monoclonal and polyclonal antibodies to placental alkaline phosphatase. *Br. J. Cancer* **49**: 123–128.

Johnson, K. A., L. Hessle, S. Vaingankar, C. Wennberg, S. Mauro, S. Narisawa, J. W. Goding, K. Sano, J. L. Millán and R. Terkeltaub (2000) Osteoblast tissue-nonspecific alkaline phosphatase antagonizes and regulates PC-1. *Am. J. Physiol. Regul. Integr. Comp. Physiol.* **279**: R1365–R1377.

Johnson, K., J. Goding, D. Van Etten, A. Sali, S. I. Hu, D. Farley, H. Krug, L. Hessle,

J. L. Millán and R. Terkeltaub (2003) Linked deficiencies in extracellular PP$_i$ and osteopontin mediate pathologic calcification associated with defective PC-1 and ANK expression. *J. Bone Miner. Res.* **18**: 994–1004.

Johnson-Pais, T. L. and R. J. Leach (1996) 1,25-Dihydroxyvitamin D3 and transforming growth factor-beta act synergistically to override extinction of liver/bone/kidney alkaline phosphatase in osteosarcoma hybrid cells. *Exp. Cell Res.* **226**: 67–74.

Joosten, C. E. and M. L. Shuler (2003) Effect of culture conditions on the degree of sialylation of a recombinant glycoprotein expressed in insect cells. *Biotechnol. Prog.* **19**: 739–749.

Joshi, L., T. R. Davis, T. S. Mattu, P. M. Rudd, R. A. Dwek, M. L. Shuler and H. A. Wood (2000) Influence of baculovirus-host cell interactions on complex N-linked glycosylation of a recombinant human protein. *Biotechnol. Prog.* **16**: 650–656.

Jourquin, G. and J. M. Kauffmann (1998) Fluorimetric determination of theophylline in serum by inhibition of bovine alkaline phosphatase in AOT based water/in oil microemulsion. *J. Pharm. Biomed. Anal.* **18**: 585–596.

Kallioniemi, O. P., M. M. Nieminen, J. Lehtinen, T. Veneskoski and T. Koivula (1987) Increased serum placental-like alkaline phosphatase activity in smokers originates from the lungs. *Eur. J. Respir. Dis.* **71**: 170–176.

Kalofonos, H. P., T. R. Pawlikowska, A. Hemingway, N. Courtenay-Luck, B. Dhokia, D. Snook, G. B. Sivolapenko, G. R. Hooker, C. G. McKenzie and P. J. Lavender (1989) Antibody guided diagnosis and therapy of brain gliomas using radiolabeled monoclonal antibodies against epidermal growth factor receptor and placental alkaline phosphatase. *J. Nucl. Med.* **30**: 1636–1645.

Kalofonos, H. P., C. Kosmas, T. R. Pawlikowska, A. Bamias, D. Snook, B. Dhokia, G. B. Sivolapenko, N. S. Courtenay-Luck and A. A. Epenetos (1990) Immunolocalization of testicular tumours using radiolabeled monoclonal antibody to placental alkaline phosphatase. *J. Nucl. Med. Allied Sci.* **34**: 294–298.

Kam, W., E. Clauser, Y. S. Kim, Y. W. Kan and W. J. Rutter (1985) Cloning, sequencing and chromosomal localization of human term placental alkaline phosphatase cDNA. *Proc. Natl. Acad. Sci. USA* **82**: 8715–8719.

Kanabrocki, E.L., L.E. Scheving, F. Halberg, R.L. Brewer and T.J.Bird (1973) Circadian variations in presumably healthy men under conditions of peace-time army reserve unit training. *Space Life Sci.* **4**: 258–270.

Kanakis, I., M. Nikolaou, D. Pectasides, C. Kiamouris and N. K. Karamanos (2004) Determination and biological relevance of serum cross-linked Type I collagen N-telopeptide and bone-specific alkaline phosphatase in breast metastatic cancer. *J. Pharm. Biomed. Anal.* **34**: 827–832.

Kang, J. O., W. A. Hudak, W. J. Crowley and B. S. Criswell (1990) Placental-type alkaline phosphatase in peritoneal fluid of women with endometriosis. *Clin. Chim. Acta* **186**: 285–294.

Kapojos, J. J., K. Poelstra, T. Borghuis, A. Van Den Berg, H. J. Baelde, P. A. Klok and W. W. Bakker (2003) Induction of glomerular alkaline phosphatase after challenge with lipopolysaccharide. *Int. J. Exp. Pathol.* **84**: 135–144.

Kash, S. F., R. S. Johnson, L. H. Tecott, J. L. Noebels, R. D. Mayfield, D. Hanahan and S. Baekkeskov (1997) Epilepsy in mice deficient in the 65-kDa isoform of glutamic acid decarboxylase. *Proc. Natl. Acad. Sci. USA* **94**: 14060–14065.

Kasono, K., K. Sato, D. C. Han, Y. Fujii, T. Tsushima and K. Shizume (1988) Stimulation of alkaline phosphatase activity by thyroid hormone in mouse osteoblast-like cells (MC3T3-E1): a possible mechanism of hyperalkaline phosphatasia in hyperthyroidism. *Bone Miner.* **4**: 355–363.

Kaufman, M. H. (1992) *The Atlas of Mouse Development*. San Diego: Academic Press.

Khan, K. N., T. Tsutsumi, K. Nakata, K. Nakao, Y. Kato and S. Nagataki (1998) Regulation of alkaline phosphatase gene expression in human hepatoma cells by bile acids. *J. Gastroenterol. Hepatol.* **13**: 643–650.

Kihn, L., A. Dinwoodie and R. A. Stinson (1991) High-molecular-weight alkaline phosphatase in serum has properties similar to the enzyme in plasma membranes of the liver. *Am. J. Clin. Pathol.* **96**: 470–478.

Kiledjian, M. and T. Kadesch (1990) Analysis of the human liver/bone/kidney alkaline phosphatase promoter *in vivo* and *in vitro*. *Nucleic Acids Res.* **18**: 957–961.

Kiledjian, M. and T. Kadesch (1991) Post-transcriptional regulation of the human liver/bone/kidney alkaline phosphatase gene. *J. Biol. Chem.* **266**: 4207–4213.

Kim, K. M. (1976) Calcification of matrix vesicles in human aortic valve and aortic media. *Fed. Proc.* **35**: 156–162.

Kim, E. E. and H. W. Wyckoff (1991) Reaction mechanism of alkaline phosphatase based on crystal structures. Two-metal ion catalysis. *J. Mol. Biol.* **218**: 449–464.

Kim, J., H. K. Kim, K. Kim, S. R. Kim and W. K. Cho (1989) Multiple forms of alkaline phosphatase in mouse preimplantation embryos. *J. Reprod. Fertil.* **86**: 65–72.

Kim, J. H., S. Meng, A. Shei and R. A. Hodin (1999) A novel Sp1-related cis element involved in intestinal alkaline phosphatase gene transcription. *Am. J. Physiol.* **276**: G800–807.

Kim, K. Y., Y. J. Cho, G. A. Jeon, P. D. Ryu and J. N. Myeong (2003) Membrane-bound alkaline phosphatase gene induces antitumor effect by G2/M arrest in etoposide phosphate-treated cancer cells. *Mol. Cell. Biochem.* **252**: 213–221.

Kim, Y. J., M. H. Lee, J. M. Wozney, J. Y. Cho and H. M. Ryoo (2004) Bone morphogenetic protein-2-induced alkaline phosphatase expression is stimulated by Dlx5 and repressed by Msx2. *J. Biol. Chem.* **279**: 50773–50780.

Kinoshita, Y., T. Okamoto, H. Mano, Y. Furuhashi, S. Goto and Y. Tomoda (1990) Establishment of hybridomas secreting monoclonal antibodies to placental alkaline phosphatase and development of an enzyme immunoassay for its determination. *Nippon Sanka Fujinka Gakkai Zasshi* **42**: 613–619.

Kirchberger, J., H. Seidel and G. Kopperschlager (1987) Interaction of procion red HE-3B and other reactive dyes with alkaline phosphatase: a study by means of kinetic, difference spectroscopic and chromatographic methods. *Biomed. Biochim. Acta* **46**: 653–663.

Kishi, F., S. Matsuura and T. Kajii (1989) Nucleotide sequence of the human liver-type alkaline phosphatase cDNA. *Nucleic Acids Res.* **17**: 2129.

Kleerekoper, M. (1996) Biochemical markers of bone remodeling. *Am. J. Med. Sci.* **312**: 270–277.

Knoll, B. J., K. N. Rothblum and M. Longley (1988) Nucleotide sequence of the human placental alkaline phosphatase gene. Evolution of the 5′-flanking region by deletion/substitution. *J. Biol. Chem.* **263**: 12020–12027.

Kobayashi, T. and J. M. Robinson (1991) A novel intracellular compartment with unusual secretory properties in human neutrophils. *J. Cell. Biol.* **113**: 743–756.

Kobayashi, T., T. Sugimoto, M. Kanzawa and K. Chihara (1998) Identification of an enhancer sequence in 5′-flanking region of 1A exon of mouse liver/bone/kidney-type alkaline phosphatase gene. *Biochem. Mol. Biol. Int.* **44**: 683–691.

Kohno, H., K. Sudo and T. Kanno (1983) Intestinal alkaline phosphatase linked to immunoglobulin G of the kappa type. *Clin. Chim. Acta* **135**: 41–48.

Koide, O., S. Iwai, K. Baba and H. Iri (1987) Identification of testicular atypical germ cells by an immunohistochemical technique for placental alkaline phosphatase. *Cancer* **60**: 1325–1330.

Kodama, H., K. Asai, T. Adachi, Y. Mori, K. Hayashi, K. Hirano and T. Stigbrand (1994) Expression of a heterodimeric (placental–ntestinal) hybrid alkaline phosphatase in KB cells. *Biochim. Biophys. Acta* **1218**: 163–172.

Kokado, A., H. Arakawa and M. Maeda (2002) Chemiluminescent assay of alkaline phosphatase using dihydroxyacetone phosphate as substrate detected with lucigenin. *Luminescence* **17**: 5–10.

Kolbus, N., W. Beuche, K. Felgenhauer and M. Mader (1996) Definition of a discontinuous immunodominant epitope of intestinal alkaline phosphatase, an autoantigen in acute bacterial infections. *Clin. Immunol. Immunopathol.* **80**: 298–306.

Komarnytsky, S., N. V. Borisjuk, L. G. Borisjuk, M. Z. Alam and I. Raskin (2000) Production of recombinant proteins in tobacco guttation fluid. *Plant Physiol.* **124**: 927–934.

Komaru, K., Y. Ishida, Y. Amaya, M. Goseki-Sone, H. Orimo and K. Oda (2005) Novel aggregate formation of a frameshift mutant protein of tissue-nonspecific alkaline phosphatase is ascribed to three cysteine residues in the C-terminal extension. Retarded secretion and proteasomal degradation. *FEBS J.* **272**: 1704–1717.

Komoda, T., M. Sonoda, M. Ikeda, S. Hokari and Y. Sakagishi (1981) Inhibition of alkaline phosphatase by bismuth. *Clin. Chim. Acta* **116**: 161–169.

Komoda, T., I. Koyama, A. Nagata, Y. Sakagishi, K. DeSchryver-Kecskemeti and D. H. Alpers (1986) Ontogenic and phylogenic studies of intestinal, hepatic and placental alkaline phosphatases. Evidence that intestinal alkaline phosphatase is a late evolutionary development. *Gastroenterology* **91**: 277–286.

Koshida, K., T. Stigbrand, E. Munck-Wikland, H. Hisazumi and B. Wahren (1990) Analysis of serum placental alkaline phosphatase activity in testicular cancer and cigarette smokers. *Urol. Res.* **18**: 169–173.

Koshida, K., A. Nishino, H. Yamamoto, T. Uchibayashi, K. Naito, H. Hisazumi, K. Hirano, Y. Hayashi, B. Wahren and L. Andersson (1991) The role of alkaline phosphatase isoenzymes as tumor markers for testicular germ cell tumors. *J. Urol.* **146**: 57–60.

Koshida, K., T. Uchibayashi, H. Yamamoto, K. Yokoyama and K. Hirano (1996) A potential use of a monoclonal antibody to placental alkaline phosphatase (PLAP) to detect lymph node metastases of seminoma. *J. Urol.* **155**: 337–341.

Koshida, K., K. Yokoyama, T. Imao, H. Konaka, K. Hirano, T. Uchibayashi and M. Namiki (1998) Immunolocalization of anti-placental alkaline phosphatase monoclonal antibody in mice with testicular tumors and lymph node metastasis. *Urol. Res.* **26**: 23–28.

Kozlenkov, A., T. Manes, M. F. Hoylaerts and J. L. Millán (2002) Function assignment to conserved residues in mammalian alkaline phosphatases. *J. Biol. Chem.* **277**: 22992–22999.

Kozlenkov, A., M. H. Le Du, P. Cuniasse, T. Ny, M. F. Hoylaerts and J. L. Millán (2004) Residues determining the binding specificity of uncompetitive inhibitors to tissue-nonspecific alkaline phosphatase. *J. Bone Miner. Res.* **19**: 1862–1872.

Kralovanszky, J., Z. Szentirmay, I. Besznyak and S. Eckhardt (1984) Placental type alkaline phosphatase in possibly premalignant alterations of human gastric mucosa. *Oncology* **41**: 189–194.

Krull, I. S., H. H. Stuting and S. C. Krzysko (1988) Conformational studies of bovine alkaline phosphatase in hydrophobic interaction and size-exclusion chromatography with linear diode array and low-angle laser light scattering detection. *J. Chromatogr.* **442**: 29–52.

Kumar, P., J. P. Nagpaul, B. Singh, R. C. Bansal and R. Sharma (1978) Nature of inhibition of rat testicular alkaline phosphatase by isatin. *Experientia* **34**: 434–435.

Kurn, N., M. G. Taylor, E. Carlton, M. Caufield and E. F. Ullman (1994) Performance characteristics of a new immunoassay for serum bone specific alkaline phosphatase. *J. Bone Miner. Res.* **9**: S189.

Kuwana, T. and S. B. Rosalki (1991) Measurement of alkaline phosphatase of intestinal origin in plasma by p-bromotetramisole inhibition. *J. Clin. Pathol.* **44**: 236–237.

Kuwana, T., O. Sugita and M. Yakata (1991) Sugar chain heterogeneity of bone and liver alkaline phosphatase in serum. *Enzyme* **45**: 63–66.

Kwong, W. H. and P. P. Tam (1984) The pattern of alkaline phosphatase activity in the developing mouse spinal cord. *J. Embryol. Exp. Morphol.* **82**: 241–251.

Kyeyune-Nyombi, E., K. H. Lau, D. J. Baylink and D. D. Strong (1991) 1,25-Dihydroxyvitamin D3 stimulates both alkaline phosphatase gene transcription and mRNA stability in human bone cells. *Arch. Biochem. Biophys.* **291**: 316–325.

Kyeyune-Nyombi, E., V. Nicolas, D. D. Strong and J. Farley (1995) Paradoxical effects of phosphate to directly regulate the level of skeletal alkaline phosphatase activity in human osteosarcoma (SaOS-2) cells and inversely regulate the level of skeletal alkaline phosphatase mRNA. *Calcif. Tissue Int.* **56**: 154–159.

Lakke, E. A., J. G. van der Veeken and E. Marani (1988) The prenatal development of alkaline phosphatase activity in the hypothalamus of the rat. *Basic Appl. Histochem.* **32**: 179–185.

Lange, P. H., J. L. Millán, T. Stigbrand, R. L. Vessella, E. Ruoslahti and W. H. Fishman (1982) Placental alkaline phosphatase as a tumor marker for seminoma. *Cancer Res.* **42**: 3244–3247.

Langman, M. J., E. Leuthold, E. B. Robson, J. Harris, J. E. Luffman and H. Harris (1966) Influence of diet on the "intestinal" component of serum alkaline phosphatase in people of different ABO blood groups and secretor status. *Nature* **212**: 41–43.

Latham, K. M. and E. J. Stanbridge (1990) Identification of the HeLa tumor-associated antigen, p75/150, as intestinal alkaline phosphatase and evidence for its transcriptional regulation. *Proc. Natl. Acad. Sci. USA* **87**: 1263–1267.

Latham, K. M. and E. J. Stanbridge (1992) Examination of the oncogenic potential of a tumor-associated antigen, intestinal alkaline phosphatase, in HeLa × fibroblast cell hybrids. *Cancer Res.* **52**: 616–622.

Lawson, G. M., J. A. Katzmann, T. K. Kimlinger and J. F. O'Brien (1985) Isolation and preliminary characterization of a monoclonal antibody that interacts preferentially with the liver isoenzyme of human alkaline phosphatase. *Clin. Chem.* **31**: 381–385.

Le Du, M. H. and J. L. Millán (2002) Structural evidence of functional divergence in human alkaline phosphatases. *J. Biol. Chem.* **277**: 49808–49814.

Le Du, M. H., T. Stigbrand, M. J. Taussig, A. Menez and E. A. Stura (2001) Crystal structure of alkaline phosphatase from human placenta at 1.8 Å resolution. Implication for a substrate specificity. *J. Biol. Chem.* **276**: 9158–9165.

Leach, R. J., Z. Schwartz, T. L. Johnson-Pais, D. D. Dean, M. Luna and B. D. Boyan (1995) Osteosarcoma hybrids can preferentially target alkaline phosphatase activity to matrix vesicles: evidence for independent membrane biogenesis. *J. Bone Miner. Res.* **10**: 1614–1624.

Leathers, V. L. and A. W. Norman (1993) Evidence for calcium mediated conformational changes in calbindin-D28K (the vitamin D-induced calcium binding protein) interactions with chick intestinal brush border membrane alkaline phosphatase as studied via photoaffinity labeling techniques. *J. Cell. Biochem.* **52**: 243–252.

Leboy, P. S., J. N. Beresford, C. Devlin and M. E. Owen (1991) Dexamethasone induction of osteoblast mRNAs in rat marrow stromal cell cultures. *J. Cell. Physiol.* **146**: 370–378.

Lepire, M. L. and C. A. Ziomek (1989) Preimplantation mouse embryos express a heat-stable alkaline phosphatase. *Biol. Reprod.* **41**: 464–473.

Leung, K. S., K. P. Fung, A. H. Sher, C. K. Li and K. M. Lee (1993) Plasma bone-specific alkaline phosphatase as an indicator of osteoblastic activity. *J. Bone Joint. Surg. Br.* **75**: 288–292.

Lewis-Jones, D. I., P. M. Johnson, A. D. Desmond and P. J. McLaughlin (1992) Germ cell alkaline phosphatase in human seminal plasma following vasectomy. *Br. J. Urol.* **69**: 418–420.

Li, J., X. Luo, Q. Wang, L. M. Zheng, I. King, T. W. Doyle and S. H. Chen (1998) Synthesis and biological evaluation of a water soluble phosphate prodrug of 3-aminopyridine-2-carboxaldehyde thiosemicarbazone (3-AP) *Bioorg. Med. Chem. Lett.* **8**: 3159–3164.

Li, J., L. M. Zheng, I. King, T. W. Doyle and S. H. Chen (2001) Syntheses and antitumor activities of potent inhibitors of ribonucleotide reductase: 3-amino-4-methylpyridine-2-carboxaldehyde-thiosemicarbazone (3-AMP), 3-aminopyridine-2-carboxaldehyde-thiosemicarbazone (3-AP) and its water-soluble prodrugs. *Curr. Med. Chem.* **8**: 121–133.

Li, T. K., L. Lumeng and R. L. Veitch (1974) Regulation of pyridoxal 5′-phosphate metabolism in liver. *Biochem. Biophys. Res. Commun.* **61**: 677–684.

Li, X., W. Wang and T. Lufkin (1997) Dicistronic LacZ and alkaline phosphatase reporter constructs permit simultaneous histological analysis of expression from multiple transgenes. *Biotechniques* **23**: 874–882.

Lia-Baldini, A. S., F. Muller, A. Taillandier, J. F. Gibrat, M. Mouchard, B. Robin, B. Simon-Bouy, J. L. Serre, A. S. Aylsworth, E. Bieth, S. Delanote, P. Freisinger, J. C. Hu, H. P. Krohn, M. E. Nunes and E. Mornet (2001) A molecular approach to dominance in hypophosphatasia. *Hum. Genet.* **109**: 99–108.

Lieverse, A. G., G. G. van Essen, G. J. Beukeveld, J. Gazendam, E. C. Dompeling, L. P. ten Kate, S. A. van Belle and J. Weits (1990) Familial increased serum intestinal alkaline phosphatase: a new variant associated with Gilbert's syndrome. *J. Clin. Pathol.* **43**: 125–128.

Lin, C. W. and W. H. Fishman (1972) L-Homoarginine. An organ-specific, uncompetitive inhibitor of human liver and bone alkaline phosphohydrolases. *J. Biol. Chem.* **247**: 3082–3087.

Litmanovitz, O. Reish, T. Dolfin, S. Arnon, R. Regev, G. Grinshpan, M. Yamazaki and K. Ozono (2002) Glu274Lys/Gly309Arg mutation of the tissue-nonspecific alkaline phosphatase gene in neonatal hypophosphatasia associated with convulsions. *J. Inherit. Metab. Dis.* **25**: 35–40.

Liu, P. P., K. S. Leung, S. M. Kumta, K. M. Lee and K. P. Fung (1996) Bone-specific alkaline phosphatase in plasma as tumour marker for osteosarcoma. *Oncology* **53**: 275–280.

Llinas, P., E. Stura, A. Menez, Z. Kiss, T. Stigbrand, J. L. Millán and M. H. Le Du (2005) Structural studies of human placental alkaline phosphatase in complex with functional ligands. *J. Mol. Biol.* **350**: 441–451.

Loose, J. H., I. Damjanov and H. Harris (1984) Identity of the neoplastic alkaline phosphatase as revealed with monoclonal antibodies to the placental form of the enzyme. *Am. J. Clin. Pathol.* **82**: 173–177.

Lorenz, B. and H. C. Schroder (2001) Mammalian intestinal alkaline phosphatase acts as highly active exopolyphosphatase. *Biochem. Biophys. Acta* **1547**: 254–261.

Low, M. G. and A. R. Prasad (1988) A phospholipase D specific for the phosphatidylinositol anchor of cell-surface proteins is abundant in plasma. *Proc. Natl. Acad. Sci. USA* **85**: 980–984.

Low, M. G. and A. R. Saltiel (1988) Structural and functional roles of glycosyl-phosphatidylinositol in membranes. *Science* **239**: 268–275.

Low, M. G., R. C. Carroll and A. C. Cox (1986) Characterization of multiple forms of phosphoinositide-specific phospholipase C purified from human platelets. *Biochem. J.* **237**: 139–145.

Lowe, M. E. (1992) Site-specific mutations in the COOH-terminus of placental alkaline phosphatase: a single amino acid change converts a phosphatidylinositol-glycan-anchored protein to a secreted protein. *J. Cell. Biol.* **116**: 799–807.

Lowe, M. E. and A. W. Strauss (1990) Expression of a Nagao-type, phosphatidylinositol–glycan anchored alkaline phosphatase in human choriocarcinomas. *Cancer Res.* **50**: 3956–3962.

Lowe, M., A. W. Strauss, R. Alpers, S. Seetharam and D. H. Alpers (1990) Molecular cloning and expression of a cDNA encoding the membrane-associated rat intestinal alkaline phosphatase. *Biochim. Biophys. Acta* **1037**: 170–177.

Lyaruu, D. M., J. H. Woltgens, A. A. Dogterom and T. J. Bervoets (1983) Studies of alkaline phosphatase inhibition by *p*-bromotetramisole in non-mineralizing and mineralizing neonatal hamster tooth germs *in vitro*. *J. Biol. Buccale* **11**: 347–353.

Lyaruu, D. M., J. H. Woltgens and T. J. Bervoets (1984) Effect of 1-*p*-bromotetramisole on mineralization of hamster tooth germs *in vitro*: a light and electron microscopic study. *J. Biol. Buccale* **12**: 287–296.

Lyaruu, D. M., J. H. Woltgens and T. J. Bervoets (1987) Effect of alkaline-phosphatase inhibition by 1-*p*-bromotetramisole on the formation of trichloroacetic acid-[^{32}P]-insoluble phosphate from inorganic [^{32}P]-phosphate and [^{32}P]-pyrophosphate in non-mineralizing and mineralizing hamster molar tooth-germs *in vitro*. *Arch. Oral. Biol.* **32**: 429–432.

Ma, L. and E. R. Kantrowitz (1994) Mutations at histidine 412 alter zinc binding and eliminate transferase activity in *Escherichia coli* alkaline phosphatase. *J. Biol. Chem.* **269**: 31614–31619.

MacGregor, G. R., B. P. Zambrowicz and P. Soriano (1995) Tissue non-specific alkaline phosphatase is expressed in both embryonic and extraembryonic lineages during mouse embryogenesis but is not required for migration of primordial germ cells. *Development* **121**: 1487–1496.

Mackintosh, J., J. Simes, D. Raghavan and B. Pearson (1990) Prostatic cancer with bone metastases: serum alkaline phosphatase (SAP) as a predictor of response and the significance of the SAP "flare". *Br. J. Urol.* **66**: 88–93.

Mader, M., N. Kolbus, D. Meihorst, A. Kohn, W. Beuche and K. Felgenhauer (1994) Human intestinal alkaline phosphatase-binding IgG in patients with severe bacterial infections. *Clin. Exp. Immunol.* **95**: 98–102.

Maekawa, M., K. Sudo and T. Kanno (1985) Characteristics of the complex between alkaline phosphatase and immunoglobulin A in human serum. *Clin. Chim. Acta* **150**: 185–195.

Maelandsmo, G. M., P. J. Ross, M. Pavliv, R. A. Meulenbroek, C. Evelegh, D. A. Muruve, F. L. Graham and R. J. Parks (2005) Use of a murine secreted alkaline phosphatase as a non-immunogenic reporter gene in mice. *J. Gene Med.* **7**: 307–315.

Magnusson, P. and J.R.Farley (2002) Differences in sialic acid residues among bone alkaline phosphatase isoforms: a physical, biochemical and immunological characterization. *Calcif. Tissue Int.* **71**: 508–518.

Magnusson, P., O. Lofman and L. Larsson (1992) Determination of alkaline phosphatase isoenzymes in serum by high-performance liquid chromatography with post-column reaction detection. *J. Chromatogr.* **576**: 79–86.

Magnusson, P., O. Lofman and L. Larsson (1993) Methodological aspects on separation and reaction conditions of bone and liver alkaline phosphatase isoform analysis by high-performance liquid chromatography. *Anal. Biochem.* **211**: 156–163.

Magnusson, P., A. Hager and L. Larsson (1995a) Serum osteocalcin and bone and liver alkaline phosphatase isoforms in healthy children and adolescents. *Pediatr. Res.* **38**: 955–961.

Magnusson, P., O. Löfman, G. Toss and L. Larsson (1995b) Determination of bone alkaline phosphatase isoforms in serum by a new high-performance liquid chromatography assay in patients with metabolic bone disease. *Acta Orthop. Scand.* **66**: 203–204.

Magnusson, P., M. Degerblad, M. Saaf, L. Larsson and M. Thoren (1997) Different responses of bone alkaline phosphatase isoforms during recombinant insulin-like growth factor-I (IGF-I) and during growth hormone therapy in adults with growth hormone deficiency. *J. Bone Miner. Res.* **12**: 210–220.

Magnusson, P., M. Degerblad, M. Sääf, L. Larsson and M. Thorén (1998a) Different isoforms of bone alkaline phosphatase exist. *J. Bone Miner. Res.* **13**: 760–761.

Magnusson, P., L. Larsson, G. Englund, B. Larsson, P. Strang and L. Selin-Sjogren (1998b) Differences of bone alkaline phosphatase isoforms in metastatic bone disease and discrepant effects of clodronate on different skeletal sites indicated by the location of pain. *Clin. Chem.* **44**: 1621–1628.

Magnusson, P., L. Larsson, M. Magnusson, M. W. Davie and C. A. Sharp (1999) Isoforms of bone alkaline phosphatase: characterization and origin in human trabecular and cortical bone. *J. Bone. Miner. Res.* **14**: (1926)–(1933)

Magnusson, P., C. A. Sharp, M. Magnusson, J. Risteli, M. W. Davie and L. Larsson

(2001) Effect of chronic renal failure on bone turnover and bone alkaline phosphatase isoforms. *Kidney Int.* **60**: 257–265.

Magnusson, P., L. Arlestig, E. Paus, S. Di Mauro, M. P. Testa, T. Stigbrand, J. R. Farley, K. Nustad and J. L. Millán (2002a) Monoclonal antibodies against tissue-nonspecific alkaline phosphatase. Report of the ISOBM TD9 workshop. *Tumor Biol.* **23**: 228–248.

Magnusson, P., C.A. Sharp CA and J.R. Farley (2002b) Different distributions of human bone alkaline phosphatase isoforms in serum and bone tissue extracts. *Clin. Chim. Acta* **325**: 59–70.

Magrini, A., N. Bottini, F. Gloria-Bottini, L. Stefanini, A. Bergamaschi, E. Cosmi and E. Bottini (2003) Enzyme polymorphisms, smoking and human reproduction. A study of human placental alkaline phosphatase. *Am. J. Hum. Biol.* **15**: 781–785.

Mahmood, A., S. Mahmood, K. DeSchryver-Kecskemeti and D. H. Alpers (1993) Characterization of proteins in rat and human intestinal surfactant-like particles. *Arch. Biochem. Biophys.* **300**: 280–286.

Mahmood, A., F. Yamagishi, R. Eliakim, K. DeSchryver-Kecskemeti, T. L. Gramlich and D. H. Alpers (1994) A possible role for rat intestinal surfactant-like particles in transepithelial triacylglycerol transport. *J. Clin. Invest.* **93**: 70–80.

Mahmood, A., M. J. Engle and D. H. Alpers (2002) Secreted intestinal surfactant-like particles interact with cell membranes and extracellular matrix proteins in rats. *J. Physiol.* **542**: 237–244.

Mahmood, A., J. S. Shao and D. H. Alpers (2003) Rat enterocytes secrete SLPs containing alkaline phosphatase and cubilin in response to corn oil feeding. *Am. J. Physiol. Gastrointest. Liver Physiol.* **285**: G433–441.

Majeska, R. J. and R. E. Wuthier (1975) Studies on matrix vesicles isolated from chick epiphyseal cartilage. Association of pyrophosphatase and ATPase activities with alkaline phosphatase. *Biochim. Biophys. Acta* **391**: 51–60.

Majeska, R. J., B. C. Nair and G. A. Rodan (1985) Glucocorticoid regulation of alkaline phosphatase in the osteoblastic osteosarcoma cell line ROS 17/2.8. *Endocrinology* **116**: 170–179.

Makiya, R. and T. Stigbrand (1992a). Placental alkaline phosphatase as the placental IgG receptor. *Clin. Chem.* **38**: 2543–2545.

Makiya, R. and T. Stigbrand (1992b). Placental alkaline phosphatase has a binding site for the human immunoglobulin-G Fc portion. *Eur. J. Biochem.* **205**: 341–345.

Makiya, R. and T. Stigbrand (1992c). Placental alkaline phosphatase is related to human IgG internalization in HEp2 cells. *Biochem. Biophys. Res. Commun.* **182**: 624–630.

Makiya, R., L. E. Thornell and T. Stigbrand (1992) Placental alkaline phosphatase, a GPI-anchored protein, is clustered in clathrin-coated vesicles. *Biochem. Biophys. Res. Commun.* **183**: 803–808.

Malik, A.S. and M.G. Low (1986) Conversion of human placental alkaline phosphatase from a high Mr form to a low Mr form during butanol extraction. An investigation of the role of endogenous phosphoinositide-specific phospholipases. *Biochem. J.* **240**: 519–527.

Malo, M. S., W. Zhang, F. Alkhoury, P. Pushpakaran, M. A. Abedrapo, M. Mozumder, E. Fleming, A. Siddique, J. W. Henderson and R. A. Hodin (2004) Thyroid hormone positively regulates the enterocyte differentiation marker intestinal alkaline phosphatase gene via an atypical response element. *Mol. Endocrinol.* **18**: 1941–1962.

Mamber, S. W., A. B. Mikkilineni, E. J. Pack, M. P. Rosser, H. Wong, Y. Ueda and S. Forenza (1995) Tubulin polymerization by paclitaxel (taxol) phosphate prodrugs after metabolic activation with alkaline phosphatase. *J. Pharmacol. Exp. Ther.* **274**: 877–883.

Manes, T., K. Glade, C. A. Ziomek and J. L. Millán (1990) Genomic structure and comparison of mouse tissue-specific alkaline phosphatase genes. *Genomics* **8**: 541–554.

Manes, T., M. F. Hoylaerts, R. Muller, F. Lottspeich, W. Holke and J. L. Millán

(1998) Genetic complexity, structure and characterization of highly active bovine intestinal alkaline phosphatases. *J. Biol. Chem.* **273**: 23353–23360.

Manivel, J. C., J. Jessurun, M. R. Wick and L. P. Dehner (1987) Placental alkaline phosphatase immunoreactivity in testicular germ-cell neoplasms. *Am. J. Surg. Pathol.* **11**: 21–29.

Mano, H., Y. Furuhashi, S. Hattori, S. Goto, Y. Tomoda and T. Ichikatai (1986a). Radioimmunodetection of human choriocarcinoma xenografts by monoclonal antibody to placental alkaline phosphatase. *Jpn. J. Cancer Res.* **77**: 160–167.

Mano, H., Y. Furuhashi, Y. Morikawa, S. E. Hattori, S. Goto and Y. Tomoda (1986b). Radioimmunoassay of placental alkaline phosphatase in ovarian cancer sera and tissues. *Obstet. Gynecol.* **68**: 759–764.

Manolagas, S. C., D. W. Burton and L. J. Deftos (1981) 1,25-Dihydroxyvitamin D3 stimulates the alkaline phosphatase activity of osteoblast-like cells. *J. Biol. Chem.* **256**: 7115–7117.

Marquez, C., M. L. Toribio, M. A. Marcos, A. de la Hera, A. Barcena, L. Pezzi and C. Martinez (1989) Expression of alkaline phosphatase in murine B lymphocytes. Correlation with B cell differentiation into Ig secretion. *J. Immunol.* **142**: 3187–3192.

Marquez, C., A. De la Hera, E. Leonardo, L. Pezzi, A. Strasser and A. C. Martinez (1990) Identity of PB76 differentiation antigen and lymphocyte alkaline phosphatase. *Eur. J. Immunol.* **20**: 947–950.

Martin, C. J. and W. J. Evans (1989) Inositol hexaphosphate and its Cu(II) coordinate complex as inhibitors of intestinal alkaline phosphatase. *Res. Commun. Chem. Pathol. Pharmacol.* **65**: 289–296.

Martin, C. J. and W. J. Evans (1991a). Inactivation of intestinal alkaline phosphatase by inositol hexaphosphate–Cu(II) coordinate complexes. *J. Inorg. Biochem.* **42**: 161–175.

Martin, C. J. and W. J. Evans (1991b). Reversible inhibition of intestinal alkaline phosphatase by inositol hexaphosphate and its Cu(II) coordinate complexes. *J. Inorg. Biochem.* **42**: 177–184.

Martin, D., N. K. Spurr and J. Trowsdale (1987a). RFLP of the human placental alkaline phosphatase gene (PLAP) *Nucleic Acids Res.* **15**: 9104.

Martin, D., D. F. Tucker, P. Gorman, D. Sheer, N. K. Spurr and J. Trowsdale (1987b). The human placental alkaline phosphatase gene and related sequences map to chromosome 2 band q37. *Ann. Hum. Genet.* **51**: 145–152.

Masuhara, K., R. Yoshikawa, K. Takaoka, K. Ono, D. C. Morris and H. C. Anderson (1991) Monoclonal antibody against human bone alkaline phosphatase. *Int. Orthop.* **15**: 61–64.

Mathieu, P., P. Voisine, A. Pepin, R. Shetty, N. Savard and F. Dagenais (2005) Calcification of human valve interstitial cells is dependent on alkaline phosphatase activity. *J. Heart Valve Dis.* **14**: 353–357.

Matsushita, M., T. Irino, T. Stigbrand, T. Nakajima and T. Komoda (1998) Changes in intestinal alkaline phosphatase isoforms in healthy subjects bearing the blood group secretor and nonsecretor. *Clin. Chim. Acta* **277**: 13–24.

Matsushita, M., T. Irino, K. Oh-le and T. Komoda (2000) Specific gel electrophoresis method detects two isoforms of human intestinal alkaline phosphatase. *Electrophoresis* **21**: 281–284.

Matsuura, S., F. Kishi and T. Kajii (1990) Characterization of a 5′-flanking region of the human liver/bone/kidney alkaline phosphatase gene: two kinds of mRNA from a single gene. *Biochem. Biophys. Res. Commun.* **168**: 993–1000.

Mayne, P. D., S. Thakrar, S. B. Rosalki, A. Y. Foo and S. Parbhoo (1987) Identification of bone and liver metastases from breast cancer by measurement of plasma alkaline phosphatase isoenzyme activity. *J. Clin. Pathol.* **40**: 398–403.

McCarthy, A. D., A. M. Cortizo, G. Gimenez Segura, L. Bruzzone and S. B. Etcheverry (1998) Non-enzymatic glycosylation of alkaline phosphatase alters its biological properties. *Mol. Cell. Biochem.* **181**: 63–69.

McComb, R. B., G. N. Bowers, Jr and S. Posen (1979) *Alkaline Phosphatase*. New York: Plenum.

McDicken, I. W., G. H. Stamp, P. J. McLaughlin and P. M. Johnson

(1983) Expression of human placental-type alkaline phosphatase in primary breast cancer. *Int. J. Cancer* **32**: 205–209.

McDicken, I. W., P. J. McLaughlin, P. M. Tromans, D. M. Luesley and P. M. Johnson (1985) Detection of placental-type alkaline phosphatase in ovarian cancer. *Br. J. Cancer* **52**: 59–64.

McDougall, K., J. Beecroft, C. Wasnidge, W. A. King and A. Hahnel (1998) Sequences and expression patterns of alkaline phosphatase isozymes in preattachment bovine embryos and the adult bovine. *Mol. Reprod. Dev.* **50**: 7–17.

McLaughlin, P. J., M. H. Cheng, M. B. Slade and P. M. Johnson (1982) Expression on cultured human tumour cells of placental trophoblast membrane antigens and placental alkaline phosphatase defined by monoclonal antibodies. *Int. J. Cancer* **30**: 21–26.

McLaughlin, P. J., H. Gee and P. M. Johnson (1983) Placental-type alkaline phosphatase in pregnancy and malignancy plasma: specific estimation using a monoclonal antibody in a solid phase enzyme immunoassay. *Clin. Chim. Acta* **130**: 199–209.

McLaughlin, P. J., P. J. Travers, I. W. McDicken and P. M. Johnson (1984) Demonstration of placental and placental-like alkaline phosphatase in non-malignant human tissue extracts using monoclonal antibodies in an enzyme immunoassay. *Clin. Chim. Acta* **137**: 341–348.

McLaughlin, P. J., P. H. Warne, G. E. Hutchinson, P. M. Johnson and D. F. Tucker (1987) Placental-type alkaline phosphatase in cervical neoplasia. *Br. J. Cancer* **55**: 197–201.

Means, R. E., T. Greenough and R. C. Desrosiers (1997) Neutralization sensitivity of cell culture-passaged simian immunodeficiency virus. *J. Virol.* **71**: 7895–7902.

Mendonca, M. S., R. J. Antoniono, K. M. Latham, E. J. Stanbridge and J. L. Redpath (1991) Characterization of intestinal alkaline phosphatase expression and the tumorigenic potential of gamma-irradiated HeLa × fibroblast cell hybrids. *Cancer Res.* **51**: 4455–4462.

Merchant-Larios, H., F. Mendlovic and A. Alvarez-Buylla (1985) Characterization of alkaline phosphatase from primordial germ cells and ontogenesis of this enzyme in the mouse. *Differentiation* **29**: 145–151.

Mestrovic, V. and M. Pavela-Vrancic (2003) Inhibition of alkaline phosphatase activity by okadaic acid, a protein phosphatase inhibitor. *Biochimie* **85**: 647–650.

Meyer, J. L. (1984) Can biological calcification occur in the presence of pyrophosphate? *Arch. Biochem. Biophys.* **231**: 1–8.

Meyer, L. J., M. A. Lafferty, M. G. Raducha, C. J. Foster, K. J. Gogolin and H. Harris (1982) Production of a monoclonal antibody to human liver alkaline phosphatase. *Clin. Chim. Acta* **126**: 109–117.

Meyer, R. E., S. J. Thompson, C. L. Addy, C. Z. Garrison and R. G. Best (1995) Maternal serum placental alkaline phosphatase level and risk for preterm delivery. *Am. J. Obstet. Gynecol.* **173**: 181–186.

Mezzano, L., M. J. Sartori, S. Lin, G. Repossi and S. P. de Fabro (2005) Placental alkaline phosphatase (PLAP) study in diabetic human placental villi infected with *Trypanosoma cruzi*. *Placenta* **26**: 85–92.

Micanovic, R., C. A. Bailey, L. Brink, L. Gerber, Y. C. Pan, J. D. Hulmes and S. Udenfriend (1988) Aspartic acid-484 of nascent placental alkaline phosphatase condenses with a phosphatidylinositol glycan to become the C-terminus of the mature enzyme. *Proc. Natl. Acad. Sci USA* **85**: 1398–1402.

Michigami, T., T. Uchihashi, A. Suzuki, K. Tachikawa, S. Nakajima and K. Ozono (2005) Common mutations F310L and T1559del in the tissue-nonspecific alkaline phosphatase gene are related to distinct phenotypes in Japanese patients with hypophosphatasia. *Eur. J. Pediatr.* **164**: 277–282.

Miki, K., T. Oda, H. Suzuki, H. Niwa, Y. Endo, S. Iino, J. Miyazaki, K. Hirano and M. Sugiura (1978) Human fetal organ alkaline phosphatases. *Clin. Chim. Acta* **85**: 115–124.

Milhiet, P. E., M. C. Giocondi, O. Baghdadi, F. Ronzon, B. Roux and C. Le Grimellec (2002) Spontaneous insertion and partitioning of alkaline phosphatase into model lipid rafts. *EMBO Rep.* **3**: 485–490.

Millán, J. L. (1986) Molecular cloning and sequence analysis of human placental alkaline phosphatase. *J. Biol. Chem.* **261**: 3112–3115.

Millán, J. L. and T. Manes (1988) Seminoma-derived Nagao isozyme is encoded by a germ-cell alkaline phosphatase gene. *Proc. Natl. Acad. Sci. USA* **85**: 3024–3028.

Millán, J. L. and T. Stigbrand (1981) "Sandwich" enzyme immunoassay for placental alkaline phosphatase. *Clin. Chem.* **27**: 2014–2018.

Millán, J. L. and T. Stigbrand (1983) Antigenic determinants of human placental and testicular placental-like alkaline phosphatases as mapped by monoclonal antibodies. *Eur. J. Biochem.* **136**: 1–7.

Millán, J. L., M. P. Whyte, L. V. Avioli and W. H. Fishman (1980) Hypophosphatasia (adult form): quantitation of serum alkaline phosphatase isoenzyme activity in a large kindred. *Clin. Chem.* **26**: 840–845.

Millán, J. L., G. Beckman, A. Jeppsson and T. Stigbrand (1982a). Genetic variants of placental alkaline phosphatase as detected by a monoclonal antibody. *Hum. Genet.* **60**: 145–149.

Millán, J. L., A. Eriksson and T. Stigbrand (1982b). A possible new locus of alkaline phosphatase expressed in human testis. *Hum. Genet.* **62**: 293–295.

Millán, J. L., T. Stigbrand, E. Ruoslahti and W. H. Fishman (1982c). Characterization and use of an allotype-specific monoclonal antibody to placental alkaline phosphatase in the study of cancer-related phosphatase polymorphism. *Cancer Res.* **42**: 2444–2449.

Millán, J. L., K. Nustad and B. Norgaard-Pedersen (1985a). Highly sensitive solid-phase immunoenzymometric assay for placental and placental-like alkaline phosphatases with a monoclonal antibody and monodisperse polymer particles. *Clin. Chem.* **31**: 54–59.

Millán, J. L., T. Stigbrand and H. Jornvall (1985b). Structural comparisons of two allelic variants of human placental alkaline phosphatase. *Int. J. Biochem.* **17**: 1033–1039.

Miller, P. D., D. T. Baran, J. P. Bilezikian, S. L. Greenspan, R. Lindsay, B. L. Riggs and N. B. Watts (1999) Practical clinical application of biochemical markers of bone turnover: Consensus of an expert panel. *J. Clin. Densitom.* **2**: 323–342.

Milligan, T. P., H. R. Park, K. Noonan and C. P. Price (1997) Assessment of the performance of a capture immunoassay for the bone isoform of alkaline phosphatase in serum. *Clin. Chim Acta* **263**: 165–175.

Mochizuki, H., M. Saito, T. Michigami, H. Ohashi, N. Koda, S. Yamaguchi and K. Ozono (2000) Severe hypercalcaemia and respiratory insufficiency associated with infantile hypophosphatasia caused by two novel mutations of the tissue-nonspecific alkaline phosphatase gene. *Eur. J. Pediatr.* **159**: 375–379.

Monod, J., J. P. Changeux and F. Jacob (1963) Allosteric proteins and cellular control systems. *J. Mol. Biol.* **6**: 306–329.

Mori, S. and M. Nagano (1985a). Electron microscopic cytochemistry of alkaline phosphatase in neurons of rats. *Arch. Histol. Jpn.* **48**: 389–397.

Mori, S. and M. Nagano (1985b). Electron-microscopic cytochemistry of alkaline-phosphatase activity in endothelium, pericytes and oligodendrocytes in the rat brain. *Histochemistry* **82**: 225–231.

Morimoto, K. and K. Inouye (1997) A sensitive enzyme immunoassay of human thyroid-stimulating hormone (TSH) using bispecific F(ab')2 fragments recognizing polymerized alkaline phosphatase and TSH. *J. Immunol. Methods* **205**: 81–90.

Mornet, E., A. Taillandier, S. Peyramaure, F. Kaper, F. Muller, R. Brenner, P. Bussiere, P. Freisinger, J. Godard, M. Le Merrer, J. F. Oury, H. Plauchu, R. Puddu, J. M. Rival, A. Superti-Furga, R. L. Touraine, J. L. Serre and B. Simon-Bouy (1998) Identification of fifteen novel mutations in the tissue-nonspecific alkaline phosphatase (TNSALP) gene in European patients with severe hypophosphatasia. *Eur. J. Hum. Genet.* **6**: 308–314.

Mornet, E., E. Stura, A. S. Lia-Baldini, T. Stigbrand, A. Menez and M. H. Le Du (2001) Structural evidence for a functional role of human tissue nonspecific alkaline phosphatase in bone mineralization. *J. Biol. Chem.* **276**: 31171–31178.

Moro, L., C. Gazzarrini, D. Crivellari, E. Galligioni, R. Talamini and B. de Bernard (1993) Biochemical markers for detecting bone metastases in patients with breast cancer. *Clin. Chem.* **39**: 131–134.

Morote, J. and J. Bellmunt (2002) Bone alkaline phosphatase serum level predicts the response to antiandrogen withdrawal. *Eur. Urol.* **41**: 257–261.

Morote, J., J. A. Lorente and G. Encabo (1996) Prostate carcinoma staging. Clinical utility of bone alkaline phosphatase in addition to prostate specific antigen. *Cancer* **78**: 2374–2378.

Morote, J., Y. I. M'Hammed, E. Martinez, S. Esquena, J. A. Lorente and A. Gelabert (2002) Increase of bone alkaline phosphatase after androgen deprivation therapy in patients with prostate cancer. *Urology* **59**: 277–280.

Morote, J., E. Trilla, S. Esquena, J. M. Abascal, R. M. Segura, R. Catalan, G. Encabo and J. Reventos (2003) Analysis of bone alkaline phosphatase as a marker for the diagnosis of osteoporosis in men under androgen ablation. *Int. J. Biol. Markers* **18**: 290–294.

Morris, D. C., K. Masuhara, K. Takaoka, K. Ono and H. C. Anderson (1992) Immunolocalization of alkaline phosphatase in osteoblasts and matrix vesicles of human fetal bone. *Bone Miner.* **19**: 287–298.

Moss, D.W. (1987) Diagnostic aspects of alkaline phosphatase and its isoenzymes. *Clin. Biochem.* **20**: 225–230.

Moss, D.W. and R.K. Edwards (1984) Improved electrophoretic resolution of bone and liver alkaline phosphatases resulting from partial digestion with neuraminidase. *Clin. Chim. Acta* **143**: 177–182.

Moss, D. W. and L. G. Whitby (1975) A simplified heat-inactivation method for investigating alkaline phosphatase isoenzymes in serum. *Clin. Chim. Acta* **61**: 63–71.

Moss, D. W. and K. B. Whitaker (1987) The physical characteristics and enzymatic modification of fetal intestinal alkaline phosphatase in amniotic fluid. *Clin. Biochem.* **20**: 9–12.

Moss, D. W., R. H. Eaton, J. K. Smith and L. G. Whitby (1967) Association of inorganic-pyrophosphatase activity with human alkaline-phosphatase preparations. *Biochem. J.* **102**: 53–57.

Mostofi, F. K. and L. H. Sobin (1977) *Histological Typing of Testis Tumours. International Histological Classification of Tumours*, Vol. 16. Geneva: World Health Organization.

Mueller, H. D., H. Leung and R. A. Stinson (1985) Different genes code for alkaline phosphatases from human fetal and adult intestine. *Biochem. Biophys. Res. Commun.* **126**: 427–433.

Mueller, W. H., D. Kleefeld, B. Khattab, J. D. Meissner and R. J. Scheibe (2000) Effects of retinoic acid on N-glycosylation and mRNA stability of the liver/bone/kidney alkaline phosphatase in neuronal cells. *J. Cell. Physiol.* **182**: 50–61.

Muensch, H. A., W. C. Maslow, F. Azama, M. Bertrand, P. Dewhurst and B. Hartman (1986) Placental-like alkaline phosphatase. Re-evaluation of the tumor marker with exclusion of smokers. *Cancer* **58**: 1689–1694.

Mulivor, R. A., V. L. Hannig and H. Harris (1978a). Developmental change in human intestinal alkaline phosphatase. *Proc. Natl. Acad. Sci. USA* **75**: 3909–3912.

Mulivor, R. A., L. I. Plotkin and H. Harris (1978b). Differential inhibition of the products of the human alkaline phosphatase loci. *Ann. Hum. Genet.* **42**: 1–13.

Muller, P., W. Henn, I. Niedermayer, R. Ketter, W. Feiden, W. I. Steudel, K. D. Zang and H. Steilen-Gimbel (1999) Deletion of chromosome 1p and loss of expression of alkaline phosphatase indicate progression of meningiomas. *Clin. Cancer Res.* **5**: 3569–3577.

Mumm, S., J. Jones, P. Finnegan and M. P. Whyte (2001) Hypophosphatasia: molecular diagnosis of Rathbun's original case. *J. Bone Miner. Res.* **16**: 1724–1727.

Mumm, S., J. Jones, P. Finnegan, P. S. Henthorn, M. N. Podgornik and

M. P. Whyte (2002) Denaturing gradient gel electrophoresis analysis of the tissue nonspecific alkaline phosphatase isoenzyme gene in hypophosphatasia. *Mol. Genet. Metab.* **75**: 143–153.

Municio, M. J. and M. L. Traba (2003) Mitochondrial alkaline phosphatase as an intracellular signal in the synthesis of 1,25(OH)2D3 and 24,25(OH)2D3 in LLC-PK1 cells. *J. Physiol. Biochem.* **59**: 287–292.

Murakami, M., J. Kuratsu, Y. Mihara, K. Matsuno and Y. Ushio (1993) Histochemical study of alkaline phosphatase in primary human brain tumors: diagnostic implications for meningiomas and neurinomas. *Neurosurgery* **32**: 180–184; discussion, 184.

Murphy, J. E., T. T. Tibbitts and E. R. Kantrowitz (1995) Mutations at positions 153 and 328 in *Escherichia coli* alkaline phosphatase provide insight towards the structure and function of mammalian and yeast alkaline phosphatases. *J. Mol. Biol.* **253**: 604–617.

Murphy, G. P., M. J. Troychak, O. E. Cobb, V. A. Bowes, R. J. Kenny, R. J. Barren, III, G. M. Kenny, H. Ragde, E. H. Holmes and R. L. Wolfert (1997) Evaluation of PSA, free PSA, PSMA and total and bone alkaline phosphatase levels compared to bone scans in the management of patients with metastatic prostate cancer. *Prostate* **33**: 141–146.

Murray, E., D. Provvedini, D. Curran, B. Catherwood, H. Sussman and S. Manolagas (1987) Characterization of a human osteoblastic osteosarcoma cell line (SAOS-2) with high bone alkaline phosphatase activity. *J. Bone Miner. Res.* **2**: 231–238.

Murshed, M., D. Harmey, J. L. Millán, M. D. McKee and G. Karsenty (2005) Broadly expressed genes accounts for the special restriction of ECM mineralization to bone. *Genes Dev.* **19**: 1093–1104.

Myers, D. L., K. J. Harmon, V. Lindner and L. Liaw (2003) Alterations of arterial physiology in osteopontin-null mice. *Arterioscler. Thromb. Vasc. Biol.* **23**: 1021–1028.

Nakagawa, H., K. Umeki, K. Yamanaka, N. Kida and S. Ohtaki (1983) Macromolecular alkaline phosphatase and an immunoglobulin G that inhibited alkaline phosphatase in a patient's serum. *Clin. Chem.* **29**: 375–378.

Nakamura, T., K. Nakamura and R. A. Stinson (1988) Release of alkaline phosphatase from human osteosarcoma cells by phosphatidylinositol phospholipase C: effect of tunicamycin. *Arch. Biochem. Biophys.* **265**: 190–196.

Nakamura, Y., K. Noda, S. Shimpo, T. Oikawa, K. Kawasaki and A. Hirashita (2004) Phosphatidylinositol-dependent bond between alkaline phosphatase and collagen fibers in the periodontal ligament of rat molars. *Histochem. Cell. Biol.* **121**: 39–45.

Nakayama, M., I. Gorai, H. Minaguchi, C. Rosenquist and P. Qvist (1998) Purification and characterization of bone-specific alkaline phosphatase from a human osteosarcoma cell line. *Calcif. Tissue Int.* **62**: 67–73.

Nakayama, T., M. Yoshida and M. Kitamura (1970) L-Leucine sensitive, heat-stable alkaline-phosphatase isoenzyme detected in a patient with pleuritis carcinomatosa. *Clin. Chim. Acta* **30**: 546–548.

Narisawa, S., J. M. Sowadski and J. L. Millán (1990) An active site mutant of human placental alkaline phosphatase with deficient enzymatic activity and preserved immunoreactivity. *Clin. Chim. Acta* **186**: 189–196.

Narisawa, S., M. C. Hofmann, C. A. Ziomek and J. L. Millán (1992) Embryonic alkaline phosphatase is expressed at M-phase in the spermatogenic lineage of the mouse. *Development* **116**: 159–165.

Narisawa, S., K. A. Smans, J. Avis, M. F. Hoylaerts and J. L. Millán (1993) Transgenic mice expressing the tumor marker germ cell alkaline phosphatase: an *in vivo* tumor model for human cancer antigens. *Proc. Natl. Acad. Sci. USA* **90**: 5081–5085.

Narisawa, S., H. Hasegawa, K. Watanabe and J. L. Millán (1994) Stage-specific expression of alkaline phosphatase during neural development in the mouse. *Dev. Dyn.* **201**: 227–235.

Narisawa, S., N. Fröhlander and J. L. Millán (1997) Inactivation of two mouse alkaline phosphatase genes and establishment of a model of infantile hypophosphatasia. *Dev. Dyn.* **208**: 432–446.

Narisawa, S., C. Wennberg and J. L. Millán (2001) Abnormal vitamin B_6 metabolism in alkaline phosphatase knock-out mice causes multiple abnormalities, but not the impaired bone mineralization. *J. Pathol.* **193**: 125–133.

Narisawa, S., L. Huang, A. Iwasaki, H. Hasegawa, D. H. Alpers and J. L. Millán (2003) Accelerated fat absorption in intestinal alkaline phosphatase knockout mice. *Mol. Cell. Biol.* **23**: 7525–7530.

Narisawa, S., D. Harmey, P. Magnusson and J. L. Millán (2005) Conserved epitopes in human and mouse tissue-nonspecific alkaline phosphatase: 2nd Report of the ISOBM TD9 workshop. *Tumor Biol.* **26**: 113–120.

Nathan, E., G. Baatrup, H. Berg, B. Lund-Hansen and S. Reinholdt (1984) Persistently increased intestinal fraction of alkaline phosphatase. *Eur. J. Pediatr.* **142**: 142–144.

Nauli, A. M., S. Zheng, Q. Yang, R. Li, R. Jandacek and P. Tso (2003) Intestinal alkaline phosphatase release is not associated with chylomicron formation. *Am. J. Physiol. Gastrointest. Liver Physiol.* **284**: G583–G587.

Nayudu, R. V. and L. de Meis (1989) Energy transduction at the catalytic site of enzymes: hydrolysis of phosphoester bonds and synthesis of pyrophosphate by alkaline phosphatase. *FEBS Lett.* **255**: 163–166.

Ng, T. B. and P. P. Tam (1986) Changes of acid and alkaline phosphatase activities in the developing mouse brain. *Biol. Neonate* **50**: 107–113.

Niedermayer, I., W. Feiden, W. Henn, H. Steilen-Gimbel, W. I. Steudel and K. D. Zang (1997) Loss of alkaline phosphatase activity in meningiomas: a rapid histochemical technique indicating progression-associated deletion of a putative tumor suppressor gene on the distal part of the short arm of chromosome 1. *J. Neuropathol. Exp. Neurol.* **56**: 879–886.

Nielsen, H.K., K. Brixen and L. Mosekilde (1990a) Diurnal rhythm in serum activity of wheat-germ lectin-precipitable alkaline phosphatase: temporal relationships with the diurnal rhythm of serum osteocalcin. *Scand. J. Clin. Lab. Invest.* **50**: 851–856.

Nielsen, O. S., A. J. Munro, W. Duncan, J. Sturgeon, M. K. Gospodarowicz, M. A. Jewett, A. Malkin and G. M. Thomas (1990b) Is placental alkaline phosphatase (PLAP) a useful marker for seminoma? *Eur. J. Cancer* **26**: 1049–1054.

Nikawa, T., K. Rokutan, K. Nanba, K. Tokuoka, S. Teshima, M. J. Engle, D. H. Alpers and K. Kishi (1998) Vitamin A up-regulates expression of bone-type alkaline phosphatase in rat small intestinal crypt cell line and fetal rat small intestine. *J. Nutr.* **128**: 1869–1877.

Nilsson, E. E., S. D. Westfall, C. McDonald, T. Lison, I. Sadler-Riggleman and M. K. Skinner (2002) An *in vivo* mouse reporter gene (human secreted alkaline phosphatase) model to monitor ovarian tumor growth and response to therapeutics. *Cancer Chemother. Pharmacol.* **49**: 93–100.

Nishihara, Y., Y. Hayashi, T. Fujii, T. Adachi, T. Stigbrand and K. Hirano (1994) The alkaline phosphatase in human plexus chorioideus. *Biochim. Biophys. Acta* **1209**: 274–278.

Noda, M., K. Yoon, G. A. Rodan and D. E. Koppel (1987a). High lateral mobility of endogenous and transfected alkaline phosphatase: a phosphatidylinositol-anchored membrane protein. *J. Cell. Biol.* **105**: 1671–1677.

Noda, M., K. Yoon, M. Thiede, R. Buenaga, M. Weiss, P. Henthorn, H. Harris and G. A. Rodan (1987b). cDNA cloning of alkaline phosphatase from rat osteosarcoma (ROS 17/2.8) cells. *J. Bone Miner. Res.* **2**: 161–164.

Noda, T. H. Tokuda, M. Yoshida, E. Yasuda, Y. Hanai, S. Takai and O. Kozawa (2005) Possible Involvement of phosphatidylinositol 3-kinase/Akt pathway in insulin-like growth factor-I-induced alkaline phosphatase activity in osteoblasts. *Horm. Metab. Res.* **37**: 270–274.

Norton, P. A., B. Conyers, Q. Gong, L. F. Steel, T. M. Block and A. S. Mehta (2005) Assays for glucosidase inhibitors with potential antiviral activities:

secreted alkaline phosphatase as a surrogate marker. *J. Virol. Methods* **124**: 167–172.

Nosjean, O., I. Koyama, M. Goseki, B. Roux and T. Komoda (1997) Human tissue non-specific alkaline phosphatases: sugar-moiety-induced enzymic and antigenic modulations and genetic aspects. *Biochem. J.* **321**: 297–303.

Nouri, A. M., N. Torabi-Pour and A. A. Dabare (2000) A new highly specific monoclonal antibody against placental alkaline phosphatase: a potential marker for the early detection of testis tumour. *BJU Int.* **86**: 894–900.

Nouwen, E. J., D. E. Pollet, J. B. Schelstraete, M. W. Eerdekens, C. Hansch, A. Van de Voorde and M. E. De Broe (1985) Human placental alkaline phosphatase in benign and malignant ovarian neoplasia. *Cancer Res.* **45**: 892–902.

Nouwen, E. J., P. G. Hendrix, S. Dauwe, M. W. Eerdekens and M. E. De Broe (1987) Tumor markers in the human ovary and its neoplasms: a comparative immunohistochemical study. *Am. J. Pathol.* **126**: 230–242.

Nouwen, E. J., N. Buyssens and M. E. De Broe (1990) Heat-stable alkaline phosphatase as a marker for human and monkey type-I pneumocytes. *Cell. Tissue. Res.* **260**: 321–335.

Nozawa, S., H. Ohta, S. Izumi, S. Hayashi, F. Tsutsui, S. Kurihara and K. Watanabe (1980) Heat-stable alkaline phosphatase in the normal female genital organ-with special reference to the histochemical heat-stability test and L-phenylalanine inhibition test. *Acta Histochem. Cytochem.* **13**: 521–530.

Nozawa, S., S. Narisawa, R. Iizuka, T. Fukasawa, T. Kohji, P. K. Nakane, K. Hirano and J. L. Millán (1989) The mechanism of placental alkaline phosphatase induction *in vitro*. *Cell. Biochem. Funct.* **7**: 227–232.

Nozawa, S., Y. Udagawa, H. Ohkura, Y. Negishi, K. Akiya, N. Inaba, H. Takamizawa, E. Kimura and Y. Terashima (1990) Serum placental alkaline phosphatase (PLAP) in gynecologic malignancies – with special reference to the combination of PLAP and CA54/61 assay. *Clin. Chim. Acta* **186**: 275–284.

Nurnberg, P., H. Thiele, D. Chandler, W. Hohne, M. L. Cunningham, H. Ritter, G. Leschik, K. Uhlmann, C. Mischung, K. Harrop, J. Goldblatt, Z. U. Borochowitz, D. Kotzot, F. Westermann, S. Mundlos, H. S. Braun, N. Laing and S. Tinschert (2001) Heterozygous mutations in ANKH, the human ortholog of the mouse progressive ankylosis gene, result in craniometaphyseal dysplasia. *Nat. Genet.* **28**: 37–41.

Nustad, K., H. P. Monrad-Hansen, E. Paus, J. L. Millán and B. Norgaard-Pedersen (1984) Evaluation of a new, sensitive radioimmunoassay for placental alkaline phosphatase in pre- and post-operative sera from the Danish testicular cancer material. *Prog. Clin. Biol. Res.* **166**: 337–348.

Nuyts, G. D., H. A. Roels, G. F. Verpooten, A. M. Bernard, R. R. Lauwerys and M. E. De Broe (1992) Intestinal-type alkaline phosphatase in urine as an indicator of mercury induced effects on the S3 segment of the proximal tubule. *Nephrol. Dial. Transplant.* **7**: 225–229.

Obzansky, D. M., B. R. Rabin, D. M. Simons, S. Y. Tseng, D. M. Severino, H. Eggelte, M. Fisher, S. Harbron, R. W. Stout and M. J. Di Paolo (1991) Sensitive, colorimetric enzyme amplification cascade for determination of alkaline phosphatase and application of the method to an immunoassay of thyrotropin. *Clin. Chem.* **37**: 1513–1518.

Oda, K., Y. Amaya, M. Fukushi-Irie, Y. Kinameri, K. Ohsuye, I. Kubota, S. Fujimura and J. Kobayashi (1999) A general method for rapid purification of soluble versions of glycosylphosphatidylinositol-anchored proteins expressed in insect cells: an application for human tissue-nonspecific alkaline phosphatase. *J. Biochem.* **126**: 694–699.

Ogose, A., T. Hotta, H. Kawashima, H. Hatano, H. Umezu, Y. Inoue and N. Endo (2001) Elevation of serum alkaline phosphatase in clear cell chondrosarcoma of bone. *Anticancer Res.* **21**: 649–655.

Ohkubo, S., J. Kimura and I. Matsuoka (2000) Ecto-alkaline phosphatase in NG108-15 cells: a key enzyme mediating P1 antagonist-sensitive ATP response. *Br. J. Pharmacol.* **131**: 1667–1672.

Okamoto, T., H. Seo, H. Mano, M. Furuhashi, S. Goto, Y. Tomoda and N. Matsui (1990) Expression of human placenta alkaline phosphatase in placenta during pregnancy. *Placenta* **11**: 319–327.

Okawa, A., I. Nakamura, S. Goto, H. Moriya, Y. Nakamura and S. Ikegawa (1998) Mutation in Npps in a mouse model of ossification of the posterior longitudinal ligament of the spine. *Nat. Genet.* **19**: 271–273.

Olafsdottir, S. and J. F. Chlebowski (1989) A hybrid *Escherichia coli* alkaline phosphatase formed on proteolysis. *J. Biol. Chem.* **264**: 4529–4535.

Olsen L., S. Bressendorff, J. T. Troelsen and J. Olsen (2005) Differentiation-dependent activation of the human intestinal alkaline phosphatase promoter by HNF-4 in intestinal cells. *Am. J. Physiol. Gastrointest. Liver Physiol.* **289**: G220–226.

Onica, D., K. Rosendahl and L. Waldenlind (1989) Inherited occurrence of a heat stable alkaline phosphatase in the absence of malignant disease. *Clin. Chim. Acta* **180**: 23–34.

Onica, D., K. Rosendahl and L. Waldenlind (1990) Further characterization of a heat-stable alkaline phosphatase with low sensitivity to L-phenylalanine. *Clin. Chim. Acta* **194**: 193–202.

Onsgard-Meyer, M., A. L. McCoy and F. G. Knox (1996) Effect of bromotetramisole on renal phosphate excretion. *Proc. Soc. Exp. Biol. Med.* **213**: 193–195.

Orditura, M., F. De Vita, A. Roscigno, S. Infusino, A. Auriemma, P. Iodice, F. Ciaramella, G. Abbate and G. Catalano (1999) Amifostine: a selective cytoprotective agent of normal tissues from chemo-radiotherapy induced toxicity (review) *Oncol. Rep.* **6**: 1357–1362.

Orimo H. and T. Shimada (2005) Regulation of the human tissue-nonspecific alkaline phosphatase gene expression by all-*trans*-retinoic acid in SaOS-2 osteosarcoma cell line. *Bone* **36**: 866–876.

Orimo, H., Z. Hayashi, A. Watanabe, T. Hirayama and T. Shimada (1994) Novel missense and frameshift mutations in the tissue-nonspecific alkaline phosphatase gene in a Japanese patient with hypophosphatasia. *Hum. Mol. Genet.* **3**: 1683–1684.

Orimo, H., E. Nakajima, Z. Hayashi, K. Kijima, A. Watanabe, H. Tenjin, T. Araki and T. Shimada (1996) First-trimester prenatal molecular diagnosis of infantile hypophosphatasia in a Japanese family *Prenat. Diagn.* **16**: 559–563. [Erratum appears in *Prenat. Diagn.* 1996; **16**: 881].

Orimo, H., M. Goseki-Sone, S. Sato and T. Shimada (1997) Detection of deletion 1154–1156 hypophosphatasia mutation using TNSALP exon amplification. *Genomics* **42**: 364–366.

Orimo, H., H. J. Girshick, M. Goseki-Sone, M. Ito, K. Oda and T. Shimada (2001) Mutational analysis and functional correlation with phenotype in German patients with childhood-type hypophosphatasia. *J. Bone Miner. Res.* **16**: 2313–2319.

Osada, S. and S. Saji (2004) The clinical significance of monitoring alkaline phosphatase level to estimate postoperative liver failure after hepatectomy. *Hepatogastroenterology* **51**: 1434–1438.

O'Sullivan, M. J. and V. Marks (1981) Methods for the preparation of enzyme–antibody conjugates for use in enzyme immunoassays. *Methods Enzymol.* **73**: 147–166.

Otto, V. I., B. K. Schar, T. Sulser and E. Hanseler (1998) Specific determination of germ cell alkaline phosphatase for early diagnosis and monitoring of seminoma: performance and limitations of different analytical techniques. *Clin. Chim. Acta* **273**: 131–147.

Ovitt, C. E., A. W. Strauss, D. H. Alpers, J. Y. Chou and I. Boime (1986) Expression of different-sized placental alkaline phosphatase mRNAs in placenta and choriocarinoma cells. *Proc. Natl. Acad. Sci. USA* **83**: 3781–3785.

Ozono, K., M. Yamagata, T. Michigami, S. Nakajima, N. Sakai, G. Cai,

K. Satomura, N. Yasui, S. Okada and M. Nakayama (1996) Identification of novel missense mutations (Phe310Leu and Gly439Arg) in a neonatal case of hypophosphatasia. *J. Clin. Endocrinol. Metab.* **81**: 4458–4461.

Paiva, J., I. Damjanov, P. H. Lange and H. Harris (1983) Immunohistochemical localization of placental-like alkaline phosphatase in testis and germ-cell tumors using monoclonal antibodies. *Am. J. Pathol.* **111**: 156–165.

Palermo, C., P. Manduca, E. Gazzerro, L. Foppiani, D. Segat and A. Barreca (2004) Potentiating role of IGFBP-2 on IGF-II-stimulated alkaline phosphatase activity in differentiating osteoblasts. *Am. J. Physiol. Endocrinol. Metab.* **286**: E648–E657.

Palmer, D. A., T. E. Edmonds and N. J. Seare (1992) Flow injection electrochemical enzyme immunoassay for theophylline using a protein A immunoreactor and p-aminophenyl phosphate–p-aminophenol as the detection system. *Analyst* **117**: 1679–1682.

Palomares, L. A., C. E. Joosten, P. R. Hughes, R. R. Granados and M. L. Shuler (2003) Novel insect cell line capable of complex N-glycosylation and sialylation of recombinant proteins. *Biotechnol. Prog.* **19**: 185–192.

Pan, C. J., A. D. Sartwell and J. Y. Chou (1991) Transcriptional regulation and the effects of sodium butyrate and glycosylation on catalytic activity of human germ cell alkaline phosphatase. *Cancer Res.* **51**: 2058–2062.

Panigrahi, K., P. D. Delmas, F. Singer, W. Ryan, O. Reiss, R. Fisher, P. D. Miller, I. Mizrahi, C. Darte, B. C. Kress and R. H. Christenson (1994) Characteristics of a two-site immunoradiometric assay for human skeletal alkaline phosphatase in serum. *Clin. Chem.* **40**: 822–828.

Paragas, V. B., Y. Z. Zhang, R. P. Haugland and V. L. Singer (1997) The ELF-97 alkaline phosphatase substrate provides a bright, photostable, fluorescent signal amplification method for FISH. *J. Histochem. Cytochem.* **45**: 345–357.

Park, C., M. E. Chamberlin, C. J. Pan and J. Y. Chou (1996) Differential expression and butyrate response of human alkaline phosphatase genes are mediated by upstream DNA elements. *Biochemistry* **35**: 9807–9814.

Pedersen, B.J., A. Schlemmer, C. Hassager and C. Christiansen (1995) Changes in the C-terminal propeptide of type I procollagen and other markers of bone formation upon five days of bed rest. *Bone* **17**: 91–95.

Pelger, R. C., A. N. G. A. Lycklama, A. H. Zwinderman and N. A. Hamdy (2002) The flare in alkaline phosphatase activity post-orchidectomy predicts which patient may benefit from early chemotherapy in metastatic prostate cancer. *Prostate* **50**: 119–124.

Pezzi, L., C. Marquez, M. L. Toribio and C. Martinez (1991) Translocation of alkaline phosphatase during the activation of B cells. *Res. Immunol.* **142**: 109–115.

Picher, M., L. H. Burch, A. J. Hirsh, J. Spychala and R. C. Boucher (2003) Ecto 5′-nucleotidase and nonspecific alkaline phosphatase. Two AMP-hydrolyzing ectoenzymes with distinct roles in human airways. *J. Biol. Chem.* **278**: 13468–13479.

Pizauro, J. M., M. A. Demenis, P. Ciancaglini and F. A. Leone (1998) Kinetic characterization of a membrane-specific ATPase from rat osseous plate and its possible significance on endochodral ossification. *Biochim. Biophys. Acta* **1368**: 108–114.

Poelstra, K., W. W. Bakker, P. A. Klok, M. J. Hardonk and D. K. Meijer (1997a). A physiologic function for alkaline phosphatase: endotoxin detoxification. *Lab. Invest.* **76**: 319–327.

Poelstra, K., W. W. Bakker, P. A. Klok, J. A. Kamps, M. J. Hardonk and D. K. Meijer (1997b). Dephosphorylation of endotoxin by alkaline phosphatase *in vivo*. *Am. J. Pathol.* **151**: 1163–1169.

Pollak, A., H. Coradello, J. Leban, E. Maxa, M. Sternberg, K. Widhalm and G. Lubec (1983) Inhibition of alkaline phosphatase activity by glucose. *Clin. Chim. Acta* **133**: 15–24.

Pollak, A., E. Schober, H. Coradello, A. Lischka, S. Levin, F. Waldhauser and

G. Lubec (1984) Influence of glucose fluctuations on alkaline phosphatase activity. *Acta Diabetol. Lat.* **21**: 123–131.

Posen, S. and H. S. Grunstein (1982) Turnover rate of skeletal alkaline phosphatase in humans. *Clin. Chem.* **28**: 153–154.

Povinelli, C. M. and B. J. Knoll (1991) Trace expression of the germ-cell alkaline phosphatase gene in human placenta. *Placenta* **12**: 663–668.

Price, C. P., T. P. Milligan and C. Darte (1997) Direct comparison of performance characteristics of two immunoassays for bone isoform of alkaline phosphatase in serum. *Clin. Chem.* **43**: 2052–2057.

Price, G. H (1980) Inhibition of alkaline phosphatase by several diuretics. *Clin. Chim. Acta* **101**: 313–319.

Quarles, L. D., D. A. Yohay, L. W. Lever, R. Caton and R. J. Wenstrup (1992) Distinct proliferative and differentiated stages of murine MC3T3-E1 cells in culture: an *in vitro* model of osteoblast development. *J. Bone Miner. Res.* **7**: 683–692.

Ramasamy, I (1991) Affinity electrophoresis of alkaline phosphatase using polyacrylamide gels. *Clin. Chim. Acta* **199**: 243–251.

Rambaldi, A., M. Terao, S. Bettoni, M. L. Tini, R. Bassan, T. Barbui and E. Garattini (1990) Expression of leukocyte alkaline phosphatase gene in normal and leukemic cells: regulation of the transcript by granulocyte colony-stimulating factor. *Blood* **76**: 2565–2571.

Rawadi, G., B. Vayssiere, F. Dunn, R. Baron and S. Roman-Roman (2003) BMP-2 controls alkaline phosphatase expression and osteoblast mineralization by a Wnt autocrine loop. *J. Bone Miner. Res.* **18**: 1842–1853.

Ray, K., J. Vockley and H. Harris (1984) Epitopes of human intestinal alkaline phosphatases, defined by monoclonal antibodies. *FEBS Lett.* **174**: 294–299.

Ray, K., M. J. Weiss, N. C. Dracopoli and H. Harris (1988) Probe 8B/E5′ detects a second RFLP at the human liver/bone/kidney alkaline phosphatase (*ALPL*) locus. *Nucleic Acids Res.* **16**: 2361.

Raymond, F. D., D. W. Moss and D. Fisher (1994) Separation of alkaline phosphatase isoforms with and without intact glycan–phosphatidylinositol anchors in aqueous polymer phase systems. *Clin. Chim. Acta* **227**: 111–120.

Reale, M. G., D. Santini, G. G. Marchei, A. Manna, A. Del Nero, V. Bianco, P. Marchei and L. Frati (1995) Skeletal alkaline phosphatase as a serum marker of bone metastases in the follow-up of patients with breast cancer. *Int. J. Biol. Markers* **10**: 42–46.

Redman, C. A., J. E. Thomas-Oates, S. Ogata, Y. Ikehara and M. A. Ferguson (1994) Structure of the glycosylphosphatidylinositol membrane anchor of human placental alkaline phosphatase. *Biochem. J.* **302**: 861–865.

Register, T. C. and R. E. Wuthier (1984) Effect of vanadate, a potent alkaline phosphatase inhibitor, on 45Ca and 32Pi uptake by matrix vesicle-enriched fractions from chicken epiphyseal cartilage. *J. Biol. Chem.* **259**: 3511–3518.

Reynolds, J. J. and G. W. Dew (1977) Comparison of the inhibition of avian and mammalian bone alkaline phosphatases by levamisole and compound R8231. *Experientia* **33**: 154–155.

Rezende, A. A., J. M. Pizauro, P. Ciancaglini and F. A. Leone (1994) Phosphodiesterase activity is a novel property of alkaline phosphatase from osseous plate. *Biochem. J.* **301**: 517–522.

Rezende, L. A., P. Ciancaglini, J. M. Pizauro and F. A. Leone (1998) Inorganic pyrophosphate-phosphohydrolytic activity associated with rat osseous plate alkaline phosphatase. *Cell. Mol. Biol. (Noisy-le-Grand)* **44**: 293–302.

Riggs, B. L (2000) Are biochemical markers for bone turnover clinically useful for monitoring therapy in individual osteoporotic patients? *Bone* **26**: 551–552.

Riklund, K. E., R. A. Makiya, B. E. Sundstrom, L. E. Thornell and T. I. Stigbrand (1990) Experimental radioimmunotherapy of HeLa tumours in nude mice with ^{131}I-labeled monoclonal antibodies. *Anticancer Res.* **10**: 379–384.

Rindi, G., V. Ricci, G. Gastaldi and C. Patrini (1995) Intestinal alkaline phosphatase can transphosphorylate thiamin to thiamin monophosphate during intestinal transport in the rat. *Arch. Physiol. Biochem.* **103**: 33–38.

Risco, F. and M. L. Traba (1994) Possible involvement of a magnesium dependent mitochondrial alkaline phosphatase in the regulation of the 25-hydroxyvitamin D3-1 alpha-and 25-hydroxyvitamin D324R-hydroxylases in LLC-PK1 cells. *Magnes. Res.* **7**: 169–178.

Riva, P., M. Marangolo, V. Tison, G. Moscatelli, G. Franceschi, A. Spinelli, G. Rosti, P. Morigi, N. Riva and D. Tirindelli (1990) Radioimmunotherapy trials in germ testicular carcinoma: a phase I study. *Int. J. Biol. Markers* **5**: 188–194.

Roberts, S. J., A. J. Stewart, P. J. Sadler and C. Farquharson (2004) Human PHOSPHO1 exhibits high specific phosphoethanolamine and phosphocholine phosphatase activities. *Biochem. J.* **382**: 59–65.

Robison, R (1923) The possible significance of hexosephosphoric esters in ossification. *Biochem. J.* **17**: 286–293.

Robson, E. B. and H. Harris (1965) Genetics of the alkaline phosphatase polymorphism of the human placenta. *Nature* **207**: 1257–1259.

Rodan, G. A. and B. A. Rodan (1983) Expression of the osteoblastic phenotype. In: Peck, W. A. (ed.), *Bone and Mineral Research*, Vol. 2. Amsterdam: Elsevier, pp. 244–285.

Rodan, S. B., G. Wesolowski, K. Yoon and G. A. Rodan (1989) Opposing effects of fibroblast growth factor and pertussis toxin on alkaline phosphatase, osteopontin, osteocalcin and Type I collagen mRNA levels in ROS 17/2.8 cells. *J. Biol. Chem.* **264**: 19934–19941.

Roelofs, H., T. Manes, T. Janszen, J. L. Millán, J. W. Oosterhuis and L. H. Looijenga (1999) Heterogeneity in alkaline phosphatase isozyme expression in human testicular germ cell tumours: An enzyme-/immunohistochemical and molecular analysis. *J. Pathol.* **189**: 236–244.

Rooseboom, M., J. N. Commandeur and N. P. Vermeulen (2004) Enzyme-catalyzed activation of anticancer prodrugs. *Pharmacol. Rev.* **56**: 53–102.

Root, C.F., Jr, J.S. Fine and K.J.Clayson (1987) Isoelectric focusing of neuraminidase-treated alkaline phosphatase isoenzymes on agarose gel. *Clin. Chem.* **33**: 830–832.

Rosalki, S. B. and A. Y. Foo (1984) Two new methods for separating and quantifying bone and liver alkaline phosphatase isoenzymes in plasma. *Clin. Chem.* **30**: 1182–1186.

Rosalki, S. B. and A. Y. Foo (1989) Lectin affinity electrophoresis of alkaline phosphatase for the differentiation of bone and hepatobiliary disease. *Electrophoresis* **10**: 604–611.

Rosenthal, I., S. Bernstein and B. Rosen (1996) Alkaline phosphatase activity in *Penicillium roqueforti* and in blue-veined cheeses. *J. Dairy Sci.* **79**: 16–19.

Ross, P.D., B.C. Kress, R.E. Parson, R.D.Wasnich, K.A. Armour and I.A.Mizrahi (2000) Serum bone alkaline phosphatase and calcaneus bone density predict fractures: a prospective study. *Osteoporos. Int.* **11**: 76–82.

Rossi Norrlund, R., D. Holback, L. Johansson, S. O. Hietala and K. R. Ahlstrom (1997) Combinations of nonlabeled, ^{125}I-labeled and anti-idiotypic antiplacental alkaline phosphatase monoclonal antibodies at experimental radioimmunotargeting. *Acta Radiol.* **38**: 1087–1093.

Rowan, R. A. and D. S. Maxwell (1981) An ultrastructural study of vascular proliferation and vascular alkaline phosphatase activity in the developing cerebral cortex of the rat. *Am. J. Anat.* **160**: 257–265.

Ruedl, C., G. Gstraunthaler and M. Moser (1989) Differential inhibitory action of the fungal toxin orellanine on alkaline phosphatase isoenzymes. *Biochim. Biophys. Acta* **991**: 280–283.

Ruoslahti, E. (2002) Drug targeting to specific vascular sites. *Drug. Discov. Today* **7**: 1138–1143.

Rutsch, F., S. Vaingankar, K. Johnson, I. Goldfine, B. Maddux, P. Schauerte, H. Kalhoff, K. Sano, W. A. Boisvert, A. Superti-Furga and R. Terkeltaub (2001) PC-1 nucleoside triphosphate pyrophosphohydrolase deficiency in idiopathic infantile arterial calcification. *Am. J. Pathol.* **158**: 543–554.

Rutsch, F., N. Ruf, S. Vaingankar, M. R. Toliat, A. Suk, W. Hohne, G. Schauer, M. Lehmann, T. Roscioli, D. Schnabel, J. T. Epplen, A. Knisely, A. Superti-Furga, J. McGill, M. Filippone, A. R. Sinaiko, H. Vallance, B. Hinrichs, W. Smith, M. Ferre, R. Terkeltaub and P. Nurnberg (2003) Mutations in *ENPP1* are associated with 'idiopathic' infantile arterial calcification. *Nat. Genet.* **34**: 379–381.

Sahin, U., F. Hartmann, P. Senter, C. Pohl, A. Engert, V. Diehl and M. Pfreundschuh (1990) Specific activation of the prodrug mitomycin phosphate by a bispecific anti-CD30/anti-alkaline phosphatase monoclonal antibody. *Cancer Res.* **50**: 6944–6948.

Sakiyama, T., J. C. Robinson and J. Y. Chou (1979) Characterization of alkaline phosphatases from human first trimester placentas. *J. Biol. Chem.* **254**: 935–938.

Sali, A., J. M. Favaloro, R. Terkeltaub and J. W. Goding (1999) Germline deletion of the nucleoside triphosphate pyrophosphohydrolase (NTPPPH) plasma cell membrane glycoprotein-a (PC-1) produces abnormal calicification of periarticular tissues. In: Vanduffel, L. and R. Lemmems (eds), *Ecto-ATPases and Related Ectoenzymes*. Shaker Publishing BV, Maastricht, The Netherlands, pp. 267–282.

Santini, V. (2001) Amifostine: chemotherapeutic and radiotherapeutic protective effects. *Expert Opin. Pharmacother.* **2**: 479–489.

Santini, V. and F. J. Giles (1999) The potential of amifostine: from cytoprotectant to therapeutic agent. *Haematologica* **84**: 1035–1042.

Sarrouilhe, D., P. Lalegerie and M. Baudry (1992) Endogenous phosphorylation and dephosphorylation of rat liver plasma membrane proteins, suggesting an 18 kDa phosphoprotein as a potential substrate for alkaline phosphatase. *Biochim. Biophys. Acta* **1118**: 116–122.

Sartori, M. J., S. Lin, F. M. Frank, E. L. Malchiodi and S. P. de Fabro (2002) Role of placental alkaline phosphatase in the interaction between human placental trophoblast and *Trypanosoma cruzi*. *Exp. Mol. Pathol.* **72**: 84–90.

Sartori, M. J., L. Mezzano, S. Lin, S. Munoz and S. P. de Fabro (2003) Role of placental alkaline phosphatase in the internalization of trypomatigotes of *Trypanosoma cruzi* into HEp2 cells. *Trop. Med. Int. Health* **8**: 832–839.

Sato, K., D. C. Han, Y. Fujii, T. Tsushima and K. Shizume (1987) Thyroid hormone stimulates alkaline phosphatase activity in cultured rat osteoblastic cells (ROS 17/2.8) through 3,5,3′-triiodo-L-thyronine nuclear receptors. *Endocrinology* **120**: 1873–1881.

Sato, N., Y. Takahashi and S. Asano (1994) Preferential usage of the bone-type leader sequence for the transcripts of liver/bone/kidney-type alkaline phosphatase gene in neutrophilic granulocytes. *Blood* **83**: 1093–1101.

Say, J. C., K. Ciuffi, R. P. Furriel, P. Ciancaglini and F. A. Leone (1991) Alkaline phosphatase from rat osseous plates: purification and biochemical characterization of a soluble form. *Biochim. Biophys. Acta* **1074**: 256–262.

Schär, B. K., V. I. Otto and E. Hanseler (1997) Simultaneous detection of all four alkaline phosphatase isoenzymes in human germ cell tumors using reverse transcription-PCR. *Cancer Res.* **57**: 3841–3846.

Scheibe, R. J., I. Moeller-Runge and W. H. Mueller (1991) Retinoic acid induces the expression of alkaline phosphatase in P19 teratocarcinoma cells. *J. Biol. Chem.* **266**: 21300–21305.

Scherer, S. S (1996) Molecular specializations at nodes and paranodes in peripheral nerve. *Microsc. Res. Tech.* **34**: 452–461.

Schlemmer, A., C. Hassager, S.B. Jensen and C. Christiansen (1992) Marked diurnal variation in urinary excretion of pyridinium cross-links in premenopausal women. *J. Clin. Endocrinol. Metab.* **74**: 476–480.

Schoenau, E., W. Boeswald, R. Wanner, K. H. Herzog, B. Boewing, H. J. Boehles and K. Stehr (1989) High-molecular-mass ("biliary") isoenzyme of alkaline phosphatase and the diagnosis of liver dysfunction in cystic fibrosis. *Clin. Chem.* **35**: 1888–1890.

Schreiber, J. H., J. A. Schisa and J. M. Wilson (1993) Recombinant retroviruses containing novel reporter genes. *Biotechniques* **14**: 818–823.

Schreiber, W. E. and L. Whitta (1986) Alkaline phosphatase isoenzymes resolved by electrophoresis on lectin-containing agarose gel. *Clin. Chem.* **32**: 1570–1573.

Schwartz, Z., G. Knight, L. D. Swain and B. D. Boyan (1988) Localization of vitamin D3-responsive alkaline phosphatase in cultured chondrocytes. *J. Biol. Chem.* **263**: 6023–6026.

Schwartz, Z., G. G. Langston, L. D. Swain and B. D. Boyan (1991) Inhibition of 1,25-(OH)2D3- and 24,25-(OH)2D3-dependent stimulation of alkaline phosphatase activity by A23187 suggests a role for calcium in the mechanism of vitamin D regulation of chondrocyte cultures. *J. Bone Miner. Res.* **6**: 709–718.

Seabrook, R. N., E. M. Bailyes, C. P. Price, K. Siddle and J. P. Luzio (1988) The distinction of bone and liver isoenzymes of alkaline phosphatase in serum using a monoclonal antibody. *Clin. Chim. Acta* **172**: 261–266.

Seargeant, L. E. and R. A. Stinson (1979) Inhibition of human alkaline phosphatases by vanadate. *Biochem. J.* **181**: 247–250.

Sengelov, H., M. H. Nielsen and N. Borregaard (1992) Separation of human neutrophil plasma membrane from intracellular vesicles containing alkaline phosphatase and NADPH oxidase activity by free flow electrophoresis. *J. Biol. Chem.* **267**: 14912–14917.

Senter, P. D., M. G. Saulnier, G. J. Schreiber, D. L. Hirschberg, J. P. Brown, I. Hellstrom and K. E. Hellstrom (1988) Anti-tumor effects of antibody–alkaline phosphatase conjugates in combination with etoposide phosphate. *Proc. Natl. Acad. Sci. USA* **85**: 4842–4846.

Senter, P. D., G. J. Schreiber, D. L. Hirschberg, S. A. Ashe, K. E. Hellstrom and I. Hellstrom (1989) Enhancement of the *in vitro* and *in vivo* antitumor activities of phosphorylated mitomycin C and etoposide derivatives by monoclonal antibody-alkaline phosphatase conjugates. *Cancer Res.* **49**: 5789–5792.

Sergi, C., E. Mornet, J. Troeger and T. Voigtlaender (2001) Perinatal hypophosphatasia: radiology, pathology and molecular biology studies in a family harboring a splicing mutation (648+1A) and a novel missense mutation (N400S) in the tissue-nonspecific alkaline phosphatase (TNSALP) gene. *Am. J. Med. Genet.* **103**: 235–240.

Serra, B., M. D. Morales, A. J. Reviejo, E. H. Hall and J. M. Pingarron (2005) Rapid and highly sensitive electrochemical determination of alkaline phosphatase using a composite tyrosinase biosensor. *Anal. Biochem.* **336**: 289–294.

Shalhevet, D., D. Krull, P. Clamp, R. Feltes, E. Atac and L. B. Schook (1993) A HindIII polymorphism at the porcine bone, liver, kidney alkaline phosphatase (*ALPL*) locus. *Anim. Genet.* **24**: 140.

Shao, J.-S., M. Engle, Q. Xie, R. E. Schmidt, S. Narisawa, J. L. Millán and D. H. Alpers (2000) Effect of tissue nonspecific alkaline phosphatase in maintenance of murine colon and stomach. *Microsc. Res. Tech.* **51**: 121–128.

She, Q. B., J. J. Mukherjee, T. Chung and Z. Kiss (2000a). Placental alkaline phosphatase, insulin and adenine nucleotides or adenosine synergistically promote long-term survival of serum-starved mouse embryo and human fetus fibroblasts. *Cell. Signal.* **12**: 659–665.

She, Q. B., J. J. Mukherjee, J. S. Huang, K. S. Crilly and Z. Kiss (2000b). Growth factor-like effects of placental alkaline phosphatase in human fetus and mouse embryo fibroblasts. *FEBS Lett.* **469**: 163–167.

Shibata, H., M. Fukushi, A. Igarashi, Y. Misumi, Y. Ikehara, Y. Ohashi and K. Oda (1998) Defective intracellular transport of tissue-nonspecific alkaline phosphatase with an Ala162→Thr mutation associated with lethal hypophosphatasia. *J. Biochem.* **123**: 968–977.

Shidoji, Y. and S. H. Kim (2004) Mutually distinctive gradients of three types of intestinal alkaline phosphatase along the longitudinal axis of the rat intestine. *Digestion* **70**: 10–15.

Shigenari, A., A. Ando, T. Baba, T. Yamamoto, Y. Katsuoka and H. Inoko (1998) Characterization of alkaline phosphatase genes expressed in seminoma by cDNA cloning. *Cancer Res.* **58**: 5079–5082.

Shinozaki, T. and K. P. Pritzker (1996) Regulation of alkaline phosphatase: implications for calcium pyrophosphate dihydrate crystal dissolution and other alkaline phosphatase functions. *J. Rheumatol.* **23**: 677–683.

Shinozaki, T., Y. Xu, T. F. Cruz and K. P. Pritzker (1995) Calcium pyrophosphate dihydrate (CPPD) crystal dissolution by alkaline phosphatase: interaction of alkaline phosphatase on CPPD crystals. *J. Rheumatol.* **22**: 117–123.

Shinozaki, T., H. Watanabe, K. Takagishi and K. P. Pritzker (1998) Allotype immunoglobulin enhances alkaline phosphatase activity: implications for the inflammatory response. *J. Lab. Clin. Med.* **132**: 320–328.

Shioi, A., Y. Nishizawa, S. Jono, H. Koyama, M. Hosoi and H. Morii (1995) Beta-glycerophosphate accelerates calcification in cultured bovine vascular smooth muscle cells. *Arterioscler. Thromb. Vasc. Biol.* **15**: 2003–2009.

Shioi, A., M. Katagi, Y. Okuno, K. Mori, S. Jono, H. Koyama and Y. Nishizawa (2002) Induction of bone-type alkaline phosphatase in human vascular smooth muscle cells: roles of tumor necrosis factor-alpha and oncostatin M derived from macrophages. *Circ. Res.* **91**: 9–16.

Shirazi, S. P., R. B. Beechey and P. J. Butterworth (1981a). Potent inhibition of membrane-bound rat intestinal alkaline phosphatase by a new series of phosphate analogues. *Biochem. J.* **194**: 797–802.

Shirazi, S. P., R. B. Beechey and P. J. Butterworth (1981b). The use of potent inhibitors of alkaline phosphatase to investigate the role of the enzyme in intestinal transport of inorganic phosphate. *Biochem. J.* **194**: 803–809.

Shoba, L. N. and J. C. Lee (2003) Inhibition of phosphatidylinositol 3-kinase and p70S6 kinase blocks osteogenic protein-1 induction of alkaline phosphatase activity in fetal rat calvaria cells. *J. Cell. Biochem.* **88**: 1247–1255.

Siddique, A., M. S. Malo, L. M. Ocuin, B. F. Hinnebusch, M. A. Abedrapo, J. W. Henderson, W. Zhang, M. Mozumder, V. W. Yang and R. A. Hodin (2003) Convergence of the thyroid hormone and gut-enriched Kruppel-like factor pathways in the context of enterocyte differentiation. *J. Gastrointest. Surg.* **7**: 1053–1061; discussion, 1061.

Sinha, P.K., A. Bianchi-Bosisio, W. Meyer-Sabellek and P.G. Righetti (1986) Resolution of alkaline phosphatase isoenzymes in serum by isoelectric focusing in immobilized pH gradients. *Clin. Chem.* **32**: 1264–1268.

Sisley, A. C., T. R. Desai, K. L. Hynes, B. L. Gewertz and P. K. Dudeja (1999) Decrease in mucosal alkaline phosphatase: a potential marker of intestinal reperfusion injury. *J. Lab. Clin. Med.* **133**: 335–341.

Skakkebaek, N. E., J. G. Berthelsen, A. Giwercman and J. Muller (1987) Carcinoma-*in-situ* of the testis: possible origin from gonocytes and precursor of all types of germ cell tumours except spermatocytoma. *Int. J. Androl.* **10**: 19–28.

Skillen, A. W. and J. Harrison (1980) Serum alkaline phosphatase of intestinal origin: detection by acrylamide gel electrophoresis and L-*p*-bromotetramisole inhibition compared. *Clin. Chem.* **26**: 786.

Skynner, M. J., D. J. Drage, W. L. Dean, S. Turner, D. J. Watt and N. D. Allen (1999) Transgenic mice ubiquitously expressing human placental alkaline phosphatase (PLAP): an additional reporter gene for use in tandem with beta-galactosidase (lacZ) *Int. J. Dev. Biol.* **43**: 85–90.

Slaughter, C. A., M. C. Coseo, M. P. Cancro and H. Harris (1981) Detection of enzyme polymorphism by using monoclonal antibodies. *Proc. Natl. Acad. Sci. USA* **78**: 1124–1128.

Slaughter, C. A., K. J. Gogolin, M. C. Coseo, L. J. Meyer, J. Lesko and H. Harris (1983) Discrimination of human placental alkaline phosphatase allelic variants by monoclonal antibodies. *Am. J. Hum. Genet.* **35**: 1–20.

Smans, K. A., M. F. Hoylaerts, H. F. Hendrickx, M. J. Goergen and M. E. De Broe (1991) Tumor-cell lysis by in-situ-activated human peripheral-blood mononuclear cells. *Int. J. Cancer* **47**: 431–438.

Smans, K. A., M. F. Hoylaerts, S. Narisawa, J. L. Millán and M. E. De Broe (1995) Bispecific antibody-mediated lysis of placental and germ cell alkaline phosphatase targeted solid tumors in immunocompetent mice. *Cancer Res.* **55**: 4383–4390.

Smans, K. A., M. B. Ingvarsson, P. Lindgren, S. Canevari, H. Walt, T. Stigbrand, T. Backstrom and J. L. Millán (1999) Bispecific antibody-mediated lysis of primary cultures of ovarian carcinoma cells using multiple target antigens. *Int. J. Cancer* **83**: 270–277.

Smith, G. P. and T. J. Peters (1981) Subcellular localization and properties of pyridoxal phosphate phosphatases of human polymorphonuclear leukocytes and their relationship to acid and alkaline phosphatase. *Biochim. Biophys. Acta* **661**: 287–294.

Smith, G. P., G. Sharp and T. J. Peters (1985) Isolation and characterization of alkaline phosphatase-containing granules (phosphasomes) from human polymorphonuclear leucocytes. *J. Cell. Sci.* **76**: 167–178.

Smith, M., M. J. Weiss, C. A. Griffin, J. C. Murray, K. H. Buetow, B. S. Emanuel, P. S. Henthorn and H. Harris (1988) Regional assignment of the gene for human liver/bone/kidney alkaline phosphatase to chromosome 1p36.1-p34. *Genomics* **2**: 139–143.

Sogabe, N., L. Mizoi, K. Asahi, I. Ezawa and M. Goseki-Sone (2004) Enhancement by lactose of intestinal alkaline phosphatase expression in rats. *Bone* **35**: 249–255.

Solter, P. F., W. E. Hoffmann, M. D. Chambers and D. J. Schaeffer (1997) CCK-8 infusion increases plasma LMW alkaline phosphatase coincident with enterohepatic circulation of bile acids. *Am. J. Physiol.* **273**: G381–G388.

Sorensen, S (1988) Wheat-germ agglutinin method for measuring bone and liver isoenzymes of alkaline phosphatase assessed in postmenopausal osteoporosis. *Clin. Chem.* **34**: 1636–1640.

Souka, A. P., F. L. Raymond, E. Mornet, L. Geerts and K. H. Nicolaides (2002) Hypophosphatasia associated with increased nuchal translucency: a report of two affected pregnancies. *Ultrasound Obstet. Gynecol.* **20**: 294–295.

Souvannavong, V., C. Lemaire, D. De Nay, S. Brown and A. Adam (1995) Expression of alkaline phosphatase by a B-cell hybridoma and its modulation during cell growth and apoptosis. *Immunol. Lett.* **47**: 163–170.

Souvannavong, V., C. Lemaire, S. Brown and A. Adam (1997) UV irradiation of a B-cell hybridoma increases expression of alkaline phosphatase: involvement in apoptosis. *Biochem. Cell. Biol.* **75**: 783–788.

Sowa, H., H. Kaji, T. Yamaguchi, T. Sugimoto and K. Chihara (2002a). Smad3 promotes alkaline phosphatase activity and mineralization of osteoblastic MC3T3-E1 cells. *J. Bone Miner. Res.* **17**: 1190–1199.

Sowa, H., H. Kaji, T. Yamaguchi, T. Sugimoto and K. Chihara (2002b). Activations of ERK1/2 and JNK by transforming growth factor beta negatively regulate Smad3-induced alkaline phosphatase activity and mineralization in mouse osteoblastic cells. *J. Biol. Chem.* **277**: 36024–36031.

Speer, M. Y., M. D. McKee, R. E. Guldberg, L. Liaw, H. Y. Yang, E. Tung, G. Karsenty and C. M. Giachelli (2002) Inactivation of the osteopontin gene enhances vascular calcification of matrix Gla protein-deficient mice: evidence for osteopontin as an inducible inhibitor of vascular calcification *in vivo*. *J. Exp. Med.* **196**: 1047–1055.

Spentchian, M., Y. Merrien, M. Herasse, Z. Dobbie, D. Glaser, S. E. Holder, S. A. Ivarsson, D. Kostiner, S. Mansour, A. Norman, J. Roth, F. Stipoljev, J. L. Taillemite, J. J. van der Smagt, J. L. Serre, B. Simon-Bouy, A. Taillandier and E. Mornet (2003) Severe hypophosphatasia: characterization of fifteen novel mutations in the ALPL gene. *Hum. Mutat.* **22**: 105–106.

Srivastava, A. K., G. Masinde, H. Yu, D. J. Baylink and S. Mohan (2005) Mapping quantitative trait loci that influence blood levels of alkaline phosphatase in MRL/MpJ and SJL/J mice. *Bone* **35**: 1086–1094.

Stec, B., K. M. Holtz and E. R. Kantrowitz (2000) A revised mechanism for the alkaline phosphatase reaction involving three metal ions. *J. Mol. Biol.* **299**: 1303–1311.

Stefaner, I., A. Stefanescu, W. Hunziker and R. Fuchs (1997) Expression of placental alkaline phosphatase does not correlate with IgG binding, internalization and transcytosis. *Biochem. J.* **327**: 585–592.

Steinherz, P. G., L. J. Steinherz, J. S. Nisselbaum and M. L. Murphy (1984) Transient, marked, unexplained elevation of serum alkaline phosphatase. *Jama* **252**: 3289–3292.

Steitz, S. A., M. Y. Speer, M. D. McKee, L. Liaw, M. Almeida, H. Yang and C. M. Giachelli (2002) Osteopontin inhibits mineral deposition and promotes regression of ectopic calcification. *Am. J. Pathol.* **161**: 2035–2046.

Stendahl, U., A. Lindgren, B. Tholander, R. Makyia and T. Stigbrand (1989) Expression of placental alkaline phosphatase in epithelial ovarian tumours. *Tumor Biol.* **10**: 126–132.

Stephens, M. C. and K. Dakshinamurti (1975) Brain lipids in pyridoxine-deficient young rats. *Neurobiology* **5**: 262–269.

Stewart, A. J., R. Schmid, C. A. Blindauer, S. J. Paisey and C. Farquharson (2003) Comparative modelling of human PHOSPHO1 reveals a new group of phosphatases within the haloacid dehalogenase superfamily. *Protein Eng.* **16**: 889–895.

Stieber, P., D. Nagel, C. Ritzke, N. Rossler, C. M. Kirsch, W. Eiermann and A. Fateh-Moghadam (1992) Significance of bone alkaline phosphatase, CA 15-3 and CEA in the detection of bone metastases during the follow-up of patients suffering from breast carcinoma. *Eur. J. Clin. Chem. Clin. Biochem.* **30**: 809–814.

Stigbrand, T., R. Jemmerson, J. L. Millán and W. H. Fishman (1987) A hidden antigenic determinant on membrane-bound human placental alkaline phosphatase. *Tumor Biol.* **8**: 34–44.

Stigbrand, T., S. O. Hietala, B. Johansson, R. Makiya, K. Riklund and L. Ekelund (1989) Tumour radioimmunolocalization in nude mice by use of antiplacental alkaline phosphatase monoclonal antibodies. *Tumor Biol.* **10**: 243–251.

Stigbrand, T., K. Riklund, B. Tholander, K. Hirano, O. Lalos and U. Stendahl (1990) Placental alkaline phosphatase (PLAP)/PLAP-like alkaline phosphatase as tumour marker in relation to CA 125 and TPA for ovarian epithelial tumours. *Eur. J. Gynaecol. Oncol.* **11**: 351–360.

Stornelli, A., M. Arauz, H. Baschard and R. L. de la Sota (2003) Unilateral and bilateral vasectomy in the dog: alkaline phosphatase as an indicator of tubular patency. *Reprod. Domest. Anim.* **38**: 1–4.

Strom, M., J. Krisinger and H. F. DeLuca (1991) Isolation of a mRNA that encodes a putative intestinal alkaline phosphatase regulated by 1,25-dihydroxyvitamin D-3. *Biochim. Biophys. Acta* **1090**: 299–304.

Studer, M., M. Terao, M. Gianni and E. Garattini (1991) Characterization of a second promoter for the mouse liver/bone/kidney-type alkaline phosphatase gene: cell and tissue specific expression. *Biochem. Biophys. Res. Commun.* **179**: 1352–1360.

Sugimoto, N., S. Iwamoto, Y. Hoshino and E. Kajii (1998) A novel missense mutation of the tissue-nonspecific alkaline phosphatase gene detected in a patient with hypophosphatasia. *J. Hum. Genet.* **43**: 160–164.

Sumikawa, K., T. Okochi and K. Adachi (1990) Differences in phosphatidate hydrolytic activity of human alkaline phosphatase isozymes. *Biochim. Biophys. Acta* **1046**: 27–31.

Sunderland, C. A., J. O. Davies and G. M. Stirrat (1984) Immunohistology of normal and ovarian cancer tissue with a monoclonal antibody to placental alkaline phosphatase. *Cancer Res.* **44**: 4496–4502.

Sussman, N. L., R. Eliakim, D. Rubin, D. H. Perlmutter, K. DeSchryver-Kecskemeti and D. H. Alpers (1989) Intestinal alkaline phosphatase is secreted bidirectionally from villous enterocytes. *Am. J. Physiol.* 257: G14–G23.

Suzuki, A., G. Palmer, J. P. Bonjour and J. Caverzasio (1999) Regulation of alkaline phosphatase activity by p38 MAP kinase in response to activation of Gi protein-coupled receptors by epinephrine in osteoblast-like cells. *Endocrinology* 140: 3177–3182.

Suzuki, A., J. Guicheux, G. Palmer, Y. Miura, Y. Oiso, J. P. Bonjour and J. Caverzasio (2002) Evidence for a role of p38 MAP kinase in expression of alkaline phosphatase during osteoblastic cell differentiation. *Bone* 30: 91–98.

Swallow, D. M., S. Povey, M. Parkar, P. W. Andrews, H. Harris, B. Pym and P. Goodfellow (1986) Mapping of the gene coding for the human liver/bone/kidney isozyme of alkaline phosphatase to chromosome 1. *Ann. Hum. Genet.* 50: 229–235.

Taillandier, A., L. Zurutuza, F. Muller, B. Simon-Bouy, J. L. Serre, L. Bird, R. Brenner, O. Boute, J. Cousin, D. Gaillard, P. H. Heidemann, B. Steinmann, M. Wallot and E. Mornet (1999) Characterization of eleven novel mutations (M45L, R119H, 544delG, G145V, H154Y, C184Y, D289V, 862+5A, 1172delC, R411X, E459K) in the tissue-nonspecific alkaline phosphatase (TNSALP) gene in patients with severe hypophosphatasia. Mutations in Brief No. 217. Online. *Hum. Mutat.* 13: 171–172.

Taillandier, A., E. Cozien, F. Muller, Y. Merrien, E. Bonnin, C. Fribourg, B. Simon-Bouy, J. L. Serre, E. Bieth, R. Brenner, M. P. Cordier, S. De Bie, F. Fellmann, P. Freisinger, V. Hesse, R. C. Hennekam, D. Josifova, L. Kerzin-Storrar, N. Leporrier, M. T. Zabot and E. Mornet (2000) Fifteen new mutations (–195C>T, L-12X, 298-2A>G, T117N, A159T, R229S, 997+2T>A, E274X, A331T, H364R, D389G, 1256delC, R433H, N461I, C472S) in the tissue-nonspecific alkaline phosphatase (TNSALP) gene in patients with hypophosphatasia. *Hum. Mutat.* 15: 293.

Taillandier, A., A.S. Lia-Baldini, M. Mouchard, B. Robin, F. Muller, B. Simon-Bouy, J.L. Serre, A. Bera-Louville, M. Bonduelle, J. Eckhardt, D. Gaillard, A.G. Myhre, S. Kortge-Jung, L. Larget-Piet, E. Malou, D. Sillence, I.K. Temple, G. Viot and E. Mornet (2001) Twelve novel mutations in the tissue-nonspecific alkaline phosphatase gene (ALPL) in patients with various forms of hypophosphatasia. *Hum. Mutat.* 18: 83–84.

Taillandier, A., S. L. Sallinen, I. Brun-Heath, P. De Mazancourt, J. L. Serre and E. Mornet (2005) Childhood hypophosphatasia due to a *de novo* missense mutation in the tissue-nonspecific alkaline phosphatase gene. *J. Clin. Endocrinol. Metab.* 90: 2436–2439.

Takahasi, T., Iwantanti, A. S. Mizuno, Y. Morishita, H. Nishio, S. Kodama and T. Matsuo (1984) The relationship between phosphoethanolamine level in serum and intractable seizure on hypophosphatasia infantile form. In: Cohn, D. V., T. Fugita, J. T. Potts and R. V. J. Talmage (eds), *Endocrine Control of Bone and Calcium Metabolism*. Amsterdam: Excerpta Medica, pp. 93–94.

Takami, N., K. Oda and Y. Ikehara (1992) Aberrant processing of alkaline phosphatase precursor caused by blocking the synthesis of glycosylphosphatidyl-inositol. *J. Biol. Chem.* 267: 1042–1047.

Takeya, M., F. G. Klier and W. H. Fishman (1984) Molecular dimensions of the SS and FF phenotypes of human placental alkaline phosphatase. Rotary shadowing and negative staining electron microscope measurements. *J. Mol. Biol.* 173: 253–264.

Tang, H. T., C. E. Lunte, H. B. Halsall and W. R. Heineman (1988) *p*-Aminophenyl phosphate: an improved substrate for electrochemical enzyme immnoassay. *Anal. Chim. Acta* 214: 187–195.

Tanimura, A., D. H. McGregor and H. C. Anderson (1986a). Calcification in atherosclerosis. I. Human studies. *J. Exp. Pathol.* 2: 261–273.

Tanimura, A., D. H. McGregor and H. C. Anderson (1986b). Calcification in atherosclerosis. II. Animal studies. *J. Exp. Pathol.* **2**: 275–297.

Tate, S. S., R. Urade, R. Micanovic, L. Gerber and S. Udenfriend (1990) Secreted alkaline phosphatase: an internal standard for expression of injected mRNAs in the Xenopus oocyte. *Faseb J.* **4**: 227–231.

Telfer, J. F. and C. D. Green (1993a). Induction of germ-cell alkaline phosphatase by butyrate and cyclic AMP in BeWo choriocarcinoma cells. *Biochem. J.* **296**: 59–65.

Telfer, J. F. and C. D. Green (1993b). Placental alkaline phosphatase activity is inversely related to cell growth rate in HeLaS3 cervical cancer cells. *FEBS Lett.* **329**: 238–244.

Tenenbaum, H. C (1987) Levamisole and inorganic pyrophosphate inhibit beta-glycerophosphate induced mineralization of bone formed *in vitro*. *Bone Miner.* **3**: 13–26.

Tenenhouse, H. S., C. R. Scriver and E. J. Vizel (1980) Alkaline phosphatase activity does not mediate phosphate transport in the renal-cortical brush-border membrane. *Biochem. J.* **190**: 473–476.

Terao, M. and B. Mintz (1987) Cloning and characterization of a cDNA coding for mouse placental alkaline phosphatase. *Proc. Natl. Acad. Sci. USA* **84**: 7051–7055.

Terao, M., D. Pravtcheva, F. H. Ruddle and B. Mintz (1988) Mapping of gene encoding mouse placental alkaline phosphatase to chromosome 4. *Somat. Cell. Mol. Genet.* **14**: 211–215.

Terao, M., M. Studer, M. Gianni and E. Garattini (1990) Isolation and characterization of the mouse liver/bone/kidney-type alkaline phosphatase gene. *Biochem. J.* **268**: 641–648.

Tesch, W., T. Vandenbos, P. Roschgr, N. Fratzl-Zelman, K. Klaushofer, W. Beertsen and P. Fratzl (2003) Orientation of mineral crystallites and mineral density during skeletal development in mice deficient in tissue nonspecific alkaline phosphatase. *J. Bone Miner. Res.* **18**: 117–125.

Thibaudeau, G., J. Drawbridge, A. W. Dollarhide, T. Haque and M. S. Steinberg (1993) Three populations of migrating amphibian embryonic cells utilize different guidance cues. *Dev. Biol.* **159**: 657–668.

Thiede, M. A., K. Yoon, E. E. Golub, M. Noda and G. A. Rodan (1988) Structure and expression of rat osteosarcoma (ROS 17/2.8) alkaline phosphatase: product of a single copy gene. *Proc. Natl. Acad. Sci. USA* **85**: 319–323.

Tholander, B., A. Taube, A. Lindgren, O. Sjoberg, U. Stendahl, A. Kiviranta, K. Hallman, L. Holm, E. Weiner and L. Tamsen (1990) Pretreatment serum levels of CA-125, carcinoembryonic antigen, tissue polypeptide antigen and placental alkaline phosphatase, in patients with ovarian carcinoma, borderline tumors or benign adnexal masses: relevance for differential diagnosis. *Gynecol. Oncol.* **39**: 16–25.

Thompson, R. Q., G. C. Barone, III, H. B. Halsall and W. R. Heineman (1991) Comparison of methods for following alkaline phosphatase catalysis: spectrophotometric versus amperometric detection. *Anal. Biochem.* **192**: 90–95.

Tian, X. J., X. H. Song, S. L. Yan, Y. X. Zhang and H. M. Zhou (2003) Study of refolding of calf intestinal alkaline phosphatase. *J. Protein Chem.* **22**: 417–422.

Tibbitts, T. T., J. E. Murphy and E. R. Kantrowitz (1996) Kinetic and structural consequences of replacing the aspartate bridge by asparagine in the catalytic metal triad of *Escherichia coli* alkaline phosphatase. *J. Mol. Biol.* **257**: 700–715.

Tibi, L., A. Collier, A. W. Patrick, B. F. Clarke and A. F. Smith (1988) Plasma alkaline phosphatase isoenzymes in diabetes mellitus. *Clin. Chim. Acta* **177**: 147–155.

Tibi, L., A. W. Patrick, P. Leslie, A. D. Toft and A. F. Smith (1989) Alkaline phosphatase isoenzymes in plasma in hyperthyroidism. *Clin. Chem.* **35**: 1427–1430.

Tietze, C. C., M. J. Becich, M. Engle, W. F. Stenson, A. Mahmood, R. Eliakim and D. H. Alpers (1992) Caco-2 cell transfection by rat intestinal alkaline phosphatase cDNA increases surfactant-like particles. *Am. J. Physiol.* **263**: G756–G766.

Tintut, Y., Z. Alfonso, T. Saini, K. Radcliff, K. Watson, K. Bostrom and L. L. Demer (2003) Multilineage potential of cells from the artery wall. *Circulation* **108**: 2505–2510.

Tobiume, H., S. Kanzaki, S. Hida, T. Ono, T. Moriwake, S. Yamauchi, H. Tanaka and Y. Seino (1997) Serum bone alkaline phosphatase isoenzyme levels in normal children and children with growth hormone (GH) deficiency: a potential marker for bone formation and response to GH therapy. *J. Clin. Endocrinol. Metab.* **82**: 2056–2061.

Toh, Y., M. Yamamoto, H. Endo, Y. Misumi and Y. Ikehara (1989) Isolation and characterization of a rat liver alkaline phosphatase gene. A single gene with two promoters. *Eur. J. Biochem.* **182**: 231–237.

Toh, Y., M. Yamamoto, H. Endo, Y. Misumi and Y. Ikehara (1990) Duplicate leader exons in the mouse liver-type alkaline phosphatase gene. *Biochem. Int.* **22**: 213–218.

Tokumitsu, S., K. Tokumitsu and W. H. Fishman (1981a). Immunocytochemical demonstration of intracytoplasmic alkaline phosphatase in HeLa TCRC-1 cells. *J. Histochem. Cytochem.* **29**: 1080–1087.

Tokumitsu, S., K. Tokumitsu and W. H. Fishman (1981b). Intracellular alkaline phosphatase activity in cultured human cancer cells. *Histochemistry* **73**: 1–13.

Tonik, S. E., A. E. Ortmeyer, J. E. Shindelman and H. H. Sussman (1983) Elevation of serum placental alkaline phosphatase levels in cigarette smokers. *Int. J. Cancer* **31**: 51–53.

Tsavaler, L., R. C. Penhallow, W. Kam and H. H. Sussman (1987) Pst I restriction fragment length polymorphism of the human placental alkaline phosphatase gene in normal placentae and tumors. *Proc. Natl. Acad. Sci. USA* **84**: 4529–4532.

Tsavaler, L., R. C. Penhallow and H. H. Sussman (1988) Pst I restriction fragment length polymorphism of human placental alkaline phosphatase gene: Mendelian segregation and localization of mutation site in the gene. *Proc. Natl. Acad. Sci. USA* **85**: 7680–7684.

Tso, P., G. Pinkston, D. M. Klurfeld and D. Kritchevsky (1984) The absorption and transport of dietary cholesterol in the presence of peanut oil or randomized peanut oil. *Lipids* **19**: 11–16.

Tsonis, P. A., W. S. Argraves and J. L. Millán (1988) A putative functional domain of human placental alkaline phosphatase predicted from sequence comparisons. *Biochem. J.* **254**: 623–624.

Tsukazaki, K., E. G. Hayman and E. Ruoslahti (1985) Effects of ricin A chain conjugates of monoclonal antibodies to human alpha-fetoprotein and placental alkaline phosphatase on antigen-producing tumor cells in culture. *Cancer Res.* **45**: 1834–1838.

Tu, Z., L. H. Ke and G. He (2004) The application of alkaline phosphatase labeled HBV probe in serum detection. *Virus Genes* **28**: 151–156.

Turner, B. M. (1986) Use of alkaline phosphatase-conjugated antibodies for detection of protein antigens on nitrocellulose filters. *Methods Enzymol.* **121**: 848–855.

Turner, R. M. and S. M. McDonnell (2003) Alkaline phosphatase in stallion semen: characterization and clinical applications. *Theriogenology* **60**: 1–10.

Tylzanowski, P., K. Verschueren, D. Huylebroeck and F. P. Luyten (2001) Smad-interacting protein 1 is a repressor of liver/bone/kidney alkaline phosphatase transcription in bone morphogenetic protein-induced osteogenic differentiation of C2C12 cells. *J. Biol. Chem.* **276**: 40001–40007.

Uchida, T., T. Shimoda, H. Miyata, T. Shikata, S. Iino, H. Suzuki, T. Oda, K. Hirano and M. Sugiura (1981) Immunoperoxidase study of alkaline phosphatase in testicular tumor. *Cancer* **48**: 1455–1462.

Uhler, M. D. and A. Abou-Chebl (1992) Cellular concentrations of protein kinase A modulate prostaglandin and cAMP induction of alkaline phosphatase. *J. Biol. Chem.* **267**: 8658–8665.

Ureña, P. and M.C. de Vernejoul (1999) Circulating biochemical markers of bone remodeling in uremic patients. *Kidney Int.* **55**: 2141–2156.

Van Belle, H (1976) Alkaline phosphatase. I. Kinetics and inhibition by levamisole of purified isoenzymes from humans. *Clin. Chem.* **22**: 972–976.

Van Belle, H., M. E. De Broe and R. J. Wieme (1977) L-p-Bromotetramisole, a new reagent for use in measuring placental or intestinal isoenzymes of alkaline phosphatase in human serum. *Clin. Chem.* **23**: 454–459.

Van den Bos T., J. Oosting, V. Everts and W. Beertsen (1995) Mineralization of alkaline phosphatase-complexed collagen implants in the rat in relation to serum inorganic phosphate. *J. Bone Miner. Res.* **10**: 616–624.

Van de Voorde, A., G. de Groote, P. de Waele, M. E. de Broe, D. Pollet, J. de Boever, D. Vandekerckhove and W. Fiers (1985a). Screening of sera and tumor extracts of cancer patients using a monoclonal antibody directed against human placental alkaline phosphatase. *Eur. J. Cancer Clin. Oncol.* **21**: 65–71.

Van de Voorde, A., R. Serreyn, J. de Boever, P. de Waele, D. Vandekerckhove and W. Fiers (1985b). The occurrence of human placental alkaline phosphatase (PLAP) in extracts of normal, benign and malignant tissues of the female genital tract. *Tumor Biol.* **6**: 545–553.

Van Hoof, V. O., L. G. Lepoutre, M. F. Hoylaerts, R. Chevigne and M. E. De Broe (1988) Improved agarose electrophoretic method for separating alkaline phosphatase isoenzymes in serum. *Clin. Chem.* **34**: 1857–1862.

Van Hoof, V. O., M. F. Hoylaerts, H. Geryl, M. Van Mullem, L. G. Lepoutre and M. E. De Broe (1990) Age and sex distribution of alkaline phosphatase isoenzymes by agarose electrophoresis. *Clin. Chem.* **36**: 875–878.

Van Hoof, V. O., A. T. Van Oosterom, L. G. Lepoutre and M. E. De Broe (1992) Alkaline phosphatase isoenzyme patterns in malignant disease. *Clin. Chem.* **38**: 2546–2551.

Van Hoof, V. O., M. Martin, P. Blockx, A. Prove, A. Van Oosterom, M. M. Couttenye, M. E. De Broe and L. G. Lepoutre (1995) Immunoradiometric method and electrophoretic system compared for quantifying bone alkaline phosphatase in serum. *Clin. Chem.* **41**: 853–857.

Van Straalen, J. P., E. Sanders, M. F. Prummel and G. T. Sanders (1991) Bone-alkaline phosphatase as indicator of bone formation. *Clin. Chim. Acta* **201**: 27–33.

Van Veen, S. Q., A. K. van Vliet, M. Wulferink, R. Brands, M. A. Boermeester and T. M. van Gulik (2005) Bovine intestinal alkaline phosphatase attenuates the inflammatory response in secondary peritonitis in mice. *Infect. Immun.* **73**: 4309–4314.

Vega-Warner, A. V., H. Gandhi, D. M. Smith and Z. Ustunol (2000) Polyclonal-antibody-based ELISA to detect milk alkaline phosphatase. *J. Agric. Food Chem.* **48**: 2087–2091.

Veisman, D. N., A. D. McCarthy and A. M. Cortizo (2005) Bone-specific alkaline phosphatase activity is inhibited by bisphosphonates: role of divalent cations. *Biol. Trace Elem. Res.* **104**: 131–140.

Venetz, W. P., C. Mangan and I. W. Siddiqi (1990) Kinetic determination of alkaline phosphatase activity based on hydrolytic cleavage of the P–F bond in monofluorophosphate and fluoride ion-selective electrode. *Anal. Biochem.* **191**: 127–132.

Vergnes, H., J. Grozdea, A. Brisson-Lougarre, G. Bourrouillou, C. Blum, J. Martin and P. Colombies (1988) An enzymatic marker in mothers of trisomy 21 children: neutrophil alkaline phosphatase. *Enzyme* **39**: 174–180.

Vergnes, H., J. Grozdea, C. Denier, G. Bourrouillou, P. Calvas, F. De La Farge, P. Valdiguie and M. Calot (2000) Lower alkaline phosphatase activity and occurrence of an abnormal hybrid intestinal/tissue non-specific isoform in Down's syndrome amniotic fluids. *Early Hum. Dev.* **58**: 17–24.

Vergote, I., M. Onsrud and K. Nustad (1987) Placental alkaline phosphatase as a tumor marker in ovarian cancer. *Obstet. Gynecol.* **69**: 228–232.

Verpooten, G. F., E. J. Nouwen, M. F. Hoylaerts, P. G. Hendrix and

M. E. de Broe (1989) Segment-specific localization of intestinal-type alkaline phosphatase in human kidney. *Kidney Int.* **36**: 617–625.

Vinet, B. and L. Zizian (1979) Enzymic assay for serum theophylline. *Clin. Chem.* **25**: 1370–1372.

Vittur, F., N. Stagni, L. Moro and B. de Bernard (1984) Alkaline phosphatase binds to collagen; a hypothesis on the mechanism of extravesicular mineralization in epiphyseal cartilage. *Experientia* **40**: 836–837.

Vockley, J., M. P. D'Souza, C. J. Foster and H. Harris (1984a). Structural analysis of human adult and fetal alkaline phosphatases by cyanogen bromide peptide mapping. *Proc. Natl. Acad. Sci. USA* **81**: 6120–6123.

Vockley, J., L. J. Meyer and H. Harris (1984b). Differentiation of human adult and fetal intestinal alkaline phosphatases with monoclonal antibodies. *Am. J. Hum. Genet.* **36**: 987–1000.

Vogelsang, H., A. Hamwi and P. Ferenci (1996) Elevated liver isoenzymes of alkaline phosphatase and disease activity in patients with Crohn's disease. *Digestion* **57**: 11–15.

Von der Maase, H., A. Giwercman, J. Muller and N. E. Skakkebaek (1987) Management of carcinoma-*in-situ* of the testis. *Int. J. Androl.* **10**: 209–220.

Vorbrodt, A. W., A. S. Lossinsky and H. M. Wisniewski (1986) Localization of alkaline phosphatase activity in endothelia of developing and mature mouse blood–brain barrier. *Dev. Neurosci.* **8**: 1–13.

Vovk, A. I., V. I. Kalchenko, S. A. Cherenok, V. P. Kukhar, O. V. Muzychka and M. O. Lozynsky (2004) Calix[4]arene methylenebisphosphonic acids as calf intestine alkaline phosphatase inhibitors. *Org. Biomol. Chem.* **2**: 3162–3166.

Wada, A., A. P. Wang, H. Isomoto, Y. Satomi, T. Takao, A. Takahashi, S. Awata, T. Nomura, Y. Fujii, S. Kohno, K. Okamoto, J. Moss, J. L. Millán and T. Hirayama (2005) Placental and intestinal alkaline phosphatases are receptors for *Aeromonas sobria* hemolysin. *Int. J. Med. Microbiol.* **294**: 427–435.

Wada, N. and J. Y. Chou (1993) Characterization of upstream activation elements essential for the expression of germ cell alkaline phosphatase in human choriocarcinoma cells. *J. Biol. Chem.* **268**: 14003–14010.

Wada, T., M. D. McKee, S. Steitz and C. M. Giachelli (1999) Calcification of vascular smooth muscle cell cultures: inhibition by osteopontin. *Circ. Res.* **84**: 166–178.

Wahren, B., P. A. Holmgren and T. Stigbrand (1979) Placental alkaline phosphatase, alphafetoprotein and carcinoembryonic antigen in testicular tumors. Tissue typing by means of cytologic smears. *Int. J. Cancer* **24**: 749–753.

Wahren, B., J. Hinkula, T. Stigbrand, A. Jeppsson, L. Andersson, P. L. Esposti, F. Edsmyr and J. L. Millán (1986) Phenotypes of placental-type alkaline phosphatase in seminoma sera as defined by monoclonal antibodies. *Int. J. Cancer* **37**: 595–600.

Walach, N. and Y. Gur (1993) Leukocyte alkaline phosphatase and carcinoembryonic antigen in lung cancer patients. *Oncology* **50**: 279–284.

Wallace, B.H., J.A. Lott, J. Griffiths and R.B. Kirkpatrick (1996) Isoforms of alkaline phosphatase determined by isoelectric focusing in patients with chronic liver disorders. *Eur. J. Clin. Chem. Clin. Biochem.* **34**: 711–720.

Wallace, P. M. and P. D. Senter (1991) In vitro and in vivo activities of monoclonal antibody-alkaline phosphatase conjugates in combination with phenol mustard phosphate. *Bioconj. Chem.* **2**: 349–352.

Wan, D. Y., F. L. Cerklewski and J. E. Leklem (1993) Increased plasma pyridoxal-5′-phosphate when alkaline phosphatase activity is reduced in moderately zinc-deficient rats. *Biol. Trace Elem. Res.* **39**: 203–210.

Wang, M., C. Orsini, D. Casanova, J. L. Millán, A. Mahfoudi and V. Thuillier (2001) MUSEAP, a novel reporter gene for the study of long-term gene expression in immunocompetent mice. *Gene* **279**: 99–108.

Warnes, T. W., P. Hine, G. Kay and A. Smith (1981) Intestinal alkaline phosphatase in bile: evidence for an enterohepatic circulation. *Gut* **22**: 493–498.

Warren, R. C., C. F. McKenzie, C. H. Rodeck, G. Moscoso, D. J. Brock and L. Barron (1985) First trimester diagnosis of hypophosphatasia with a monoclonal antibody to the liver/bone/kidney isoenzyme of alkaline phosphatase. *Lancet* **ii**: 856–858.

Watanabe, H., H. Tokuyama, H. Ohta, Y. Satomura, T. Okai, A. Ooi, M. Mai and N. Sawabu (1990) Expression of placental alkaline phosphatase in gastric and colorectal cancers. An immunohistochemical study using the prepared monoclonal antibody. *Cancer* **66**: 2575–2582.

Watanabe, H., M. Hashimoto-Uoshima, M. Goseki-Sone, H. Orimo and I. Ishikawa (2001) A novel point mutation (C571T) in the tissue-non-specific alkaline phosphatase gene in a case of adult-type hypophosphatasia. *Oral Dis.* **7**: 331–335.

Watanabe, H., H. Takinami, M. Goseki-Sone, H. Orimo, R. Hamatani and I. Ishikawa (2005) Characterization of the mutant (A115V) tissue-nonspecific alkaline phosphatase gene from adult-type hypophosphatasia. *Biochem. Biophys. Res. Commun.* **327**: 124–129 (2005).

Watanabe, S., T. Watanabe, W. B. Li, B. W. Soong and J. Y. Chou (1989) Expression of the germ cell alkaline phosphatase gene in human choriocarcinoma cells. *J. Biol. Chem.* **264**: 12611–12619.

Watanabe, T., N. Wada, E. E. Kim, H. W. Wyckoff and J. Y. Chou (1991) Mutation of a single amino acid converts germ cell alkaline phosphatase to placental alkaline phosphatase. *J. Biol. Chem.* **266**: 21174–21178.

Watanabe, T., N. Wada and J. Y. Chou (1992) Structural and functional analysis of human germ cell alkaline phosphatase by site-specific mutagenesis. *Biochemistry* **31**: 3051–3058.

Watson, K. E., K. Bostrom, R. Ravindranath, T. Lam, B. Norton and L. L. Demer (1994) TGF-beta 1 and 25-hydroxycholesterol stimulate osteoblast-like vascular cells to calcify. *J. Clin. Invest.* **93**: 2106–2113.

Waymire, K. G., J. D. Mahuren, J. M. Jaje, T. R. Guilarte, S. P. Coburn and G. R. MacGregor (1995) Mice lacking tissue non-specific alkaline phosphatase die from seizures due to defective metabolism of vitamin B-6. *Nat. Genet.* **11**: 45–51.

Wei, S. C. and G. J. Doellgast (1981) Immunochemical studies of human placental-type variants of alkaline phosphatase. Structural differences between the "Nagao isoenzyme" and the placental "D-variant". *Eur. J. Biochem.* **118**: 39–45.

Weijers, R. N (1980) Properties and clinical significance of an alkaline phosphatase–lipoprotein-X complex. *Clin. Chem.* **26**: 609–612.

Weijers, R. N., H. Kruijswijk and J. M. Baak (1980) Purification of an alkaline phosphatase–lipoprotein-X complex. *Clin. Chem.* **26**: 604–608.

Weiss, M. J., P. S. Henthorn, M. A. Lafferty, C. Slaughter, M. Raducha and H. Harris (1986) Isolation and characterization of a cDNA encoding a human liver/bone/kidney-type alkaline phosphatase. *Proc. Natl. Acad. Sci. USA* **83**: 7182–7186.

Weiss, M. J., R. S. Spielman and H. Harris (1987) A high-frequency RFLP at the human liver/bone/kidney-type alkaline phosphatase locus. *Nucleic Acids Res.* **15**: 860.

Weiss, M. J., D. E. Cole, K. Ray, M. P. Whyte, M. A. Lafferty, R. A. Mulivor and H. Harris (1988) A missense mutation in the human liver/bone/kidney alkaline phosphatase gene causing a lethal form of hypophosphatasia. *Proc. Natl. Acad. Sci. USA* **85**: 7666–7669.

Weissig, H., A. Schildge, M. F. Hoylaerts, M. Iqbal and J. L. Millán (1993) Cloning and expression of the bovine intestinal alkaline phosphatase gene: biochemical characterization of the recombinant enzyme. *Biochem. J.* **290**: 503–508.

Wenham, P. R., B. Chapman and A. F. Smith (1983) Two macromolecular complexes between alkaline phosphatase and immunoglobulin A in a patient's serum. *Clin. Chem.* **29**: 1845–1849.

Weninger, M., R. A. Stinson, H. Plenk, Jr, P. Bock and A. Pollak (1989) Biochemical and morphological effects of human hepatic alkaline phosphatase in a neonate with hypophosphatasia. *Acta Paediatr. Scand. Suppl.* **360**: 154–160.

Wenkert, D., M. N. Podgornik, S. P. Coburn, L. M. Ryan, S. Mumm and M. P. Whyte (2002) Dietary phosphate restriction therapy for hypophosphatasia: preliminary observations. *J. Bone Miner. Res.* **7**: S1; S384.

Wennberg, C., A. Kivela and P. A. Holmgren (1995) Placental and germ cell alkaline phosphatase RFLPs and haplotypes associated with spontaneous abortion. *Hum. Hered.* **45**: 272–277.

Wennberg, C., L. Hessle, P. Lundberg, S. Mauro, S. Narisawa, U. H. Lerner and J. L. Millán (2000) Functional characterization of osteoblasts and osteoclasts from alkaline phosphatase knockout mice. *J. Bone Miner. Res.* **15**: 1879–1888.

Wennberg, C., A. Kozlenkov, S. Di Mauro, N. Fröhlander, L. Beckman, M. F. Hoylaerts and J. L. Millán (2002) Structure, genomic DNA typing and kinetic characterization of the D allozyme of placental alkaline phosphatase (PLAP/ALPP) *Hum. Mutat.* **19**: 258–267.

Whisnant, A. R., S. E. Johnston and S. D. Gilman (2000) Capillary electrophoretic analysis of alkaline phosphatase inhibition by theophylline. *Electrophoresis* **21**: 1341–1348.

Whyte, M. P (1994) Hypophosphatasia and the role of alkaline phosphatase in skeletal mineralization. *Endocr. Rev.* **15**: 439–461.

Whyte, M. P. Hypophosphatasia (1995) In: Scriver, C. R., A. L. Beaudet, W. S. Sly and D. Valle (eds), *The Metabolic and Molecular Bases of Inherited Disease.* New York: McGraw-Hill, pp. 4095–4112.

Whyte, M. P., S. L. Teitelbaum, W. A. Murphy, M. A. Bergfeld and L. V. Avioli (1979) Adult hypophosphatasia. Clinical, laboratory and genetic investigation of a large kindred with review of the literature. *Medicine* **58**: 329–347.

Whyte, M. P., W. A. Murphy and M. D. Fallon (1982a). Adult hypophosphatasia with chondrocalcinosis and arthropathy. Variable penetrance of hypophosphatasemia in a large Oklahoma kindred. *Am. J. Med.* **72**: 631–641.

Whyte, M. P., R. Valdes, Jr, L. M. Ryan and W. H. McAlister (1982b). Infantile hypophosphatasia: enzyme replacement therapy by intravenous infusion of alkaline phosphatase-rich plasma from patients with Paget's bone disease. *J. Pediatr.* **101**: 379–386.

Whyte, M. P., W. H. McAlister, L. S. Patton, H. L. Magill, M. D. Fallon, W. B. Lorentz, Jr and H. G. Herrod (1984) Enzyme replacement therapy for infantile hypophosphatasia attempted by intravenous infusions of alkaline phosphatase-rich Paget's plasma: results in three additional patients. *J. Pediatr.* **105**: 926–933.

Whyte, M. P., J. D. Mahuren, L. A. Vrabel and S. P. Coburn (1985) Markedly increased circulating pyridoxal-5'-phosphate levels in hypophosphatasia. Alkaline phosphatase acts in vitamin B_6 metabolism. *J. Clin. Invest.* **76**: 752–756.

Whyte, M. P., J. D. Mahuren, K. N. Fedde, F. S. Cole, E. R. McCabe and S. P. Coburn (1988) Perinatal hypophosphatasia: tissue levels of vitamin B_6 are unremarkable despite markedly increased circulating concentrations of pyridoxal-5'-phosphate. Evidence for an ectoenzyme role for tissue-nonspecific alkaline phosphatase. *J. Clin. Invest.* **81**: 1234–1239.

Whyte, M. P., M. Landt, L. M. Ryan, R. A. Mulivor, P. S. Henthorn, K. N. Fedde, J. D. Mahuren and S. P. Coburn (1995) Alkaline phosphatase: placental and tissue-nonspecific isoenzymes hydrolyze phosphoethanolamine, inorganic pyrophosphate and pyridoxal 5'-phosphate. Substrate accumulation in carriers of hypophosphatasia corrects during pregnancy. *J. Clin. Invest.* **95**: 1440–1445.

Whyte, M. P., J. Kurtzberg, W. H. McAlister, S. Mumm, M. N. Podgornik, S. P. Coburn, L. M. Ryan, C. R. Miller, G. S. Gottesman, A. K. Smith, J. Douville, B. Waters-Pick, R. D. Armstrong and P. L. Martin (2003) Marrow cell transplantation for infantile hypophosphatasia. *J. Bone Miner. Res.* **18**: 624–636.

Wick, M. R., P. E. Swanson and J. C. Manivel (1987) Placental-like alkaline phosphatase reactivity in human tumors: an immunohistochemical study of 520 cases. *Hum. Pathol.* **18**: 946–954.

Wiedemann, A. L., S. C. Charney, A. M. Barger, D. J. Schaeffer and B. E. Kitchell (2005) Assessment of corticosteroid-induced alkaline phosphatase as a prognostic indicator in canine lymphoma. *J. Small Anim. Pract.* **46**: 185–190.

Wilcox, F. H. and B. A. Taylor (1981) Genetics of the Akp-2 locus for alkaline phosphatase of liver, kidney, bone and placenta in the mouse. Linkage with the Ahd-1 locus on chromosome 4. *J. Hered.* **72**: 387–390.

Wilcox, F. H., L. Hirschhorn, B. A. Taylor, J. E. Womack and T. H. Roderick (1979) Genetic variation in alkaline phosphatase of the house mouse (*Mus musculus*) with emphasis on a manganese-requiring isozyme. *Biochem. Genet.* **17**: 1093–1107.

Williams, D. G., P. G. Byfield and D. W. Moss (1985) Inhibition of human alkaline phosphatase isoenzymes by the affinity reagent reactive yellow 13. *Enzyme* **33**: 70–74.

Wilson, P. D., G. P. Smith and T. J. Peters (1983) Pyridoxal 5′-phosphate: a possible physiological substrate for alkaline phosphatase in human neutrophils. *Histochem. J.* **15**: 257–264.

Wisdom, G. B (2004) Conjugation of antibodies to alkaline phosphatase. In Burns, R. (ed.), *Methods in Molecular Biology. Vol. 295. Immunochemical Protocols*, 3rd edn. Totowa, NJ: Humana Press, pp. 123–126.

Witters, I., P. Moerman, E. Mornet and J. P. Fryns (2004) Positive maternal serum triple test screening in severe early onset hypophosphatasia. *Prenat. Diagn.* **24**: 494–497.

Woitge, H.W., M.J. Seibel and R. Ziegler (1996) Comparison of total and bone-specific alkaline phosphatase in patients with nonskeletal disorders or metabolic bone diseases. *Clin. Chem.* **42**: 1796–1804.

Woitge, H.W., C. Scheidt-Nave, C. Kissling, G. Leidig-Bruckner, K. Meyer, A.Grauer, S.H. Scharla, R. Ziegler and M.J. Seibel (1998) Seasonal variation of biochemical indexes of bone turnover: results of a population-based study. *J. Clin. Endocrinol. Metab.* **83**: 68–75.

Wolf, P. (1990) High-molecular-weight alkaline phosphatase and alkaline phosphatase lipoprotein X complex in cholestasis and hepatic malignancy. *Arch. Pathol. Lab. Med.* **114**: 577–579.

Wolf, P. L. (1994) Clinical significance of serum high-molecular-mass alkaline phosphatase, alkaline phosphatase–lipoprotein-X complex and intestinal variant alkaline phosphatase. *J Clin. Lab. Anal.* **8**: 172–176.

Wolf, P. L. (1995) The significance of transient hyperphosphatasemia of infancy and childhood to the clinician and clinical pathologist. *Arch. Pathol. Lab. Med.* **119**: 774–775.

Wolff, J. M., T. Ittel, W. Boeckmann, T. Reinike, F. K. Habib and G. Jakse (1996) Skeletal alkaline phosphatase in the metastatic workup of patients with prostate cancer. *Eur. Urol.* **30**: 302–306.

Wolff, J. M., T. H. Ittel, W. Boeckmann, F. K. Habib and G. Jakse (1997) Metastatic workup of patients with prostate cancer employing skeletal alkaline phosphatase. *Anticancer Res.* **17**: 2995–2997.

Wolff, J. M., T. H. Ittel, H. Borchers, O. Boekels and G. Jakse (1999) Metastatic workup of patients with prostate cancer employing alkaline phosphatase and skeletal alkaline phosphatase. *Anticancer Res.* **19**: 2653–2655.

Woltgens, J. H., D. M. Lyaruu and A. L. Bronckers (1985) Effect of 1-*p*-bromotetramisole on phosphatases in neonatal hamster bone and tooth germs at different pH. *J. Biol. Buccale* **13**: 3–10.

Wong, Y. W. and M. G. Low (1992) Phospholipase resistance of the glycosyl-phosphatidylinositol membrane anchor on human alkaline phosphatase. *Clin. Chem.* **38**: 2517–2525.

Wray, L. K. and H. Harris (1984) Use of monoclonal antibodies to assign phenotypes to placental alkaline phosphatase from cultured human cell lines derived from tumors. *Am. J. Hum. Genet.* **36**: 471–476.

Wu, L. N., B. R. Genge, G. C. Lloyd and R. E. Wuthier (1991) Collagen-binding proteins in collagenase-released matrix vesicles from cartilage. Interaction between matrix vesicle proteins and different types of collagen. *J. Biol. Chem.* **266**: 1195–1203.

Xie, Q. and D. H. Alpers (2000) The two isozymes of rat intestinal alkaline phosphatase are products of two distinct genes. *Physiol. Genomics* **3**: 1–8.

Xie, Q. M., Y. Zhang, S. Mahmood and D. H. Alpers (1997) Rat intestinal alkaline phosphatase II messenger RNA is present in duodenal crypt and villus cells. *Gastroenterology* **112**: 376–386.

Xu, Q., Z. Lu and X. Zhang (2002) A novel role of alkaline phosphatase in protection from immunological liver injury in mice. *Liver* **22**: 8–14.

Xu, X. and E. R. Kantrowitz (1992) The importance of aspartate 327 for catalysis and zinc binding in *Escherichia coli* alkaline phosphatase. *J. Biol. Chem.* **267**: 16244–16251.

Xu, X. and E. R. Kantrowitz (1993) Binding of magnesium in a mutant *Escherichia coli* alkaline phosphatase changes the rate-determining step in the reaction mechanism. *Biochemistry* **32**: 10683–10691.

Xu, X., X. Q. Qin, E. R. Kantrowitz (1994) Probing the role of histidine-372 in zinc binding and the catalytic mechanism of *Escherichia coli* alkaline phosphatase by site-specific mutagenesis. *Biochemistry* **33**: 2279–2284.

Xu, Y., T. Cruz, P. T. Cheng and K. P. Pritzker (1991) Effects of pyrophosphatase on dissolution of calcium pyrophosphate dihydrate crystals. *J. Rheumatol.* **18**: 66–71.

Xue, S. and P. N. Rao (1981) Sodium butyrate blocks HeLa cells preferentially in early G1 phase of the cell cycle. *J. Cell. Sci.* **51**: 163–171.

Yamagishi, F., M. J. Becich, B. A. Evans, T. Komoda and D. H. Alpers (1994a). Clearance of surfactant-like particle proteins from circulation in rats. *Am. J. Physiol.* **266**: G596–G605.

Yamagishi, F., T. Komoda and D. H. Alpers (1994b). Secretion and distribution of rat intestinal surfactant-like particles after fat feeding. *Am. J. Physiol.* **266**: G944–G952.

Yamaguchi, R. and N. Shimozato (1978) Diagnosis of placental function by prediction curves for heat-stable alkaline phosphatase (HSAP) Tohoku. *J. Exp. Med.* **124**: 73–82.

Yamamoto, K., T. Awogi, K. Okuyama and N. Takahashi (2003) Nuclear localization of alkaline phosphatase in cultured human cancer cells. *Med. Electron. Microsc.* **36**: 47–51.

Yan, S., Y. Liu, X. Tian, Y. Zhang and H. Zhou (2003) Effect of extraneous zinc on calf intestinal alkaline phosphatase. *J. Protein Chem.* **22**: 371–375.

Yeh, K., M. Yeh, P. R. Holt and D. H. Alpers (1994) Development and hormonal modulation of postnatal expression of intestinal alkaline phosphatase mRNA species and their encoded isoenzymes. *Biochem. J.* **301**: 893–899.

Yokota, Y. (1978) Purification and characterization of alkaline phosphatase from cultured rat ascites hepatoma cells. *J. Biochem.* **83**: 1293–1298.

Yoon, K., M. A. Thiede and G. A. Rodan (1988) Alkaline phosphatase as a reporter enzyme. *Gene* **66**: 11–17.

Young, G. P., S. Friedman, S. T. Yedlin and D. H. Allers (1981) Effect of fat feeding on intestinal alkaline phosphatase activity in tissue and serum. *Am. J. Physiol.* **241**: G461–G468.

Yuhas, J. M. (1980) Active versus passive absorption kinetics as the basis for selective protection of normal tissues by S-2-(3-aminopropylamino)ethylphosphorothioic acid. *Cancer Res.* **40**: 1519–1524.

Yusa, N., K. Watanabe, S. Yoshida, N. Shirafuji, S. Shimomura, K. Tani, S. Asano and N. Sato (2000) Transcription factor Sp3 activates the liver/bone/kidney-type alkaline phosphatase promoter in hematopoietic cells. *J. Leukoc. Biol.* **68**: 772–777.

Zackson, S. L. and M. S. Steinberg (1988) A molecular marker for cell guidance information in the axolotl embryo. *Dev. Biol.* **127**: 435–442.

Zernik, J., M. A. Thiede, K. Twarog, M. L. Stover, G. A. Rodan, W. B. Upholt and D. W. Rowe (1990) Cloning and analysis of the 5′ region of the rat bone/liver/kidney/placenta alkaline phosphatase gene. A dual-function promoter. *Matrix* **10**: 38–47.

Zernik, J., B. Kream and K. Twarog (1991) Tissue-specific and dexamethasone-inducible expression of alkaline phosphatase from alternative promoters of the rat bone/liver/kidney/placenta gene. *Biochem. Biophys. Res. Commun.* **176**: 1149–1156.

Zhang, F., M. W. Wolff, D. Williams, K. Busch, S. C. Lang, D. W. Murhammer and R. J. Linhardt (2001) Affinity purification of secreted alkaline phosphatase produced by baculovirus expression vector system. *Appl. Biochem. Biotechnol.* **90**: 125–136.

Zhang, F., D. W. Murhammer and R. J. Linhardt (2002a). Enzyme kinetics and glycan structural characterization of secreted alkaline phosphatase prepared using the baculovirus expression vector system. *Appl. Biochem. Biotechnol.* **101**: 197–210.

Zhang, Y., J. S. Shao, Q. M. Xie and D. H. Alpers (1996) Immunolocalization of alkaline phosphatase and surfactant-like particle proteins in rat duodenum during fat absorption. *Gastroenterology* **110**: 478–488.

Zhang, Y. X., Y. Zhu, H. W. Xi, Y. L. Liu and H. M. Zhou (2002b). Refolding and reactivation of calf intestinal alkaline phosphatase with excess magnesium ions. *Int. J. Biochem. Cell. Biol.* **34**: 1241–1247.

Zhang, Y. X., X. H. Song, S. L. Yan and H. M. Zhou (2003) The unfolding intermediate state of calf intestinal alkaline phosphatase during denaturation in guanidine solutions. *J. Protein Chem.* **22**: 405–409.

Zhou, H., S. S. Manji, D. M. Findlay, T. J. Martin, J. K. Heath and K. W. Ng (1994) Novel action of retinoic acid. Stabilization of newly synthesized alkaline phosphatase transcripts. *J. Biol. Chem.* **269**: 22433–22439.

Ziomek, C. A., M. L. Lepire and I. Torres (1990) A highly fluorescent simultaneous azo dye technique for demonstration of nonspecific alkaline phosphatase activity. *J. Histochem. Cytochem.* **38**: 437–442.

Zsengeller, Z. K., C. Halbert, A. D. Miller, S. E. Wert, J. A. Whitsett and C. J. Bachurski (1999) Keratinocyte growth factor stimulates transduction of the respiratory epithelium by retroviral vectors. *Hum. Gene Ther.* **10**: 341–353.

Zurutuza, L., F. Muller, J. F. Gibrat, A. Taillandier, B. Simon-Bouy, J. L. Serre and E. Mornet (1999) Correlations of genotype and phenotype in hypophosphatasia. *Hum. Mol. Genet.* **8**: 1039–1046.

Index

a

ABO genotype 159
absorption 155
active site
– calcium binding site 111
– see also protein structure
adult IAP, surfactant-like particles 17
$Akp2^{-/-}$ see knockout mice
$Akp3^{-/-}$ see knockout mice
$Akp5^{-/-}$ see knockout mice
Akp6 159
Akp-ps1 8 ff.
allelic variants 82
allosteric behavior 80
allosterism see enzymatic properties
allotype-specific 91
allozymes 50 ff.
amifostine 223 f.
ank/ank mice see mutant mice
ankylosis 144
anticancer drugs 223
antigenic domain see epitope maps
3-AP 224
AP pseudogene, Akp-ps1 6 ff.
apnea 152
ascites 194
ATPase 87
autosomal dominant 123
autosomal recessive 123

b

bIAP see intestinal alkaline phosphatase, bovine 9
bile salts 178
biochemical marker, bone turnover 167
biological roles see functions
bispecific monoclonal antibody see immunolocalization/immunotherapy
bone isoform 99
bone metastasis
– osteoblastic 173
– osteolytic 173
bone mineralization 107, 141
bone nodules 138
bovine AP 9
breast cancer 173 ff.
bromotetramisole 79

c

calcification 148
calcium 107
calcium pyrophosphate dihydrate 108, 231
calcium site see protein structure
calcium-binding site (M4) 111
calf IAP, hereogeneity 60
carcinoma-in-situ see neoplastic expression
catalytic inhibition 67 ff.
– competitive 69 ff.
– efficiency 31, 83
– GCAP 71
– mechanism 29
– noncompetitive 69 ff.
– PLAP 71
– TNAP 76
– uncompetitive 71 ff.
cell therapy 229
chemiluminescence 243
chemotherapeutic drugs 217
cholestasis 175
chondrocalcinosis 231
choriocarcinoma 23, 189
CIS 191
clearance 167
clinical conditions 183
clinical usefulness 167
collagen 14, 20 f., 41 f., 111, 148 ff., 161, 175, 180, 228, 230
competitive inhibition 68

Mammalian Alkaline Phosphatases: From Biology to Applications in Medicine and Biotechnology. J. L. Millán
Copyright © 2006 WILEY-VCH Verlag GmbH & Co. KGaA, Weinheim
ISBN: 3-527-31079-7

competitive inhibitor 39 ff.
computer modeling 75
cooperativity 80
CPPD *see* calcium pyrophosphate dihydrate
cross-reactivity 100
crown domain *see* protein structure

d

3-D structure, Glu429 43
demetalated 35
dental abnormalities 136
detoxification 127
disulphide bonds *see* protein structure
docking 78
double-knockout mice 144
duodenum 17, 45
dysgerminomas 195

e

EAP
– *Akp5* 6 ff.
– *see also* embryonic alkaline phosphatase
ECAP *see Escherichia coli* alkaline phosphatase
ECM *see* extracellular matrix
ectopic calcification 234
ectopic expression 188
EIA *see* enzyme immunoassay
electrophoresis 168
ELISA *see* enzyme-linked immunosorbent assay
embryonal carcinomas 189
embryonic alkaline phosphatase
– developmental expression 15 f.
– gene structure 8
– knockout mice 160 ff.
– nomenclature 6
– secreted 217 f.
embryos 134
endotoxin 177, 232
Enpp1$^{-/-}$ mice *see* knockout mice
enterocytes 154
enzymatic cascade 220
enzymatic properties 67 ff.
– allosteric behavior 80 ff.
– catalytic efficiency 83 ff.
– catalytic inhibition 67 ff.
– enzyme superfamily 88 f.
– substrate specificity 85 ff.
enzyme immunoassay *see* methodologies
enzyme replacement therapy 227
enzyme-linked immunosorbent assays 100, 218
– *see also* methodologies

epiphyses 134
epitope maps 91 ff.
epitope-tagged 123
epitopes 11, 91
– GCAP 98
– IAP 98
– PLAP 98
– TNAP 98
Escherichia coli alkaline phosphatase 25 ff., 56, 73 ff., 209
etoposide 224
eutopic expression 188
evolution 8
expression
– *ALPP* 22
– *ALPP2* 22
– bone 14
– cartilage 13
– developmental 13 ff.
– endothelial cells 14
– GCAP 22
– health and disease 167 ff.
– leukocytes 14
– matrix vesicles 14
– neural tube 13
– primordial germ cells 13
– skeleton 14
– spinal cord 13
– testis 14
– TNAP gene 13
– TSAP gene 15, 22
expression systems
– *Escherichia coli* 212
– insect cells 210
– mammalian 209
– plants 212
– yeast cells 212
expressivity 123
extracellular matrix 41, 148 ff.
extracellular PP$_i$ 145

f

fat absorption 127
fat feeding 156
fetal disorders 181
fibrillar collagen 149
food industry 237 f.
Fourier Transform infrared imaging spectroscopy 142
functions
– IAP 126
– TNAP 107 ff., 131 ff.
– TSAPs 126 ff., 154 ff.

g

GCAP *see* germ cell alkaline phosphatase
gene nomenclature 5 f.
gene regulation 19 ff.
- downstream promoter 19
- 1,25-(OH)₂D3 20
- TNAP gene 19
- upstream promoter 19
gene structure 5 ff.
gene therapy vectors 217
genetic background 135
genetic polymorphism 49 ff.
- D-allele 49
- allelic variants 50
- common alleles 49
- electrophoretic polymorphism 49
- rate alleles 49
genomic DNA 10 ff., 172
genomic organization 5 ff.
- chromosome 5 ff.
- exons 5 ff.
- homology 9
- introns 5 ff.
- mRNA 7 ff.
- promoter 7 ff.
germ cell alkaline phosphatase
- *ALPP2* 6 ff.
- developmental expression 15 f.
- epitopes 91 ff.
- gene regulation 22 ff.
- gene structure 7 f.
- genetic polymorphism 91 ff.
- ovarian cancer 194 ff.
- protein structure 25 ff.
- testicular cancer 189 ff.
gestation length 161
glycans 56
glycation of APs 61
N-glycosylation 55
glycosylation sites 54
- N-glycosylation 55
- O-glycosylation 55
- *see also* post-translational modifications
glycosylphosphatidylinositol 58 ff., 155, 167, 169, 176 ff., 183, 266
GPI *see* glycosylphosphatidylinositol
GPI anchor 56
- GPI-PLC 58 ff., 155, 167, 169, 176, 209
- GPI-PLD 58 ff., 156, 167, 176 ff., 183, 266
- modeled structure 59
- structure 58
- *see also* post-translational modifications

GPI-PLC *see* GPI-specific phospholipase C
GPI-PLD *see* GPI-specific phospholipase D
GPI-specific phospholipase C 58 ff., 155, 167, 169, 176, 209
GPI-specific phospholipase D 58 ff., 156, 167, 176 ff., 183, 266
- N-butanol 60
- in serum 59

h

heterodimers 81
high performance liquid chromatography (HPLC) 247 f.
- *see also* methodologies
high-fat feeding 158
L-homoarginine 40 ff., 67
homodimer 82
hydroxyapatite 108, 141 ff., 234 ff.
Hyp mice *see* mutant mice
hyperostosis 144
hyperphosphatasia 185 ff.
hypophosphatasia 107
- clinical presentation 107 f.
- expressivity 123 f.
- mouse models 131 ff.
- penetrance 123 f.
- prenatal diagnosis 172
- pyridoxal-5′-phosphate 108
- pyrophosphate 108
- TNAP mutations 109 ff.
- treatment 227 ff.
hypophosphatemic 126

i

IAP 5 ff.
- adult IAP 16
- *Akp3* 6 ff.
- *ALPI* 6 ff.
- fetal IAP 16
- receptors 127
- secretion 155
- secretor status 127
- *see also* intestinal alkaline phosphatase
IAPI, *Alpi* 6 ff.
IAPII, *Alpi2* 6 ff.
IgG transport 128
immunoassays 171
immunocompetent 200
immunoglobulin 184
immunoglobulin complexes 184 f.
immunohistochemistry 190
immunolocalization 197 ff.
immunoradiometric (IRMA) assays 101

– *see also* methodologies
immunoreactivity 169
immunotargeting 195, 200
immunotherapy 197 ff.
impaired myelination 140
infantile hypophosphatasia 107, 132
inhibition 42, 67
inhibition constants *see* catalytic inhibition
inhibition properties 72
inhibitors
 – adenosine triphosphate 70, 88
 – bisphosphonates 69, 231
 – bromotetramisole 79, 126, 244
 – cyclic AMP 88
 – l-homoarginine 40 ff., 67, 72, 77 ff., 168 ff., 214, 244
 – inorganic phosphate 40, 61, 69 f., 72
 – l-leucine 50 f., 60, 67, 71 ff., 81 f., 186 f., 193, 244
 – l-leucinamide 74
 – l-leu-gly-gly 186, 244
 – levamisole 65, 67, 72, 77 ff., 87, 126, 136, 168 f., 180, 214, 232, 235, 244, 248
 – l-phenylalanine 40 f., 45, 47, 50, 60, 67, 72 ff., 94, 132, 169, 186, 244, 248
 – l-phe-gly-gly 186, 244
 – orthovanadate 70, 231
 – phenylphosphonate 61, 70 f.
 – phosphonates 126
 – tetramisole 67, 78, 136
 – theophylline 67, 72, 77 ff., 220
 – l-tryptophan 60, 67, 186
 – *see also* catalytic inhibition
inorganic phosphate 1, 29, 40, 61, 69 ff., 84, 87, 107 f., 126 f., 136, 141, 144, 148 ff., 178, 229 f.
inorganic pyrophosphate *see* physiological substrates
intestinal alkaline phosphatase
 – adult 16
 – *ALPI* 6 ff.
 – bovine 9
 – developmental expression 16 ff.
 – epitopes 98 f.
 – fat feeding 154 ff., 182 ff.
 – fetal 16
 – gene regulation 24
 – gene structure 7 ff.
 – knockout mice 154 ff.
 – other pathologies 182 ff.
 – protein structure 25 ff.
 – tumoral expression 204 f.
intracellular PP_i 146

introns *see* gene structure
IRMA *see* immunoradiometric assays

j
jejunum 45

k
Kasahara isozyme *see* neoplastic expression
k_{cat} 31, 69
kidney 126
kinetic parameters 32 f., 76
K_m 31, 68
knockout mice 132
 – $Akp2^{-/-}$ 8, 131 ff.
 – $Akp3^{-/-}$ 8, 154 ff.
 – $Akp5^{-/-}$ 8 ff., 160 ff.
 – $Enpp1^{-/-}$ 144 ff.

l
levamisole 77
ligands 33, 37
lipopolysaccharide 127
lipoprotein-X 175
liver 125
liver isoform 99

m
MAbs *see* monoclonal antibodies
malignant trophoblast 16
mammalian cells 209
matrix vesicles 14, 59, 64 f., 108, 136, 139 ff., 234 f.
metal ions *see* protein structure
metalated 81
methodologies 241 ff.
 – amperometric 241 ff.
 – chemiluminescent 243 f.
 – electrophoretic 245
 – heat inactivation 244
 – heat stability 244
 – high-performance liquid chromatographic 247 f.
 – histochemical 254
 – HPLC 247 f.
 – immunoassays 249 ff.
 – inhibition 244
 – lectins 245 f.
 – luminometric 242
 – mRNA 253
 – spectrophotometric 241 ff.
micro-computed tomography 142
missense mutation 109
mitomycin 224

monoclonal antibodies 91
- allotype-specific 93
- polymorphism 93
- see also epitope maps, immunolocalization, immunotherapy, and methodologies
monomer-monomer interface see protein structure
mouse models 124
mutagenesis 30 ff.
mutant mice
- ank/ank 143 ff., 230, 235
- Hyp 126, 230
mutation, noncatalytic peripheral binding site 49
MV see matrix vesicles

n

Nagao isozyme see neoplastic expression
neoplastic expression 187 ff.
new AP locus, Akp6 6 ff.
p-nitrophenylphosphate (pNPP) 39
nomenclature 6 ff.
noncompetitive inhibition 68
nonsecretor 160
nonsmoker 193
NPP1 143
nucleosidetriphosphate pyrophosphohydrolase 139
nuclotidetriphosphate pyrophosphohydrolase 108
nude mice 199

o

obesity 159
1,25-$(OH)_2D3$ 20
oligonucleotide probes 219
oligosaccharides 55
OPN see osteopontin
ortholog 160
osteoblasts 136
osteomalacia 107
osteopontin 134, 146 ff., 235 f.
osteoporosis 171, 174
osteosarcomas 173
other tumors 196
ovarian cancer 188, 194 ff., 204, 254

p

paclitaxol 224
Paget's disease 178
pasteurization 237 f.
PEA see phosphoethanolamine
penetrance 123

peripheral-binding site, see protein structure
PGC see primordial germ cells
phenotypes 98
phenotypic abnormalities
- $Akp2^{-/-}$ 131 ff.
- $Akp3^{-/-}$ 154 ff.
- $Akp5^{-/-}$ 160 ff.
- see also knockout mice
L-phenylalanine see inhibitors
phosphate 35
phosphocholine 151
phosphodiesterase 87
phosphoethanolamine 89, 108
- hypophosphatasia 86
phosphoprotein phosphatase 88
O-phosphorylethanolamine phospholyase 125, 152
phosphoserine, see protein structure, active site; see also substrates
physiological substrates
- inorganic pyrophosphate 86 f., 108, 123 f., 135, 139 ff.
- pyridoxal-5'-phosphate 70, 86, 108, 123 ff., 135, 139 ff., 152, 228
P_i see inorganic phosphate
Pichia Pastoris see recombinant AP expression
placenta 16
placental alkaline phosphatase
- allelic polymorphism 11 f., 49 ff., 91 ff.
- ALPP 6 ff.
- breast 16
- cell division 129
- crystal structure 25 f.
- endometrium 16
- epitopes 91 ff.
- expression 15 f.
- gene regulation 22 f.
- gene structure 7 f.
- GPI anchor 56 ff.
- immunolocalization/immunotherapy 197 ff.
- other tumors 196 f.
- ovarian cancer 194 ff.
- oviduct 16
- placenta 16
- pregnancy 180 ff.
- protein structure 25 ff.
- secreted 213 ff.
- testicular cancer 189 ff.
- variants 181
PLAP see placental alkaline phosphatase
PLAP/GCAP, smokers 192
plasma membrane 64

PLP *see* pyridoxal-5-phosphate
pNPP *see* p-nitrophenylphosphate
POMP *see* prodrug
post-translational modifications
– glycosylation sites 54 ff.
– GPI-anchor 56 ff.
– nonenzymatic glycation 61 f.
– quaternary structure 62 f.
PP$_i$ *see* inorganic pyrophosphate
pregnancy
– TNAP expression 5
– PLAP expression 180 ff.
– primordial germ cells 15
prodrug 223 ff.
– amifostine 223
– 3-aminopyridine-2-carboxaldehyde thiosemicarbazone 224
– etoposide 224
– mitomycin 224
– mitomycin phosphate 225
– paclitaxel phosphate 226
– phenol mustard phosphate, POMP 224, 226
prostate cancer 173
protein structure 25 ff.
– active site 28 ff.
– Ca^{2+} ion 47
– calcium site 36
– catalytic mechanism 28
– crown domain 41 ff.
– crystallographic coordinates 25
– dimer 25
– disulfide bonds 37 ff.
– α-helix 25
– homology 26
– model 26
– monomer-monomer interface 46 ff.
– monomers 25
– noncatalytic 36, 46
– noncatalytic peripheral binding site 46
– peripheral binding site 47 ff.
– quaternary structure 62
– β-strand 25
– N-terminal arm 38 ff.
– tetrameric 62
– uncompetitive inhibition 41
puberty 171
pyridoxal 140
pyridoxal-5'-phosphate *see* physiological substrates
pyridoxine-5'-phosphate 70
pyrophosphatase, hypophosphatasia 86

r
radiolabeled 202
recombinant APs 209 ff.
recombinant AP expression 209 ff.
– baculovirus 210 ff.
– *in vitro* reporter 213 ff.
– *in vivo* reporter 215 ff.
– *Escherichia coli* 212
– mammalian cells 209 ff.
– *Pichia pastoris* 212
Regan isozyme *see* neoplastic expression
reporter 213
reporter gene 215
restriction fragment length polymorphisms
– *Akp2* 11
– *ALPL* 11
– *ALPP* 11
– *ALPP2* 11
retinoic acid (RA) 21
reverse transcriptase polymerase chain reaction *see* methodologies, mRNA
RFLP *see* restriction fragment length polymorphisms
rickets 107
RT-PCR *see* methodologies, mRNA

s
SEAP *see* placental alkaline phosphatase, secreted
secondary ossification centers 134
secretor status 159
seizures 132
seminoma 189
sialic acid residues 169
single nucleotide primer extension 51, 195 ff., 254
– *see also* methodologies, mRNA
site-directed mutagenesis 95
SLP *see* surfactant-like particle
smoker 193
SNuPE *see* single nucleotide primer extension
spinal cord 134
spontaneous abortions 181
starvation 154
subcellular localization 64 ff.
substrates
– adenosine monophosphate 47, 70, 87, 152
– adenosine diphosphate 87
– adenosine-3'-phosphate-5'-phosphosulfate
adenosine triphosphate 244

- p-aminophenylphosphate 241
- bis(p-nitrophenyl)phosphate 87 f.
- 5-bromo-4-chloro-3-indolyl phosphate 218
- chemiluminescent 243
- cyclic AMP 87 f.
- dihydroxyacetone phosphate 244
- 1,2-dioxetane phosphate 243
- disodium 3-(4-methoxyspiro[1,2-dioxetane-3,2′(5′-chloro)-tricyclo-[3.3.1.1.]decan]-4-yl)-phenyl phosphate 214, 243
- β-glycerophosphate 21, 86 f., 136, 144, 149, 229
- ELF-97, 255
- flavin adenine dinucleotide-3′-phosphate 220 f., 242 f.
- fructose-6-phosphate 87
- glucose-1-phosphate 87
- glucose-6-phosphate 87
- 3-indoxyl phosphate 242
- inhibition 244
- inorganic pyrophosphate *see* physiological substrates
- lipopolysaccharide 127, 232
- menadiol diphosphate 254 ff.
- 4-methylumbelliferyl phosphate 218
- monofluorophosphate 86, 242
- α-naphthylphosphate 86
- naphtol AS-MX phosphate 254 ff.
- p-nitrophenylphosphate 39 f., 47, 61, 70, 73, 86 f., 98, 123 f., 220, 241 ff.
- phosphatidates 86,
- phosphocholine 151
- phosphoethanolamine 86, 108, 125, 135, 151 f., 227 f.
- phosphoprotein 1, 88, 221
- phosphoserine 86, 152
- phosphothreonine 86
- phosphotyrosine 86
- pyridoxal-5′-phosphate *see* physiological substrates
- riboflavin phosphates 242
substrate specificities 85
superfamily of enzymes 88
surfactant-like particles (SLPs) 17, 65, 127, 154 ff.
syncytiotrophoblast 180

t
N-terminal arm *see* protein structure
testicular cancer 189
testicular germ cell tumors 93, 189 ff., 202 f., 254

testis 190
tetramisole 67
TGCTs *see* testicular germ cell tumors
theophylline 67
therapeutic targets 148
therapeutic uses 227 ff.
- CPPD disease 231
- ectopic calcification 234 ff.
- endotoxin 231 ff.
- hypophosphatasia 227 ff.
three-dimensional-structure
- active site cleft 45
- Glu429 45
- Mg atom 45
- phosphoserine 45
- Tyr367 44
- Zn atom 42, 45
tissue-nonspecific alkaline phosphatase (TNAP)
- *Akp2* 6 ff.
- *ALPI* 6 ff.
- *ALPL* 6 ff.
- bone formation 167 ff.
- bone metastasis 99, 173 ff.
- breast cancer 173 ff.
- cholestasis 99, 125, 175 ff., 232
- developmental expression 13 f.
- epitopes 99 ff.
- gene regulation 19 ff.
- gene structure 5 ff.
- hypophosphatasia 107 ff.
- hyperphosphatasemia 185
- isoforms 170
- knockout mice 131 ff.
- liver metastasis 177
- membrane-bound 60
- nomenclature 5 f.
- nonskeletal tissues 124 ff., 139 ff., 152 ff.
- osteosarcoma 19 ff., 86, 100 ff., 171 ff., 213, 239, 248, 251
- other pathologies 178 ff.
- polymorphism 11, 52 ff.
- prostate cancer 173 ff., 189, 214
- skeletal mineralization 107 f., 131 ff., 141 ff.
tissue-specific alkaline phosphatases (TSAP)
- cervix 15
- developmental expression 15 ff.
- gene regulation 22 ff.
- gene structure 5 ff.
- gonads 15
- lung 15
- nomenclature 5 f.

– PGCs 15
– testis 15
– thymus 15
– type I pneumocytes 15
TNAP *see* tissue-nonspecific alkaline phosphatase
trafficking 123
transfection 24
transgenic mouse 149, 200
transmission electron microscopy 142
treatment 227
TSAPs *see* tissue-specific alkaline phosphatases
tumor markers *see* neoplastic expression
tumoral expression 204
tumors, other 196

u
uncompetitive inhibition 71
uncompetitive inhibitor 40 ff.

v
vascular calcification 234
veterinary medicine 239
veterinary uses 239 f.
vitamin B_6 139
V_{max} 68

w
wheat germ agglutinin (WGA) *see* methodologies, lectins